'Speculum Britanniae'
Regional Study, Antiquarianism, and Science
in Britain to 1700

Local history has been studied in Britain for at least 500 years. In this comprehensive study Stan Mendyk examines many of the first county and regional histories compiled in Britain (focusing especially on England) up to about 1700.

Mendyk considers first the precedent set by the ancients, such as Strabo and Ptolemy, in firmly establishing chorographical (ie topographical-historical) investigation, and then within that framework explores the work of a number of British historians: Bede, Gildas, William of Worcester, John Leland, William Camden, and William Lambarde. He offers summaries of the contents of their works, an identification of the authors and their connections with one another, and explanations of the means used to collect information. He notes trends within the works, such as the infusion of a patriotic element into history and the role of insularity in shaping them.

Placing these works in a broader context, Mendyk discusses some of the changes in the study of history during this period and assesses their impact on local historiography: the conflict between ancients and moderns, the development of natural history and geo-history as fields of study, and the overall development of historiography in the Tudor-Stuart period. Political and social upheaval each play a part, as do a number of other factors: the voyages of discovery, the work of Sir Francis Bacon and the Royal Society in promoting the new science, and the continuing interest in such non-scientific endeavours as alchemy and astrology.

Mendyk presents an approach that integrates history, geography, historiography, and science to expand the boundaries of conventional interpretations of local history, and in so doing illuminates a remarkable body of early work in the field.

STAN A. E. MENDYK has taught history at the University of New Brunswick in Saint John. He now lives in Oshawa, Ontario.

Map of Britain, preceding the title-page, from William Camden *Britannia*, 1607 edition

STAN A.E. MENDYK

'Speculum Britanniae'

REGIONAL STUDY,
ANTIQUARIANISM,
AND SCIENCE
IN BRITAIN
TO 1700

UNIVERSITY OF TORONTO PRESS
Toronto Buffalo London

©University of Toronto Press 1989
Toronto Buffalo London
Printed in Canada
Reprinted in 2018
ISBN 0-8020-5744-6
ISBN 978-1-4875-8082-7 (paper)

Printed on acid-free paper

Canadian Cataloguing in Publication Data

Mendyk, Stanley G.
  'Speculum Britanniae'

Bibliography: p.
Includes index.
ISBN 0-8020-5744-6

1. Great Britain – History, Local – Historiography.
2. Great Britain – Intellectual life – 16th century.
3. Great Britain – Intellectual life – 17th century.
4. Science – Great Britain – History. I. Title.

DA1.M46 1988   941.05   C88-094168-5

FOR ROSALIA AND LAUDIS

... the most flourishing and excellent, most renowmed and famous Isle of the whole world: so rich in commodities, so beautifull in situation, so resplendent in all glorie, that if the most Omnipotent had fashioned the world round like a ring, as hee did like a globe, it might haue bene most worthily the only gemme therein.

William Camden on Britain, from *Remaines concerning Britaine*, 1614

> Great Brittaine, shadow of the starry Sphears,
> Selfe-viewing Beauties true presented Grace
> In Thetis Myrrhour, on this Orbe appeares,
> In Worth excelling, as extoll'd in Place;
> Like the rich Croisade on th' Imperiall Ball,
> As much adorning as surmounting all.

Richard Zouche, from *The Dove, or passages of Cosmography*, 1613

Contents

PREFACE ix

ILLUSTRATIONS

1 Introduction: 'Where the Choir Was ... ' 3
2 'A Hole Worlde of Thinges Very Memorable' 38
3 Speculum Britanniae 57
4 Removing the 'Eclipse from the Sunne' 82
5 'Rapidly Thinning Wisps and Patches' 102
6 'Men Wake As from Deep Sleep' 114
7 Metamorphosis 136
8 'To Serve the Commonwealth of Learning' 146
9 'This Secret Call' 170
10 Remedying 'Chief Defects' 185
11 'Learning So Much Neglected' 193

12 'Industrious Searchers into the History of Nature' 213

13 Conclusion: 'Rust of Old Monuments' 229

NOTES 247

SELECT BIBLIOGRAPHY 331

INDEX 345

Preface

WHEN FIRST SEARCHING for a topic for my PH D dissertation in history, upon which this book is based, I tried to find one that would enable me to indulge in my love for local history, the history of science, historiography, and antiquarianism. This was no easy task. At first glance these areas of inquiry may appear to be mutually exclusive; to synthesize them required a certain amount of guidance. I turned to the masters of this technique – K. Theodore Hoppen, Michael Hunter, Joseph M. Levine, Stuart Piggott, Roy Porter, Barbara Shapiro, Charles Webster, and others – for valuable insights into the method, as well as for actual grist for the mill. Of course, I bear full responsibility for my interpretation of their work, and for accuracy of textual material as well as any infelicities of style. The immediate inspiration for the topic was Frank Emery's short but incisive article, 'English Regional Studies from Aubrey to Defoe' (*Geographical Journal* 124). While Emery focused on work done in England in the second half of the seventeenth century, I was stimulated to consider the similarities and differences between forms of regional study then and in the preceding decades, and to investigate the broader social, political, and intellectual climate in which that study developed.

My main concern, as it turned out, was to reveal the connective tissue among these factors, and among the practitioners of such studies themselves. Many of the latter were from the ranks of the country gentry, in whom recent historiography has taken a keen interest. I make no claim, therefore, that the figures discussed or the issues presented here are new to the student of local history, amateur or professional alike, though no doubt this is the case in some instances. But my major goals have been to bring together much scattered information about the writers and their works, in order to make the connections and contrasts mentioned above, and to expand the parameters of the subject and the issues involved in it. The result, I hope, will be to stimulate, or provoke, further inquiry into regional study.

Much has already been accomplished, especially through the emergence and growth of fine departments of local history in Britain, which have done much to make more generally known the role of the chorographer, or local historian-antiquary, in the development of British historiography in the sixteenth and seventeenth centuries. Still, a significant corpus of work in intellectual history has studiously ignored or overlooked the contributions of local and regional chorography. Post-World War II interest in English historiography of the period has been advanced by such able historians as D.C. Douglas, Frank Smith Fussner, Fritz Levy, and others – to whom this study owes much – but not even these scholars have entirely recognized the worth of the chorographers as serious historians in their own right, who expanded the nature of historical investigation, linked it firmly with topography and geography – if not entirely with geology – and made a *speculum*, or mirror, into a *theatrum* of British history, evoking the *genius loci*. There is much about British history that is 'peculiar,' distinctive, particular. Insularity, in part, made the theatre, as did a unique ecclesiological experiment in the Reformation – the church established, and challenged, by law. So, too, did the first modern revolution, of the mid-seventeenth century, and a quite inimitable structure and dynamic of imperialism. But one feature equally distinctive was a sense of regionalism, of 'country,' which was never lost sight of, from the Tudor age to the present, by even the most generalizing historians of the most, and earliest, centralized polity in the Western world. What French historians of the past couple of generations have attempted to do in the Annales school, British historians, from the early chorographers on down, have been doing consistently for centuries. Political history has been seen in regional specificity. And thanks to *mentalité* – here the Annales school has made its contribution – the new social history of recent British historians has tended to find focus and depth in county and regional investigation. This book, then, goes back to the *fons et origo* of this kind of history.

It may appear to the reader as he or she passes, in this work, from the world of the chorographer into that of the natural historian or philosopher, that the efforts of the former are denigrated when compared to those of the latter. In a sense, however, the chorographer – as antiquary, topographer, land-owner and administrator, justice of the peace, and the like – may also be considered a virtuoso in his own right. Used in this sense, the term *virtuoso* is applied to a man of merit, capable of varied tasks requiring wide learning. Thus, while regional study is shown to have become more realistic in the second half of the seventeenth century, among many of its practitioners, at least, this is not to disparage the earlier efforts of the chorographers, nor to demean the value of chorography to English historiography of the Tudor-Stuart era.

When one begins writing about regional study itself, a difficulty one soon encounters is that the only thing that holds the details together is the geographical entity – county or region – around which such studies were organized. But if one digs deeper beneath the surface, one uncovers a host of interlocking links, or even a common tradition, which links the chorographer with the natural historian, the superstitious figure with the 'scientist,' the country practitioner with the urban academic, and so on. In some instances the links are not readily visible, or are tenuous. Occasionally new influences, such as the thought of Sir Francis Bacon, have a great impact and mark a new direction for regional study. But even then the debt owed originally to the old inheritance cannot be denied entirely; as A. Rupert Hall points out, in a different context, even the revolutionary thinkers in various fields of human inquiry 'show some deep attachment or other to an older order of thought which seems almost inexplicable to a latter age.'[1]

It is hoped that to the 'natural scientists' of the seventeenth century – if one may use that somewhat anachronistic term – principally the physicists, secondly the naturalists, has been added a coterie of geohistorians whose names rarely appear among the 'giants' or 'virtuosi' – as Houghton, Westfall, and other recent historians of science call them – of the Royal Society. It is left to the reader to decide if the battle between the ancients and the moderns is worth pursuing: if the answer is positive – I have not yet reached that decision myself, perhaps influenced by Michael Hunter's admonition to the contrary found in his excellent review of Spiller's book on Meric Casaubon[2] – then perhaps my study will make its own contribution to the issue by establishing the role of the regional writers in any such conflict.

The intellectual history of the early-modern period, and not just in Britain itself, was for some time the preserve of political theorists, and while this is not necessarily pernicious, they still disproportionately dominate it. Even the better of the genre often fail to appreciate the diversity and pluralism of past theorists, whose collective endeavours, more than the individual and idiosyncratic contributions of the masters, gave much of the form and shape to past and present interpretations of society, and of politics. J.G.A. Pocock, a historian, shows what can be done with very small changes in the development of theory in explaining which ideas change institutions and even events. Of late, what has been true of political theorists in the field of intellectual history has been true of historians of science. While they are genuine historians, they suffer from a tendency to explain the past of science in reference to present developments. They also deal with a highly internal-ized causation. The same criticisms are, incidentally, applicable to historians of law. What the history of science demands (and legal history, too) is synthesis of the whole history of ideas. Recently, some scholars have aimed at making that synthesis for the history of science: Thomas Kuhn is the notable pioneer, along with Herbert Butterfield. With this study I hope to make at least a small contribution to that development, both to the history of science and to the history of history.

Another of my aims has been to point out the contribution of Baconian science, or at least of one of its branches, to the scientific revolution as manifested in England, if not in the rest of Britain. Although lately recognition of this contribution has been growing, generally scholars – especially those who accept the Kuhnian division of early-modern science into two branches, the Baconian as opposed to the 'mathematical' – tend to undervalue the importance of the non-mathematical sciences. Hall states that the Baconian sciences – for example, the earth sciences – which were 'most vaunted by contemporary propagandists, most notable for their strong "mixed" elements coming from the crafts,' were last to make 'their critical transition to modernism,'[3] following the lead of the Newtonian mathematical sciences. This leaves Baconism and descriptive science primarily notable for leaving 'only the thin legacy of empiricism, *fiat experimentum*, and the story of Francis Bacon catching his death of cold through stuffing a hen with snow.'[4] There is, of course, some truth to all this. But what is most disturbing are the wider implications that natural history – certainly the 'encyclopedic' kind – saw little or no advance before Buffon and von Humboldt, and that Baconism by itself, before its merging with Newtonian science in the nineteenth century, contributed little to the rise of science. Although Hall is careful to state that Baconism 'made important contributions to modern science,' in the next breath he deflates this statement with the words: 'If Galileo, Descartes, Newton (and others) had never heard the name of Bacon, it would have affected history little, before 1700.'[5] The attitude is typical: because Bacon's version of science, supposedly, did not operate on the same high level or 'high road' as did that of Newton, whatever that means, it should be held as somehow inferior to Newton's, and it is therefore farther than Newton's from our modern 'ideal' of science – although there is little agreement as to just what that modern ideal is. But Baconism, as a philosophy of science, if not as a science in its own right, had a pronounced effect on many British scholars before 1700, who recognized its contribution to human knowledge as a whole, and, arguably, to the scientific branch of that knowledge. The perception, then, of the method, or the exact field of study, is as important as the thing itself. If science was to blossom, it was as important in the past as it is today that it have its popularizers – and the Baconian brand was seen by many contemporaries as just that, a science. Their outlook is further justified when one considers that many of the so-called pioneers of science of the eighteenth and nineteenth centuries in fact owed much to their Baconian predecessors. This is particularly the case where natural history is concerned.

Although Bacon envisioned one all-inclusive indivisible sphere of knowledge, paradoxically his followers, while easily passing from one field of interest to another, began to lay the basis for the subdivisions of scientific subject matter that evolved into our own. Kuhn cautiously promotes a search for such natural divisions as may have existed at a particular time, and encourages investigation of the content of these evolving fields of science, which is made difficult because they were in a constant state of flux. At the same time he warns against reading too many of our own ideas about science into past history. Even with due regard, it is not too far-fetched, for example, to call John Aubrey a pioneer of 'archaeology,' Edward Lhuyd a forerunner of the botanists, and Robert Plot an early palaeontologist. If not all these men were conscious of their roles in introducing a certain amount of specialization into science, surely that does not automatically eradicate the existence of the process. If in name, at least, everything within these categories was lumped by contemporaries under the general designation of, say, natural history, or even natural philosophy, it is exactly because to these men – living as they did during the scientific revolution – the possibilities still seemed endless. Although they could look to ancient science for some hints, newly developing scientific techniques and the invention or development of scientific instruments made many shy away from utilizing terminology that might prove too restrictive where some avenues of inquiry were concerned. But if not in name – and this was the case in more than a few instances – then in fact, new branches of science – Baconian tempered, at that – were being developed. Generally, however, for our purposes, one need not be too preoccupied with the question of science as a single enterprise versus science partitioned into different streams.

If Baconism, as Kuhn noted, added little to the development of the classical sciences, it nevertheless spawned a number of new scientific fields, often rooted in prior crafts. During our period of study these remained underdeveloped, lacking 'a body of consistent theory capable of producing refined predictions.'[6] But a start was made; experimentation, collection, and observation of material was required in the first stage, and only then could one hope to arrive at sound generalizations or theories. And, in some ways at least, the Baconian scientist of the seventeenth century could borrow from the mathematical tradition. On more than one occasion we find him using models and instruments of measurement, for example, which he applied to his own use, and which ultimately derived from the mathematical.

In conclusion, in tracing the history of British regional study one theme dominates all others. During this period Britons sought to uncover and to display to the whole world the human and natural resources of their native land. The work of foreigners is considered here only if it left a clear mark upon regional study in the British Isles. Throughout this study quotations have been introduced wherever possible to convey something of the atmosphere and intent of the original sources; punctuation, capitalization, and original spelling have been retained in *nearly* all instances as in the originals. Also, short titles only are provided, except where a fuller title is instructive.

In acknowledging assistance in the production of this book, I first have to go back to my PH D dissertation committee in the Department of History at McMaster University, for, as mentioned before, this book is a much expanded and revised version of my dissertation. Thanks go to the late Professor James Daly, my dissertation supervisor, whose patience, encouragement, and skilful direction, even – or especially – in the months of his failing health, will never be forgotten. Thanks also go to the other committee members: to the late Professor Edith Wightman, for chairing it; to Professors D.J. Russo and R.E. Morton, for their searching and constructive criticisms; and especially to Professor David Barrett, who as second reader added his discerning and experienced historical insight, and who as much as anyone else directed my attention to the questions and issues that needed addressing, and then suggested the appropriate means of doing so. The uncompromisingly high standard of scholarship maintained by the faculty of the Department of History of McMaster University must always claim my gratitude.

In addition I am indebted to the staff of the following institutions for their knowledgeable and courteous help: the Bodleian Library, the British Library, the Mills Library at McMaster University, the Lambeth Library, the PRO, and various other British, Canadian, and American repositories, and English county historical societies and record offices where I conducted my research. I thank the staff of the Thomas Fisher Rare Book Library, University of Toronto, for their assistance and for permission to reproduce in this publication illustrative material from the library's collections. Thanks also go to the following, either for their aid in producing this work or for their generous support of my scholarly efforts in general: Professors Michael Cherniavsky and Royce MacGillivray, of the University of Waterloo; Nancy Lockey, Helen Miklaszewski, Rosemary Trowbridge, and Sharon Gordon for their typing; and especially Mrs Trowbridge and Ms Gordon for their research assistance and moral support, which kept me going even when it appeared that I had reached a dead end. The advice of Lina Procunier and of the reviewers associated with the Canadian Federation for the Humanities – some of which is reflected in remarks made in this preface – also proved invaluable. In fact, this book has been published with the help of a grant from the Canadian Federation for the Humanities, using funds provided by the Social Sciences and Humanities Research Council of Canada.

Of course, the ultimate responsibility for the content and organization of the finished book is mine alone, especially in view of the fact that some of the people mentioned above were involved only in early drafts, or in the dissertation upon which the book is based.

This book is lovingly dedicated to my grandmother, Rosalia Mendyk, to my mother, Laudis Grabarski, and to my uncle, Ted Mendyk, and the rest of my family, whose sympathetic interest in my work has helped me in my academic career. To them, I owe a personal and special debt. Finally, I extend my thanks to the administration and faculty of the University of New Brunswick, Saint John Campus, who aided me in my research efforts when I served there as assistant professor of history.

Portions of this book, in whole or in part or in somewhat different form, have already appeared or are scheduled to appear in print elsewhere. I therefore wish to thank the editors of the following publications for their permission to reproduce such material here: *Sixteenth Century Journal*, regarding the section on early British chorography; *Journal of the Royal Society of Antiquaries of Ireland*, regarding Gerard Boate and *Irelands Naturall History*; *Scottish Geographical Magazine*, regarding Scottish natural historians; *Notes and Records of the Royal Society of London*, regarding Robert Plot; *Omnibus*, regarding early British chorographer-genealogists; and *Old Cornwall*, regarding John Norden.

Portrait of William Camden, from *Camden's Britannia*, 1695 edition

Portrait of Francis Bacon, facing title-page, from Bacon's
Sylva Sylvarum, 1631 edition

Roman coins, from Camden's *Britannia*, 1607 edition

Title page from John Speed *The History of Great Britaine*, 1627 edition

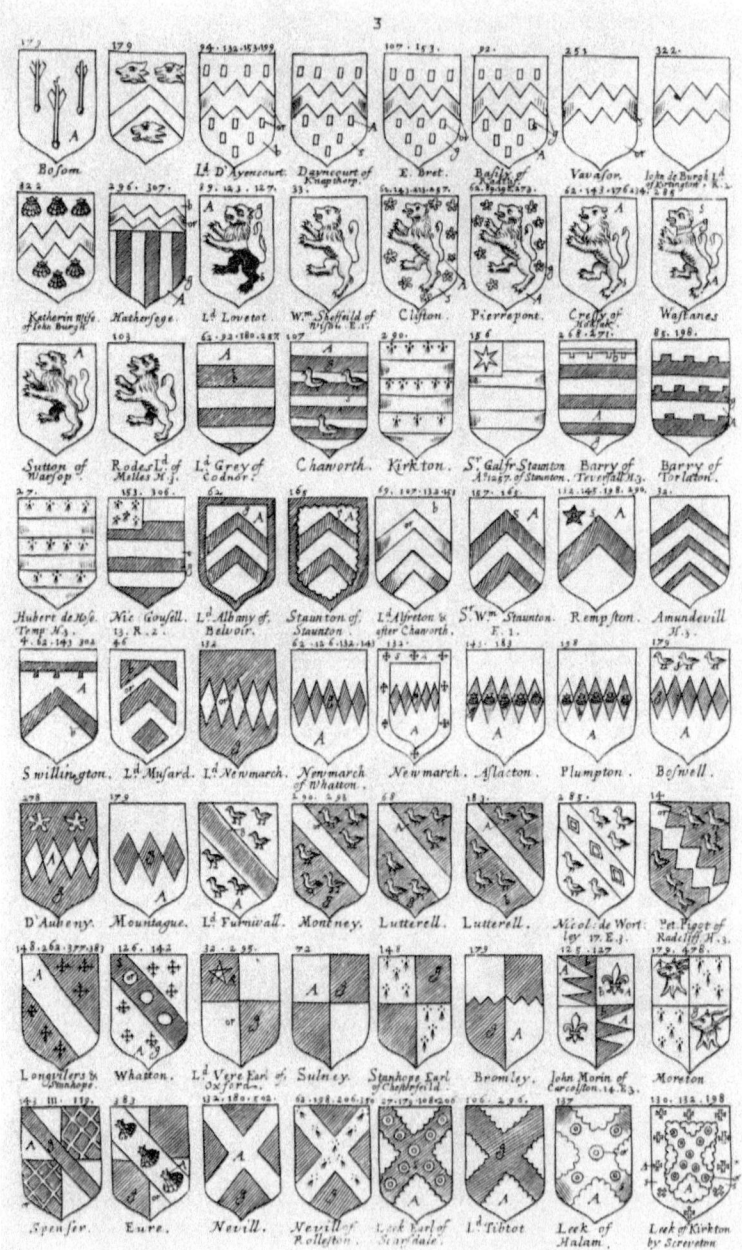

'Index of Armes,' from Robert Thoroton
The Antiquities of Nottinghamshire, 1677 edition

566 | Edward II. THE SVCCESSIONS OF Monarch 48. | Book

EDWARD THE SECOND, LORD OF IRELAND, AND DVKE OF AQVI-TAINE, &c. THE FORTIE-EIGHTH MO-NARCH OF ENGLAND, HIS RAIGNE, ACTS AND ISSVE.

CHAPTER XI.

That the *Minde* is not deriued from *Parents*, certainely the second *Edward* (called of *Caernaruon*) might (if nothing else) abundantly shew, being of a most valiant, wise and fortunate father, an vnlike sonne; yet not to begin our description of his courses, with preiudice of his person, wee will so temper our stile, that by his owne actions sincerely related, rather then by any verball censures the man may bee iudged. This cannot bee denyed, that whereas from the Conquest till his time, *England*, though it endured (by Gods iust iudgements) many bitter, sad and heauie stormes through some headinesse, ambition, or other sickneßes of minde in the Princes thereof, yet had the Men to sway and gouerne her, and those distempers were as the perturbations incident to vigorous dispositions; whereas vnder this *Edward*, who could neither get nor keepe, it seemed to endure the leuities of a Childe, though his yeares, being about twentie and three, might haue exempted him from so great infancy of iudgement, as his raigne discouered.

(2) Neuer came Prince to the crowne with more generall applause then hee: so great hopes of doing well, his victorious Father, *Edward* of *Winchester* had left vpon him, besides the right of succession, whose last warning and terrible adiurations you haue heard; with the vtter contempt and breach whereof, to the destruction of himselfe, and his friends, hee in a manner auspicated his gouernment.

(3) After that *Edward* had in his best manner proui<i>d</i>ed for the affaires of *Scotland*, where (at *Dom frees*) many of the Scottish Lords did their homage to him, as they had to his Father; the first taske which hee gaue of his future behauiours at home,

Page from Speed's *The History of Great Britaine*, 1627 edition

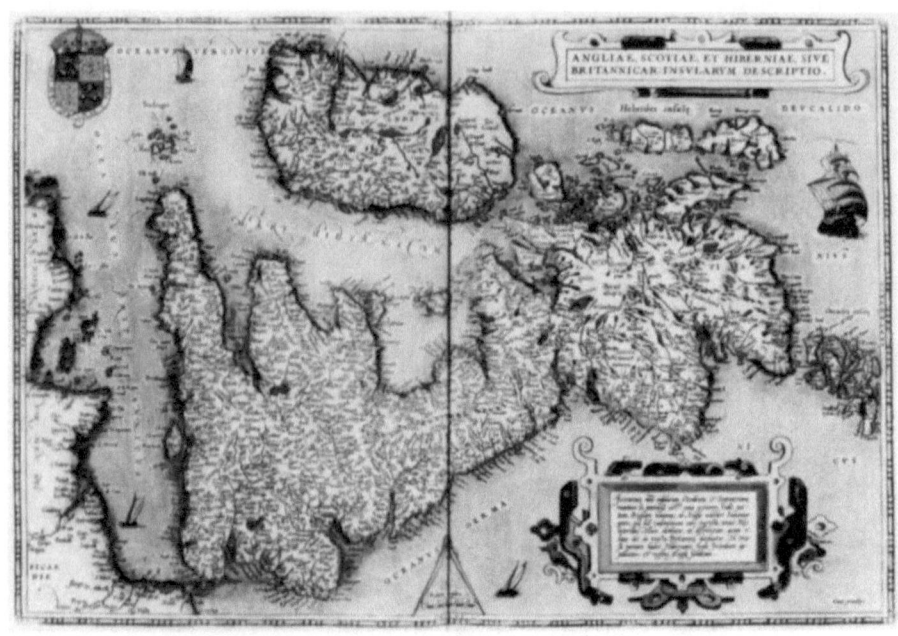

Map of Britain and Ireland, from Abraham Ortelius
Theatrum orbis terrarum, 1589 edition

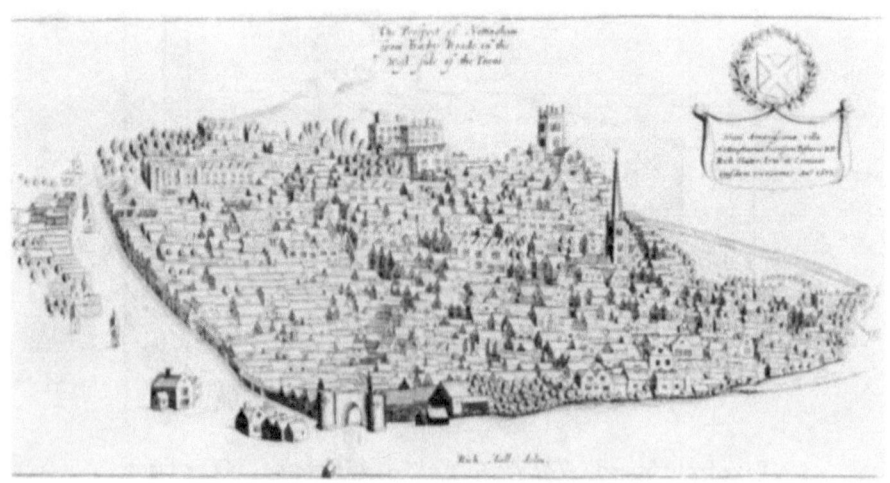

Prospect of Nottingham, from Thoroton's
The Antiquities of Nottinghamshire, 1677 edition

Map of Devon from Camden's *Britannia*, 1607 edition

'Formed Stones,' from Robert Plot
The Natural History of Staffordshire, 1686 edition

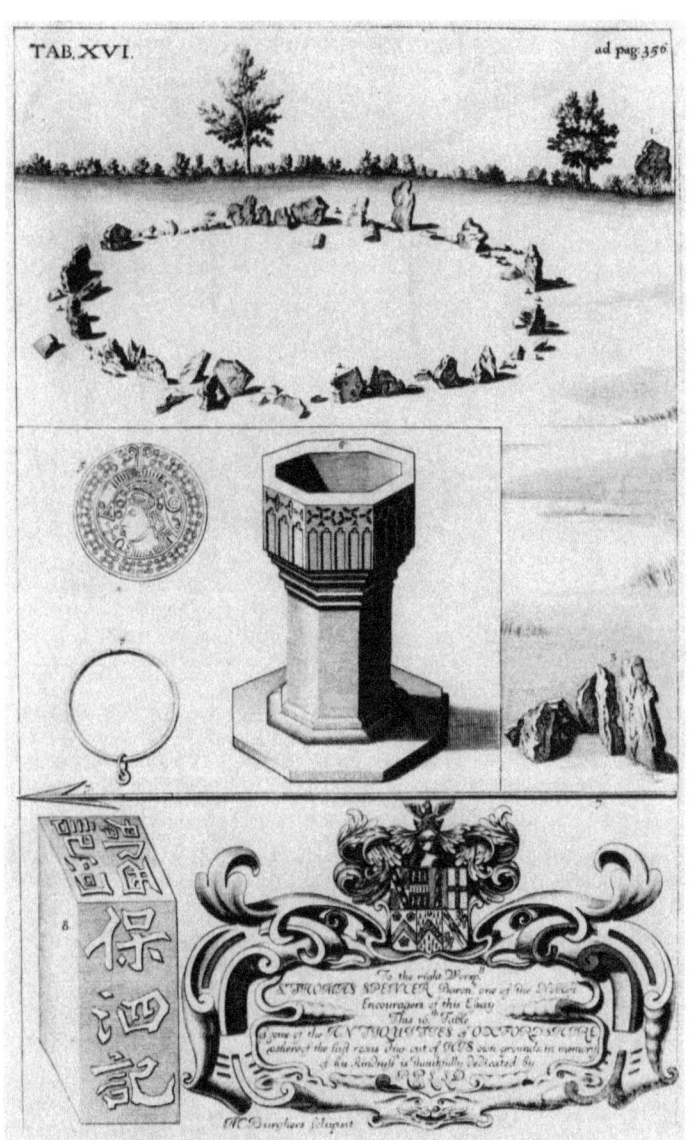

'Antiquities of Oxfordshire,' from Plot's
The Natural History of Oxfordshire, 1677 edition

THE SECOND BOOKE OF THE

3 *Hieracium 6. Clusii.*
Clusius his 6. Haukeweede.

4 *Hieracium 7. Clusii.*
Clusius his 7. Haukeweede.

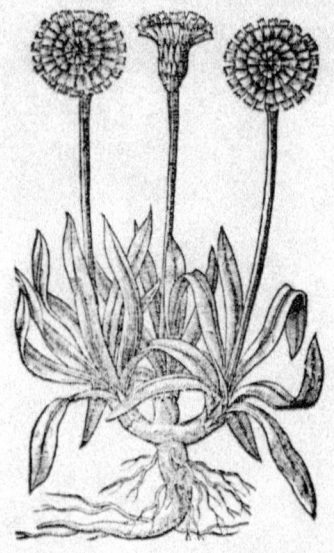

✱ *The place.*

These kindes of Haukeweede according to the report of *Clusius* do growe in Hungarie and Austrich, and in the graffie drie hils, and herbie and barraine Alpish mountaines and such like places: notwithstanding if my memorie faile me not I haue seene them in sundrie places of England, which I meane God willing better to obserue heerafter, as oportunitie shall serue me.

✱ *The time.*

He saith they flower from Maie to August, at what time the seede is ripe.

✱ *The names.*

The authour himselfe hath not saide more then heere is set downe as touching the names, so that it shall suffice what hath nowe beene saide, referring the handling thereof to a further consideration.

✱ *The nature and vertues.*

I finde not any thing at all set downe either of their natures or vertues, and therefore I forbeare to saie any thing else of them as a thing not necessarie to write any experiment vpon my owne conceit and imagination.

Of Lettuce. Chap.34.

✱ *The kindes.*

THere are according to the opinion of the auncients, of Lettuce two sorts, the one wilde or of the fielde, the other tame or of the garden: but time with the industrie of later writers haue founde out others both wilde and tame, as also artificiall, which I purpose to laie downe.

1 *Lettuce*

'SPECULUM BRITANNIAE'
Regional Study, Antiquarianism, and Science
in Britain to 1700

CHAPTER ONE

*Introduction:
'Where the Choir Was ...'*

> Oh, Kitty, how nice it would be if we could only get through into Looking-glass House! I'm sure it's got, oh! Such beautiful things in it! Let's pretend the glass has got all soft like gauze, so that we can get through. Why it's turning into a sort of mist now, I declare! It'll be easy enough to get through --.
> Lewis Carroll *Alice in Wonderland*

> Hauing thus briefly touched the generall, I purpose to proceede to the particular descriptions of this our Britania: wherein (immitating the artificial Painter, who beginneth alwaies at the head, the principall part of the bodie:) I thought it not unfit to begin my *Speculum* ['looking-glass' of] *Britaniae* with Myddlesex.
> John Norden *Speculum Britanniae ... Middlesex*

WHEN ALICE ENTERED the 'Looking-glass House' she uncovered a whole new world of unusual, interesting, and wonderful creatures and things. In some instances, as she soon realized, these were more real than fantastic, and from such encounters and experiences Alice discovered a thing or two about herself and about the real world she came from. In any case, this thrilling story, set amid a colourful background, has fuelled the imagination of and delighted many an avid reader of Lewis Carroll.

In a not entirely different fashion, for centuries previous to the publication of *Alice in Wonderland* the imagination of Britons had already been stirred by another sort of story-teller – the local historian. In most instances this story-teller was a more serious chap, although in its infancy historical accounts often straddled the fine line between myth and reality, occasionally slipping into the former. However, whenever such a story-teller wove together an accurate account by linking local historical study with topography, antiquarianism, and, later on, science, his tale was transformed – as if by magic – into a *speculum* or 'looking-glass' into the *theatrum* of British history. Such works, if written in a lively fashion, served not only to inform but also to entertain the British public, and, when rescued from obscurity, still do so today. Individually and collectively they serve as looking glasses into the British past, detailing the nation's glories, and providing a reflection of the cumulative wealth of a land and its people. Not only, then, do these writings contribute to a fuller understanding of British history, but they also reflect the outlook of past writers and generations, and, dare one say, they contribute to a better understanding of the make-up of the soul of the British today.

This book is concerned mainly with perceptions. It investigates how a group of historically minded Britons, in particular Englishmen of the early-modern era, perceived the wealth and glories of their respective regions of Britain, and why they gave concrete form to these perceptions in writing – thus leaving us with a lasting image of the land, and of the people and their way of life. These writings may be divided into two broad categories: those based mainly on literary remains and dealing largely with human activities, and those focusing on the natural historical environment, based mainly on non-literary sources; antiquarianism often permeates both categories.[1] The focus of the present study is England as the exemplar, although of course the other nations of the British Isles have their own rich traditions; examples of these are treated here, but in general the English situation typifies the whole, and in many instances the English writers were in the forefront of historical – civil and natural – and antiquarian pursuits. The tie between humanism and science is only one issue out of several that this book touches upon. Other backdrops include the general development of English historiography of the Tudor-Stuart period, the development of natural philosophy-cum-science, and even the battle between the ancients and moderns, though the value of this last issue is left to the discretion of the reader. Contrary to what some purists of historiography or science may think, there is no denying that the regional writers, (a term I shall use to refer to the coterie of historians, antiquaries, or scientists who formulated the corpus of regional description) were embroiled in the front lines of issues such as these. But the single most important concern here is bringing to light, or at least making clearer, the work of the individual writers, the field as a whole, and its contribution to a better understanding of the British past.

The first thing one notices is the extent of the antiquarian spirit as embodied in the works of the regional writers, to the degree that such a spirit can be distinguished from a growing historical consciousness, as some historians call it. The antiquarian spirit, like history itself, has inflamed the passions and

Introduction: 'Where the Choir Was ...' 5

stirred the minds of Britons for more than five hundred years, and is much alive today. There are many reasons, some merely practical ones, why people have been drawn to such activity. Some people have simply sought the truth behind the myriad legends and myths of British history, legends and myths that are cherished to this day for their ideals of heroism. Others have been inspired by the mere sight of ancient ruins, which dot the British countryside and town alike, and which symbolize the glorious past of the nation and the inner strength of its people. Inestimable is the number of men and women who, like the seventeenth-century antiquary John Aubrey, while gazing at such forms of decaying grandeur, have been moved by the relics of the past to reflect upon days of yore: 'Where the Choir was, now grass grows, where anciently were buried Kings and great men.'[2] Still others had more immediate political, social, or scientific considerations. For many the prime motive for the reconstruction of a past society or its environment was the desire to draw comparisons with the present in order to educate others. And in some instances the study of 'fragments' or 'memorials' of the past came to be transcribed into, or ancillary to, the investigation of the fragments or memorials of not the human past but that of the earth itself. Whatever the reason, underlying many of these researches was personal curiosity about either the long-forgotten or the sometimes present state of a relic, a ruin, a society, or the natural environment of a place or region; a curiosity that often found an outlet in antiquarian-historical pursuits, and that invariably led to great pleasure on the part of the investigator.[3]

Their efforts in uncovering the Braudelian-coined 'material life' or 'material civilization' of a people have not gone unnoticed by future generations, even if some of these efforts have not been documented in detail. That industrious eighteenth-century historian-antiquary Richard Gough, for example, effectively enummerated the efforts of the 'earliest' antiquary, John Leland, whom he called 'our British Pausanius,' and of others who contributed much even to our current vision of the British past.[4] From Polybius – 'the first who promises a distinct description of the Britannic isles, and the tin-works there' – on, the history, the antiquities, and the natural resources and constitution of the land itself have provided fodder for the researches and speculations of many.[5] Before the Renaissance, however, much of this type of literature, if one is to believe Gough, was 'so stuffed with fiction and [the] marvellous, that one is almost deterred from receiving the little real information they [these works] give.'[6] Still, this was not entirely the situation. Gervase of Tilbury (1152?–1223?), for one, author of *Otia Imperialia* (1209–14), demonstrated in his work a critical, rationalistic approach that, indeed, has been described as 'the earliest representative ... of realistic trend in the XII–XIII centuries chorographical [ie, topographical-historical-antiquarian] treatise.'⁷ Gervase was ahead of his time. Further development, on a widespread scale, had to await the growth in historical consciousness, combined with the refinement of more critical methods and tools of historical and antiquarian evaluation, and, eventually, with the events, ideas, and instruments developed during the scientific revolution, if not the Renaissance and humanism. This allowed for a more critical and accurate assessment of both human and earth history. By the end of the seventeenth century the intellectual climate had clearly changed, to the extent that fewer people were, as Sir Francis Bacon had characterized them: 'Credulous ... and ready to impute Accidents, and Naturall Operations, to Witch-crafts,' or to other seemingly supernatural occurrences.⁸

The seventeenth century truly was a time of enlightenment and experiment, not just in the fields of science and technology, but also in travel and discovery, in politics, and in religion. A.N. Whitehead, calling this the 'century of genius,' stated that 'A brief, and sufficiently accurate, description of the intellectual life of the European races during the succeeding two centuries and a quarter up to our own times, is that they have been living on the accumulated capital of ideas provided for them by the genius of the seventeenth century.'⁹ David Douglas, meanwhile, brought attention ot the amazing flowering of historical study that then took place.¹⁰ The fire had been lit in the previous hundred years or so, to a large degree by the chorographers – men like William Camden, William Lambarde, John Leland – and by others. As Douglas put it: 'The shires of England were one after another receiving their historians.'¹¹ By the time of Queen Elizabeth the groundwork for modern English historical scholarship was being established.¹² Leading the way were the regional writers, who contributed not only to historical research but also, in some instances, to antiquarian and geographical study.¹³ Generally they observed local antiquities and conditions firsthand, by touring either on horseback or on foot, then compiled accounts that express the pride each of them had in his county or region. William Lambarde is usually regarded as the first chorographer of a single county. His *A Perambulation of Kent* (1576) set a standard for others to follow. Lambarde's wish was that at least one man would describe each and every English county, or 'country,' using contemporary terminology. Originally he had envisioned undertaking the perambulation of all the counties (or 'shires') on his own, but was unsuccessful in this goal. Once William Camden published his massive *Britannia* in 1586, however, a national context for a concentration on localities was available; now, 'a man could attend to his own bailiwick.'¹⁴ The 'literary call' echoed throughout Britain. The response, however, was not exactly uniform; only in a few cases was the pattern of Camden's book closely

Introduction: 'Where the Choir Was ...' 7

followed.[15] While Camden's interests lay mainly in reconstructing Roman Britain – a task requiring an extensive knowledge of ancient language and history – for the most part the form adopted by the regional writers for their descriptions was one that they themselves created for their specific purposes. Little attempt was made at theoretical definition, certainly not before the mid-seventeenth century. The formulation was often a practical one, composed of many disciplines, each of them chosen for its ability to illustrate an aspect of the British scene. Just as the German chorographers had already begun, through similar efforts, to construct a great edifice of scholarship meant to survive them as a monument to their civic spirit, so, too, were the British chorographers engaged.[16]

In the task at hand, it is essential first to establish the nature and extent of the tradition from which such work sprang. After an examination of the chorographic literature, we will find that as many of the regional writers began applying newly developed historical-antiquarian and, especially, scientific methods to their own research, regional study, here including county study, eventually became more realistic and practical, and usually focused more on the natural than on the human history of a place. By the second half of the seventeenth century, such study was dominated by the production of these natural histories. Traditional chorographies were also being compiled and published, but their importance to regional study on the whole was diminished, for reasons we shall identify later. I have made no attempt to include here the work of everyone who was involved in the field, because such an effort is well beyond the scope of this book, if not beyond the capabilities of any one author. My objective, however, has been to concentrate on the activities of enough of those involved in regional study to provide as representative a sampling as possible, taking into consideration the diverse nature of many of the works.

Before proceeding further, the terminology employed in this book requires closer scrutiny. The works described here have been referred to by historians, geographers, and others as 'local histories,' 'chorographies,' 'natural histories,' 'topographical-historical studies,' 'regional geographies,' and so on. Sometimes they are simply called 'county histories,' even when the geographical unit being described in them is a region encompassing more than a single county. Because our main concern is to consider regional study in all its manifestations (ie, those centred on any one of or combination of history, geography, antiquarianism, science, etc), for practical and rational purposes the major distinction made here is between the two general types mentioned above: 'civil' history, often in the form of chorography, and 'natural' history. If all these types can sometimes be lumped indiscriminately under the general category of 'local history,' it is clear then that this term is broad enough to include the study not merely of single local communities, but also larger areas such as counties or regions.[17]

In some ways chorography arose out of antiquarianism, which itself began in Italy during the Renaissance. In the fifteenth century Flavio Biondo, in his *Roma Instaurata*, topographically and historically surveyed nearly the entire Italian peninsula, basing his work firmly on the archaeological remains of ancient Roman culture. As source material he drew upon the Latin literature of antiquity, chronicles, maps, inscriptions, historical and geographical accounts, and just about any other pertinent piece of information he could find, thus establishing a method for others to follow. By the sixteenth century antiquarian study had reached a state of popularity that enables it to be described as a movement, extending throughout Europe.[18]

In England, interest in antiquity was stimulated by the search for documentary support in the political-theological strife associated with the rise of Protestantism. The search was on for precedents for the Anglican Church, and in the process there was engendered in the English an increased interest in their past.[19] At the same time the dissolution of the monasteries following the Henrician revolt released a large amount of manuscript material from the monastic repositories, which proved useful to historically conscious men. The dissolution had other important and related effects. Ruins, such as those of decaying or destroyed monastic houses 'proved to be peculiarly fertile in stimulating consciousness of the past and in promoting historical activity.' Margaret Aston thus points out that in the next few decades antiquarian enthusiasm resulted in strong national defences of monastic institutions by Catholics and Protestants alike – sentiments echoed in the seventeenth century by William Dugdale, William Lithgow, and John Aubrey. These men were seemingly haunted by the serene beauty and by the 'echoes of "reverend history"' sounded by such desolate, ruined witnesses to the glorious past.[20] In its early stages the English antiquarian movement was chiefly manifested in the form of chorographic descriptions and in the copying of, and general interest in, public records. Sometimes the two went hand in hand. Others, however, focused their antiquarian efforts on the support of the vital element of the so-called 'Tudor myth,' the historicity of King Arthur as an ancestor of Henry VII. Enmeshed with this was support for the legends of earlier Trojan and British heroes, and for the original upholders of these, including Geoffrey of Monmouth.[21]

History was still largely narrative, and was based on written records such as chronicles; the antiquary employed these, but relied to a much greater degree on coins, ruins, relics, and other scraps of non-literary sources of information, and was less concerned with causes or personalities: 'The antiquary rescued history from the sceptics even though he did not write it. His preference for the original documents, his ingenuity in discovering forgeries, his skill in collecting and classifying the evidence, and above all, his unbounded love for learning are the antiquary's contribution to the "ethics" of the historian.'[22] By the mid-seventeenth century the term *antiquary* developed new connotations, and was more widely used. It was applied to the searching out of ancient manuscripts, to Anglo-Saxon studies, to studies of the Roman occupation in Britain, to studies of genealogic and heraldic topics, and so on. The list was almost endless, especially when one considers similar activity on the Continent. For a time the Elizabethan College (or Society) of Antiquaries, founded in 1585 and lasting into the early seventeenth century, served as a focal point for discussion of antiquarian subjects.[23] Other centres were soon established, including the library of Sir Thomas Bodley, opened in 1602, which served as storehouses of information. Bodley's library was situated at Oxford, which quickly became a clearing-house for antiquarian and historical research, remaining so for the rest of the century; into the 1700s, men like Sir Robert Sibbald in Scotland or Edward Lhuyd in Wales still kept up a stream of correspondence with colleagues stationed in Oxford, when they themselves were not there. By the 1660s, at least, more and more antiquaries originated from, or involved themselves in, the scientific circles of the time, and increasingly they applied what we would today call interpretive and analytical scientific methods – or forerunners of these – to their work. Many, therefore, appreciated the value of archaeological evidence as an autonomous province of historical inquiry, or as existing entirely outside of that province. Antiquarianism, it is clear, was in some form present in both chorography and in the new type of regional study that was being conducted by the 1660s. Certainly, writers in both periods included written sources from archaeological contexts in their studies. Though Momigliano gave them the credit for developing the use of non-literary source material, however, by the eighteenth century the antiquaries had lost any such monopoly, real or imaginary, over it – particularly as the more learned historians appropriated the antiquaries' method of checking literary by non-literary evidence.

May McKisack and T.D. Kendrick, despite their contributions to the field as a whole, are but two scholars who have failed to trace adequately British, or at least English, antiquarianism to its medieval roots. Essentially, McKisack goes back no further than the sixteenth-century antiquary John Leland – thus following in the footsteps of Gough; though acknowledging that earlier attempts have been made to examine the past 'through the intelligent use of archives and visible monuments,' Kendrick characterizes the antiquarianism of medieval England as 'fanciful and prone to fiction,' according to one recent observer.[24] In such scenarios the continuity from medieval to early-modern times is underestimated, and the role of the chronicler is usually described as uncritical or ahistorical.[25] Only lately has the pendulum begun to swing in the opposite direction, with more credit being given to 'a vast area of medieval antiquarian study [neglected by Kendrick] – local history.'[26] Richard Southern has described the antiquarian efforts of the monastic 'historians' of the Anglo-Norman period, which were designed to provide evidence of the continuous history of their houses from Anglo-Saxon times and thereby undermine the impact of the Norman invasion.[27] Unlike Southern, however, Antonia Gransden points out that the later Middle Ages was also a time of important local historical-antiquarian research, conducted mainly by monks, and includes such examples as Gervase of Canterbury's architectural history of Canterbury Cathedral and Matthew Paris's studies of the antiquities of St Albans.[28] One can also go back the other way in history, to Bede, Gildas, and Nennius, as we shall in the next chapter, to find traces of regional description.[29]

Arnaldo Momigliano was among the first to attribute the precedents of modern historical method to the antiquarian 'revolution in historical method,' which gave rise to careful separation between original and derivative sources, and wherein non-literary evidence such as inscriptions and coins was recognized as of value in defending the veracity of historical tradition – the latter often associated at the time with the ancient Greek and Roman historians.[30] The distinction between original and derivative sources became the 'common patrimony of historical research only in the late seventeenth century,' according to Momigliano, and only after the antiquaries, to which group one may add the chorographers, 'showed how to use non-literary evidence,' and 'made people reflect on the difference between collecting facts and interpreting facts.' He further elaborated the distinction between the antiquary and the historian:

(1) historians write in a chronological order; antiquaries write in a systematic order; (2) historians produce those facts which serve to illustrate or explain a certain situation; antiquaries collect all the items that are connected with a certain subject, whether they help to solve a problem or not. The subject-matter contributes to the distinction between historians and antiquaries only in so far as certain subjects (such as political institutions, religion, private life) have traditionally been considered more suitable for systematic description than for a chronological account. When a man writes in chronological order, but without explaining the facts, we call him a chronicler; when a man collects all the facts available to him but does not order them systematically, we set him aside as muddle-headed.[31]

Introduction: 'Where the Choir Was ...'

Momigliano traced antiquarian thought back to the ancient Greeks, for whom history was chiefly political history, and what was left 'was the province of learned curiosity – which the antiquarians could easily take over and explore systematically.'[32] By the seventeenth century the antiquaries were considered, by the Artes Historicae, at least, as either non-historians or, at best, 'imperfect' historians, to whom fell the task of salvaging the relics of the past that were too fragmented to be useful to the students of proper history.[33] At least the antiquaries had a monopoly over the study of classical Greece and Rome – if not over the non-classical and post-classical world, since 'the authority of the ancient historians was such that nobody was yet seriously thinking of replacing them.'[34] In the second half of the seventeenth century this situation changed, partly because the antiquaries, through the use of non-literary evidence, presented a case for the need for new histories on Greece and Rome. Their example was not lost on a new generation of scientifically minded men who sought to expand the parameters of antiquarian investigation, which still concentrated on a type of 'archaeology' not unknown to Plato. This type encompassed genealogy, lists of local magistrates, and so on; less 'scientific,' if you will, with little concern for, say, the exact measurement of artefacts.[35] The use of non-literary evidence, in part, encouraged men towards the production of a 'Book of Nature'; that is, ancient and other authorities that were found in literature were given a back seat to the collection of natural historical material in some quarters, so that the figure of the antiquary and that of the natural historian were sometimes one and the same.

Momigliano argued that the antiquary was in the forefront in 'formulating the rules for the proper interpretation of non-literary evidence' by the late seventeenth and the eighteenth centuries, producing treatises on numismatics, diplomatics, epigraphy, and iconography, of value to both himself and the historian, and in many cases was turning 'himself into a historian,' or helping historians to write 'histories of a new kind.'[36] In this last regard Momigliano made no reference to natural history, but it is clear that the direction given to the natural historians by Sir Francis Bacon, and the example established by the antiquaries, were influential on the work of the natural historians. In general, the distinction between the antiquary and the historian was becoming more and more obscure, and not merely with respect to the study of ancient history. Both the antiquary and the historian sought reliable evidence as well as factual truth. The distinction between the two, however, was never totally eradicated. Many antiquaries, unlike the 'philosophic' historians of the eighteenth century, focused too much on detail and were 'unable to reflect on principles.'[37] Still, the road was paved for the historians of today to recognize that 'the traditional subject of antiquarian research can be transformed into chapters of the history of civilization with all the necessary apparatus of erudition,' even if the image of the antiquary is somewhat tarnished.[38]

If antiquarianism was ever combined at an earlier date with history, then it was in the figure of the chorographer. In fact, one can go back to the medieval monastic accounts. These relied upon documentary evidence, such as that contained in the archives of monastic libraries, papal and royal documents, bulls and charters, and so on, and some even utilized such evidence as inscriptions on tombs. In this respect the antiquarian element was merged with the historical to produce something more than a one-dimensional chronicle of events, and was often designed to commemorate illustrious monks and benefactors. Here we find the beginnings of one motive behind chorographic research, namely the substantiation of one's position, legal or otherwise, through documentation, such as the setting out of one's family background or pedigree. There is the example of the monastic antiquary Thomas Burton:

Burton wanted to record the abbey's acquisition of property. The benefactors' pedigrees are inextricably connected with the descent of the properties which they had given... Two of the pedigrees were included not from the commemorative motive, but as part of Burton's record of a legal dispute. They are the two versions of the pedigree of a bondman, Richard de Aldwyne, who rebelled against Abbot Robert de Beverley (1356–67). Aldwyne argued that he belonged to the king, not the abbot, and produced a pedigree to prove it. But the abbot produced a pedigree showing that Richard descended from one of his bondmen.[39]

Like other antiquaries through the ages, Burton admired the pristine beauty of the representation of the crucifixion in the lay brothers' church, which, as Gransden notes, he attributed to the sculptor's piety – 'he would only carve on Fridays when fasting on bread and water.' Burton also anticipates the future chorographers in his instructive observations on topography-cum-surveying and on changes resulting from silting and erosion. He used charter evidence to trace the original boundary of the manor of Meaux, and described the effects of two mid-fourteenth-century floods. These are vividly depicted; they destroyed an abbey church at Ravenser Odd, a small port, washing away bodies and bones from graves – 'a horrible sight' – and eventually forcing the villagers to flee the site for good, as Ravenser Odd sank into the harbour and became a danger to shipping.[40] In this particular account Burton was careful to distinguish the above settlement from Old Ravenser itself, a manor that was still in existence in his day. In this he was a forerunner of the late-sixteenth-century chorographer John Norden, the surveyor who went beyond the norm in trying to visualize a site as it appeared in the past, not just the present.

Other fifteenth-century figures also exhibited an interest in what seem to be timeless qualities of antiquarian investigation. John Rous took an abiding interest in the value of pictures to the antiquary and the historian, and claimed to have anticipated Bernard of Breydenbach, who was accompanied on a trip to the Holy Land (1483) by an artist who provided a pictorial record of the journey.[41] This being the case, Rous also anticipated the work of Sir William Dugdale, who took an artist with him on his seventeenth-century chorographic tours through the English counties.[42]

It was not long before the narrow interests of the monks were extended into the wider vision of the secular antiquaries. The study of the origin and historical development of monastic houses was soon eclipsed by the study of the antiquarian-historical-topographical features of a place in general – although in the case of Burton, at least, there is evidence of this earlier. What the monks and the secular antiquaries had in common, however, is that, aside from other motives, 'curiosity was often a motive for studying the past. This was history [largely] for its own sake ...'[43] Within a hundred years or so interest in antiquities had snowballed, so that by the time the Society of Antiquaries was founded by, among others, Sir Robert Cotton and the chorographer William Camden in 1586, greater thought was being given to the facilities available for antiquarian research. The chorographer Richard Carew, for one, thought it disgraceful that England lagged behind the Continent when it came to establishing academies for such study.[44] Cotton's immense collection, however, was available to scholars, and Sir Robert was keen in promoting researches in topography and in ecclesiastical antiquities 'in the tradition of Leland and [Matthew] Parker.'[45] Throughout the seventeenth century the Cottonian library remained in private hands, where, in 1667, a servant kept the virtuoso Anthony à Wood locked up for nine hours a day when Wood was conducting research, supplying him with the required manuscripts. The Bodleian Library, though it at first catered largely to students of theology and the classics, held a purpose also for the antiquary or historian, and rivalled Cambridge University Library in resources for the historian and antiquary. Manuscripts at both places were guarded jealously against all 'foreigners.' When in London, the scholar had recourse to the Royal Library, to those of the Inns of Court and the Heralds' Office, and to the primarily theological repositories at Westminster and Lambeth. Outside the metropolis, Oxford, and Cambridge, often the only sources were the libraries of private collectors, the Cottonian being but the best known.

Investigators were often hampered by a lack of catalogues, or were forced to examine sale catalogues for the record of a book's existence; these also served as bibliographies. To the extent that records such as those at Westminster or London could be called 'public,' these were usually widely dispersed among various locations, uncatalogued or poorly catalogued, and otherwise in a disordered state. Still, access to them was in many instances easier than to the closed private collections, especially if one had some influence, personal or financial, with the clerk in charge. Sir William Dugdale, a prominent official, rarely had any such difficulty, and could be frequently located in the Tower of London investigating the rolls for his *Warwickshire* or his *Monasticon Anglicanum*.

The 'public' records of the sixteenth and seventeenth centuries were of two broad types: the 'arcana imperii' – those dealing with 'matters of estate and crown only' – often regarded as private muniments of the king, his courts, and government; and the various legal and financial records – concerning the rights and interests of the crown and subject relating to land, tenures, titles, bureaucratic precedents, and acts of Parliament – of which some were the private muniments of the king, his courts and government.[46] But the fact is that, in truth, one cannot really speak of 'record offices' in that time – not even the Tower or Rolls Chapel. 'Muniments' is the proper word. Likewise, as indicated above, it is improper to refer to 'public' records or archives. First, they were not 'public,' since the records of proceedings, etc, were the property of the clerks and could be consulted only by attorneys upon payment of a fee. It is obvious, then, that even at the best of times research was no easy matter for the chorographer or antiquary of the time.

According to Gransden, T.D. Kendrick not only underrated the role of local history – or chorography – in antiquarian study, but also, by viewing the history of the latter from the sixteenth century onwards as 'a battle between the British History and the new criticism of the Renaissance,' he overrated the effect of the Renaissance on historiography in England.[47] But overrated or not, the Renaissance, or rather humanism, did have its part to play. It certainly had an enormous impact on historical thought in England, especially during the Tudor period. And since history was, like antiquarianism, a vital element of chorography, the effects of Renaissance movements on both of these, indirectly at least, changed the course of the development of chorography. History and chorography often constituted a mixed entity, to the extent that William Wotton, in his introduction to William Lambarde's *A Perambulation of Kent*, called the work 'in substance a historie.' In the field of chorography, such distinctions, and Momigliano's separation of history and antiquarianism, are often blurred where individual studies are concerned.

Introduction: 'Where the Choir Was ...' 15

It is easy to overrate the case – as it is to understate it – for the impact of the Renaissance. For instance, although the new voyages of discovery and other speculative ventures put a premium on precise information (maps, charts, histories, and descriptions) of new lands, as Fussner has stated, and had the side effect of turning some interested parties to the investigation of parts of Britain, if only for sake of comparison, the process may have been a slower one than previously thought.[48] If printing brought a new sense of communication among a growing middle class, the fact remains that the overwhelming bulk of the population throughout Europe remained illiterate, with little access to the new currents. If the humanist historians were the first to break with the 'theological world-history of the medieval chroniclers,' much of the new history became embroiled in the religious polemical disputations associated with the Reformation, often forgoing historical veracity for the sake of defending the writer's religion.[49] Likewise, 'the rise of historical scepticism during the Renaissance was painfully slow,' – for example, with regard to the historicity of Brutus and Arthur; in England it did not emerge victorious until 'late in Elizabeth's reign.'[50] A basic problem was that until that time history, which was regarded as a branch of literature by most ancient Greek and Roman scholars, retained its strong rhetorical tradition among the humanists, despite their achievements in textual criticism, Greek studies, and Roman archaeology. But by the late sixteenth century literary and scholarly traditions of humanism began to go their separate ways, as classical studies became highly specialized.[51] In the case of chorography, there are plenty of examples of rhetoric and invective, although as time progressed these started giving way to a more objective, if drier, stating of the facts. By the middle of the seventeenth century most men were able to discard legends and myths, and to separate history as literature from history as truth. The rhetorical history generated in the Renaissance had lost ground as historical appeals to documentary evidence and factual analysis became more widespread. The educated classes as well were becoming more critical, more familiar with historical facts and historical argument. Then there was the printing press. Though still limited, its audience was growing steadily, and took greater interest in the national past. Both the past and a sense of tradition were placed in a much larger historical perspective by the successes of the printing press.[52]

Chorographers were not alone in eschewing theories of history. The humanists as a whole were more concerned with the use of history as a teacher by example and as a means of presenting ancient moral philosophy in a viable form. Because they accepted the cyclical theory of the 'ancients,' that history repeats itself, imitation of ancient scholars was all that was thought necessary:

16 'Speculum Britanniae'

Imitation was not only a question of literary form, of organization, and choice of subject matter. It was predicted on the assumption that the events of the past can be treated in the same way as those of the present because the factors which determined the actions and thoughts of men in the past are identical with those which determine them in the present. If the humanists were receptive to any general theory of history, it was the cyclical theory of the ancients, which postulated that history was repetetive and that everything that had happened before would happen again.[53]

Although the chorographers – not all of whom could be considered humanists in the strict sense of the word – displayed the type of pride in their regions akin to the civic spirit of the Italian humanists, for example, and although their works were designed as monuments to the past and, sometimes, guides to the future, as a group they were far less concerned with literary style and with delegating research to others. Some of this lack of concern may be attributed to their greater interest in antiquities *per se*, rather than in the use of historical accounts solely as servants to politics; their interest lay where careful scrutiny of the evidence counted for more than embellishment, or the concealment of the facts, where investigation was enjoyable for its own sake, and where there was more potential for new and exciting discoveries, rather than the simple plodding through of ancient texts. Naturally, the British scene, not the Greek or Roman, was the basis of their study; even William Camden's *Britannia* was concerned with the rule of the Romans only insofar as it pertained to Britain itself. Yet all such interest and curiosity ultimately can be traced to a humanistic background, which produced philologists, antiquaries and chorographers, and historians. There 'developed a critical method for the evaluation of historical material long before the adoption of such a method became a basis for the study of history in the nineteenth century.'[54]

One should be careful, however, not to confuse modern historical consciousness, or historicism, with modern historical scholarship.[55] Revisionist historians have attributed the rise of historicism to the study of Roman law in sixteenth-century France instead of to the pre-romantic or romantic movements of late eighteenth-century France, a view that 'correctly attributes a growing awareness of historical and cultural relativity in the sixteenth century to developments in historical scholarship'; however, at the same time, according to one observer, 'it incorrectly identifies this sense of relativity with historicism, thus obscuring the true nature of historical consciousness in the French Renaissance.'[56] For John Pocock, for example, the French legal theorists occupied a middle ground between Momigliano's antiquaries and

Introduction: 'Where the Choir Was ...' 17

the modern historians, having been neither narrative historians nor antiquary-like collectors of facts about the past: 'The nature of their subject forced them to consider questions of the relevance of past to present and even (if in a rudimentary form) of historical development.'[57] For Pocock, historicism evolved from historical scholarship, which in turn had evolved from legal scholarship.[58] According to Zachary Sayre Schiffman, while the rise of historical scholarship in sixteenth-century France heightened the awareness of individuality, the form of which was incompatible with the notion of development, and thus not conducive to the emergence of historicism, yet it helped pave the way for historicism. It did so 'by broadening the historical horizon to include all human creations. The idea of individuality both encouraged the expansion and restricted the range of historical consciousness in the French Renaissance.'[59]

In the course of his discussion Schiffman alludes to the work of Estienne Pasquier, 'the greatest of the sixteenth-century [French] historians.'[60] In some ways Pasquier, though much more erudite and famous today than many of the English chorographers, and though his *Recherches* covered the whole of France, reminds one of the latter. This is evident in his avowed purpose to 'study the antiquities of France,' an enterprise that included 'the scanning of many old books.'[61] Trying to steer clear of rhetoric, Pasquier was committed to the values of quantitative erudition: 'Whatever might be appropriate for an understanding of nature, human culture was accessible only through concrete investigation of specific texts and monuments.'[62] Like his British counterparts, Pasquier attempted to define the place he was treating, an entire country, as a distinct entity, distinct in his case, from Rome, conceiving France as an individuality, 'not by virtue of its development but rather by virtue of the individuality of its component parts, each of which he treated separately.'[63] His treatment, though analytical in appearance, did not require his topics – the origins of France, her chief political institutions, customs, laws, exemplary historical figures, etc – to build upon one another to present a systematic description, so that 'the order of parts could be reversed without changing the whole.'[64] In exploring the laws, languages, customs, and institutions of France, while getting away from a Renaissance-like near-total emphasis on political or military history, and while avoiding the mere narrative of history, Pasquier was able to produce a 'hybrid work, partly historical and partly antiquarian.'[65] In his study, like those of his British counterparts, one can see this 'tendency for the awareness of individuality to grow exponentially.'[66] Indeed, it was more the 'workingman's' scholar and not the 'historical syncretist,' who – if one accepts the revisionist theory – set the stage for 'historicist insights which the syncretists overlooked in their search for universal norms.'⁶⁷ The chorographer as historian-antiquary, like his successor the Baconian regional natural historian, may have contributed more to technique and knowledge than is generally recognized.

Antiquarian investigation, like historical, was not entirely dispassionate. Pasquier's background in law stimulated his interest in his nation's past in the first place, and helped shape his ideas about it. In England by the early seventeenth century, strong political motives were at work, determining the vision of the ancient constitution. Questions involving the antiquity of Parliament and the Commons's place in it encouraged, and in turn were affected by, antiquarian as well as historical determinants. Sir Henry Spelman, not normally thought of as a chorographer, is the epitome of the talented antiquary, the discoverer of the *feudum* (property given in recompense by way of pay or reward for services in the field) and of its importance to the English past. Of the chorographers, William Dugdale and William Lambarde are most closely associated with the 'politicization' of antiquarian research. In this context we shall return to Lambarde later. As for the royalist Dugdale, in the 1670s and 1680s he was largely responsible for a political controversy by claiming that the Commons did not extend back earlier than the reign of Henry III, while emphasizing the birth of the Commons as mainly a tactical power phenomenon.⁶⁸ Much antiquarian study was directed at substantiating some political or legal case or another, as in the Middle Ages. Often this was a positive development. In other instances the reverse was probably closer to the truth. In Cotton's case, politicization of his antiquarian efforts, especially within the context of his official duties, eventually compromised his scholarship, and his writings increasingly began to reflect his changing political attitudes.⁶⁹

For many chorographers, as for Pasquier, their interests and their methods of pursuing them arose mainly out of their formal training and other everyday activities; that is, out of their legal background more than out of humanism and its associated tradition of historical writing – although these cannot always be easily separated. In the end, some contributed to legal history, following in the footsteps of Flavio Biondo, who wrote on Roman offices. As in the case of the legal humanists, not to mention the later figure of Gibbon, much of their inspiration came from antiquarian pursuits such as a reflection upon Roman ruins.⁷⁰ But there were differences from Continental models. For example, in France comparative historical studies were accentuated by the dualism of the law. The awareness of the French 'of the unique characteristics of past societies made them the first modern contextualists.'⁷¹ In England, even if the common law tradition is taken as inhibiting understanding of the history of the law and of feudalism, it at least contributed to the 'essential historical task of research in original records.'⁷²

Introduction: 'Where the Choir Was ...' 19

The study of common law certainly was affected by the rise of the printing press, and the publication of the first modern legal textbooks appeared in England in the early seventeenth century. Before this, law students – including some who would go on to become chorographers – had to rely on a growing number of printed and manuscript sources, such as the critically edited and accurate abridgements of Sir Robert Brooke and Sir Anthony Fitzherbert, Thomas Littleton's *Tenures*, Thomas Phaer's *Presidentes*, and William Rastell's *Entrees*, which constituted the main printed tools for the study of land law in the late sixteenth century.[73] The Inns of Court in London tended to act as clearing-houses for legal works published in the later Tudor period.[74] It was there that many of the lawyer-chorographers studied before entering the legal profession – at Gray's Inn, Inner Temple, Lincoln's Inn, and Middle Temple. Gray's Inn generally contained the highest percentage of the sons of the nobility and gentry, the others catering largely to the sons of the lesser gentry.[75] At the Inns of Court one could complete the process of a humanistic education, one that came to incorporate a wider range of social and cultural activities, training in politics, and the opportunity of establishing contacts for future employment, possibly in the central courts and departments of state.[76] Most of our chorographers, however, returned to their home turf throughout Britain to practice law, to serve as justices of the peace, or simply to settle into that comfortable life-style generally associated with the country gentry. Much of their training at the Inns was of a practical nature, designed to allow a country gentleman to run his holdings efficiently. But in addition to teaching law, the Inns had become 'a city of small, independent schools specializing in subjects such as science, medicine, mathematics, geography, astronomy, cosmography, engineering, mining, and ordnance,' a type of education also perfectly suited towards involvement in chorography.[77] 'The study of conveyancing, writs, the forms of action, and legal processes was not alien to the city's practical, professional climate,' and, in fact, even interested some law students in scientific problems.[78] It was also at the Inns that the law student became familiar with land law, real-property law, and title search. For many, their chief concern was not the constitutional issue related to the Crown-versus-Parliament political dispute, nor even the administration of local government, but rather *title search*. For the real-property barrister title search loomed large in the late Tudor and early Stuart period, though it was of less importance to him by the later part of the seventeenth century. By then it was coming into the hands of the attorneys, because preservation in good order of title-record was much improved by landholders. Secondly, there were a good deal fewer clouded titles as monastic lands shook down in descent, and less wild conveyancing, as destructible forms of settlement were repeatedly destroyed, so that by the end of the century the field was left to Orlando Bridgman's 'strict settlement.'[79]

A general humanistic education and a training in law were perfect recipes for formulating an interest in antiquities, be they Roman ruins or simply the antiquity of a political, legal, or religious institution or of a family lineage. One example is William Lambarde, the chorographer of Kent. Lambarde entered Lincoln's Inn to study common law in August 1556 at the tender age of nineteen, and was called to the bar in 1567. While at the Inn he studied Anglo-Saxon and other ancient manuscripts and charters, one result being the publication of his *Archaionomia* (1568), a major early effort in the translation of Anglo-Saxon laws into Latin. This study contained sections on the customs of the Anglo-Saxons, their invasions, their kingdoms, and the laws of William I and Edward I.[80] Lambarde made heavy use of manuscripts in the library of Matthew Parker. He also sought the support of a wealthy patron, dedicating the work to Sir William Cordell, a bencher of Lincoln's Inn, master of the rolls, and distinguished in the interpretation of the law. It is noteworthy that, in explaining the motives behind this scholarly endeavour, and in explicating the continuity between the laws and customs of the Anglo-Saxon past and his own day, Lambarde stressed the importance for Englishmen of understanding their Anglo-Saxon past in order to understand their own society.[81] Like Pasquier in relation to France, Lambarde took pains in establishing the uniqueness of English society and culture, which he demonstrated by reference to her own distinct institutions, laws, customs, and so on, as they evolved through history, with the Norman invasion making nary a ripple in the waters of continuity, in law at least. The Tudor break from the papacy and the resultant establishment of a truly Anglican church, part of a growing feeling of national self-identification, provided Lambarde with added inspiration: 'Lambarde, who was of that English movement, not only lauded the kingdom's Anglo-Saxon heritage but he also frequently expressed vivid contempt for other national groups.'[82] This attitude is evident in his lament over the loss of Calais to the French, and in his interest in the etymology of the English language – although he agreed with Erasmus that spoken English sounded like a dog's barking: 'baw, baw, baw.'[83] Ever the typical chorographer-antiquary, 'he was unconcerned about the why and how of an event and was interested usually in discovering only what had occurred and when,' and as a lawyer 'he was "precedent-minded," not "historical-minded."'[84]

In August 1579 Lambarde was named to a new Kentish commission of the peace, and the necessity of acquiring an intimate familiarity with his new jurisdiction further propelled him into chorography. His duties included the gamut, from maintenance of law and order to road repair. Lambarde's antiquarian knowledge also stood him in good stead, as in the situation when he was asked to arbitrate a dispute involving the beacon watch, because it was well known that he had written a section on the beacons in the *Perambulation*.[85] His knowledge of ancient laws and customs served him well when it came to gaol deliveries.[86] His publication of the *Archeion: or, a Discourse upon the High Courts of Justice in England* (completed in 1591 but not published until 1635) further entitled him to the designation of 'prince of legal antiquaries.'[87] In this work the origins of Parliament are traced back to the Anglo-Saxon witan, with the author arriving at the conclusion that in his own day the king in Parliament could only act with the consent of Parliament. Lambarde is but one of the many men who combined a legal training with an interest in antiquarianism and history – even if he was more 'precedent-minded' than most – and with chorography in general. In his writings the legal framework tends to prevail over the topographical by a considerable margin. But this does not take anything away from the right to call Lambarde 'chorographer'; by the early seventeenth century the idea of chorography was extended beyond the ancient Strabolean or Ptolemaic one to include a diverse range of topics.

Chorography, then, is a key term that requires further explanation. Here the prefaces and introductions of the chorographic works are of assistance. They often set out an author's purpose and contain methodological discussions that attempt to give some articulation and coherence to the form of the work. Many of these discussions are variations on the first chapter of book 1 of Ptolemy's *Geography*, in which the limits of geography are determined.[88] For example, in the opening chapter of *A Geographicall and Anthologicall description of all the Empires and Kingdomes* (1607), entitled 'De Geographia,' Robert Stafforde referred the reader to Ptolemy: 'Geographie is an Imitation of the picture of the whole earth, with those things which are annexed thereunto. Ptol. Lib. 1 cap. 1.'[89] British and Continental chorographers were well aware of the mathematical nature of Ptolemaic geography, the purpose of which was to fix positions and to establish relationships of places on the earth's surface. It dealt with the entire earth, whereas chorography treated small portions of it and was less concerned with determining mathematical relationships than with exact description of the place, or region. Its purpose was to render a 'true likeness,' and so required the talents of an artist, one capable of 'painting the landscape' in words. Chorography's locally descriptive nature is embodied in the following definition, by the magus John Dee:

Chorographie seemeth to be an underling, and a twig, of Geographie: and yet nevertheless, is in practise manifolde, and in use very ample. This teacheth Analogically to describe a small portion or circuite of ground, with the contentes: not regarding what commensuration it hath to the whole, or any parcell, without it, contained. But in the territory or parcell of ground which it taketh in hand to make description of, it leaveth out (or undescribed) no notable, or odde thing, above the ground visible. Yea and sometimes, of thinges under ground, geveth some peculier marke: or warning: as of Mettall mines, Cole pittes, Stone quarries, etc. Thus, a Dukedome, a Shiere, a Lordship, or lesse, may be described distinctly. But marveilous pleasant, and profitable it is, in the exhibiting to our eye, and commensuration, the plat of a Citie, Towne, Forte, or Pallace, in true Symmetry: not approaching to any of them: and out of Gunne shot, etc.[90]

Arthur Hopton, meanwhile, recognized that chorography basically amounted to what we would call topography today:

Topographie (with some called Corography) is an Arte, whereby wee be taught to describe any particular place, without relation unto the whole, delivering all things of note contained therein, as ports, villages, rivers, not omitting the smallest: also to describe the platforme [plan] of houses, buildings, monuments, or any such particular thing: and therefore a Topographicall description ought to expresse every particular, which caused me the rather to call this instrument the *Topographicall Glass*, as being most apt to describe any monument, Tower, or Castle, any Mannour, country, or kingdome, so do we briefly describe England thus.[91]

Other chorographers offered variations of the above definitions, some of which did make a distinction between topography and chorography; William Pemble and Peter Heylyn saw the former as focusing on a smaller unit, for example, the town as opposed to a province.[92] As will be seen later in this book, Strabo's version of chorography – which was less a cartographical subject than it was for Ptolemy, and more concerned with history – was added to that of Ptolemy, by the British chorographers. Sometimes the two remained distinct; in most such instances the Strabolean was favoured, as it allowed for the description and inclusion of topics – eg, genealogy, antiquities, customs, etc – not allowed for in Ptolemy's more mathematical construct. And even then, the early-modern exponents of it allowed themselves plenty of licence to go beyond most any topic dreamt of by Strabo. But the 'ancients' had set the original guidelines for regional study, through chorography, to which were added the antiquarian and historical elements developed in the Renaissance.

The men who investigated neither the town nor the region but the whole

Introduction: 'Where the Choir Was ...'

world were generally known in Britain as 'cosmographers,' rather than as 'geographers.' Most of their information was derived from classical authorities such as Pliny or Tacitus, or was merely second-hand. There was little antiquarian motivation in their work, which settled more on mathematical and cartographical Ptolemaic-like subjects than did most of the chorography produced in Britain. According to Dee, cosmography 'wholly and perfectly marketh description of the heavon'ly and also the elemental part of the world.'[93] Cosmographies were part of a tradition imported from Germany.[94] In England, it seems, rare was the man who indulged in both cosmography and chorography.[95] Thomas Blundeville clarified these two terms in *His Exercises* (1594), in the course of a dialogue with an inquisitive pupil:

What is Cosmography?
Cosmography is the description of the whole world, that is to say, of heaven and earth, and all that is contained therein ...
What is Chorographie?
It is the description of some particular place, as Region, Ile, citie, or such like portion of the earth severed by it selfe fro the rest.[96]

To summarize, then, the chorographers went beyond mathematical and cartographical survey to include just about anything of antiquarian or historical relevance in their studies. Some, however, preferred to leave accounts of great events and of persons to the 'civil' histories, incorporating only material of limited local significance. Though William Camden did treat in detail many significant historical events in his *Britannia* – an instance of chorography's encompassing an area as large as an entire country – he declined to pursue the subject of Edward II's murder in Berkeley Castle because: 'I had rather you should seeke in Historians, than looke for it at my hands.'[97] Camden saw his work as chorography rather than as history; or, at least, he stated, 'neither is it any part of my meaning now to write an Historie, but a Topographie.'[98] The question of exactly how much historical information could or should be included in a chorography weighed upon the minds of many. According to Lambarde, his purpose 'specially' was 'to write a Topographie, or description of places, and no Chronographie, or storie of times (although I must now and then vse both, since the one can not fully bee performed without enterlacing the other) and for that also I shall haue iust occasion hereafter in the particulars of this Shyre, to disclose many of the same ...'[99]

As the seventeenth century approached its third decade, the topographical-historical element in many chorographic accounts seemed to be largely superseded by a growing emphasis on the incorporation of genealogical and heraldic information, and the etymology of place-names. This trend away from the strictly topographical, however, in fact existed in British chorography right from the earliest, as we shall see. Thus the early-modern chorographer had to be part historian, part genealogist, and part antiquary as well as, decreasingly, part topographer. As historian he would turn to written records for his information; as antiquary, to other physical remains, such as ruins and coins.

The *Britannia* and Lambarde's *Perambulation* opened up new ground for others to follow, the latter, for example, influencing Lambarde's 'loving friend' John Stow to write his famous *Survay of London* (1598).[100] The chorographies produced then were, in a sense, the forerunners of the Victoria County series. When well-written such a work could:

[Unfold] the rolls of family attachment, family possessions, and family distinctions. It leads you to the venerable pile of ruin, ivyed with time, and verging to destruction; an aweful picture for declining years! It shews you the beautiful retreats of the rich and noble, in which are deposited the learning and the labours of past ages. It accounts for the remnants of antiquities found among us; and, in numberless instances, it expands the mind, amuses the understanding, and is often useful in the division of property. In fine, its displays of the natural productions of the earth, and of the arts, are given for the amusement and utility of the present and after times.[101]

Chorography dominated regional study at least until 1656, when Sir William Dugdale published his *Antiquities of Warwickshire*, a large illustrated folio of more than eight hundred pages.[102] But by that time it was already on the wane. The original College of Antiquaries was no longer in existence, and the debilitating effects of the English civil war put a damper on the kind of research necessitated by chorography. Although *Warwickshire*, in its prodigious scholarship, which was perhaps unprecedented in the field of chorography of all time, set a standard of learning for later scholars, a new type of regional study was in the ascendancy. Other works, particularly after the Restoration (1660), devoted much more attention to natural history than did the chorographic writings. For example, John Aubrey of Wiltshire began to compile a study of that shire in 1659 (completed in 1685), which ignored many, though not all, of the conventions followed by the chorographers. His interests reflected the scientific concerns of Sir Francis Bacon. Hence Aubrey wanted to compile a soil map of all of England as well as a county land-use map – his marginal note for the latter is considered the earliest record of such a design. As Emery and others have made clear, regional study was used here in the new context of soil or 'land' studies carried out by many members of the Royal Society after 1660. Others active in this or related activities include Dr Robert Plot, the first keeper of the Ashmolean Museum at Oxford. Plot began collecting material for his *Natural History of Oxfordshire* (1677), exchanged information with Aubrey, and like the latter was less concerned with chronicles, legends, etymologies, and the like than with the county's natural history. In essence, a growing scientific attitude was being injected into regional study, as human history was being overshadowed by natural history and as antiquarianism of the type derived from the Renaissance evolved into a new type based upon scientific methods of inquiry. Most enlightened observers of the time were careful to distinguish the 'natural histories' from the 'chorographies,' which, though still being produced, began to represent more the armchair type of work – often designed more for sheer entertainment than for disseminating knowledge. A new generation of regional writers undertook studies of the weather and climate, fossils, subterranean as well as superterranean earth formations and configurations, the flora and fauna, and so on. 'Scientific antiquarianism' was being advanced because men were conscious of the similarities between fragments of the natural and of the man-made past. Some of the chorographers had already taken note of natural phenomena, Carew coming to mind especially, but generally natural history played a little role in their studies. But as antiquaries they led the way in showing men how to use non-literary evidence, and the ties go deeper than this.[103] The results produced by the employment of scientific antiquarianism in itself went a long way in encouraging others – in particular the historians alluded to by Momigliano – in devising rules for the proper interpretation of non-literary evidence, and in one way at least in bringing history and antiquarianism closer together.[104]

The lead in the new direction came from several quarters, not least of which was the Royal Society; as Lytton Strachey observed in his essay on John Aubrey, the foundation of this illustrious body marked 'the beginning of the modern world.'[105] In its aim of improving natural knowledge, the society helped set the foundations of the empirical method of modern science.[106] It fervently promoted, or claimed to promote, the ideas of Sir Francis Bacon, one of the first to put natural history on the same plane as civil history. Bacon thought it was necessary to expunge the mass of unrelated material up till then usually found in 'natural historical' study. For Bacon it was 'evil' to waste time 'in investigating and handling the first principles of things and the highest generalities of nature; whereas utility and the means of working result entirely from things intermediate. Hence it is that men cease not from abstracting nature till they come to potential and uninformed matter, nor on the other hand from dissecting nature till they reach the atom; things which, even if true, can do but little for the welfare of mankind.'[107]

Bacon's concern for the purification of natural history can readily be appreciated if one looks at just one example of the type circulating in his day. Thomas Hill's *A Contemplation of Mysteries* (1571) contains accounts of the properties of comets, atmospheric phenomena, earthquakes, etc. But the author had a tendency to form conjectural opinion about strange and apparently unexplainable occurrences, legends, and other topics that seem to us outside the realm of natural history. At one point Hill stated that 'In the yeare of our Lorde. 1553, was heard a woefull crye, Saying woe, woe, twise together: there was heard also the sounds of Belles, and the noyse of Trumpets: and the same tyme there happened at Duringia, that the trees, and herbs, sweate bloud ...'[108]

But Britain did not lack its share of good, recognizably natural-historical productions of quality. William Bourne's *A Booke called the Treasure for traveilers* (1578), a manual for the use of voyagers and wayfarers, contained a number of natural-historical topics, including 'the naturall causes of Sands in the Sea and rivers, and the cause of the marish ground, and Cliffes by the Sea Coasts, and rockes in the Sea.' Bourne also took note of fluvial erosion along river banks, commented upon the effect of earthquakes, and speculated on the origin of sea cliffs.[109] And there are still other, better known examples. William Turner, for one, is sometimes considered the first scientific naturalist; his main interest was in plants and their medicinal properties. Related works include books on husbandry and farming, for example, Fitzherbert's *Boke of Husbandrie* 1523), and Gerard's noteworthy *Herball* (1597). Continental naturalists were also adding knowledge to the field, men such as Georg Bauer Agricola, Andrea Cesalpino, Conrad Gesner, and others. Nevertheless, as Bacon realized, much work remained to be done; there was room for improvement, as the Royal Society well knew.[110] It promoted field studies, supplemented by analyses of the responses to its widely distributed queries on husbandry, natural resources, etc. The techniques it established for the documentation of the natural history of Britain remained in effect well into the next century, and were applied to parish as well as county and regional study.[111] Overall, the activities of the society were 'far closer to modern fieldwork' than those of the chorographers.[112] Yet in some respects, especially in the area of antiquarian thought, elements of the earlier tradition in regional study were retained in one form or another.[113] Most of these traditions were modified by the society through its promotion of the new ideas and techniques associated with the scientific revolution.

'Science' was now being differentiated from knowledge in general, 'whether that knowledge is to be derived, as Aristotle had taught, by straight deductive logic, with the geometry of Euclid as a model; or whether, as Bacon was the first to apprehend, it must gradually evolve, using observation and experiment, by refining and clarifying ... partial truths.'[114] As the seventeenth century progressed, regional study increasingly became the domain of those able to discard the Ptolemaic and Aristotelian systems that had dominated Western thought through most of the Renaissance. But though the process had begun, not all men were yet able to distinguish science from all knowledge. Because antiquarian material remains such as ruins or megalithic sites could still be equated with natural science under the category of knowledge, for this very reason men of science studied antiquities as well as natural history and almost anything else that held out a curiosity value – for the general population at the time of the Restoration took a great interest in the unusual and the macabre. In view of this fact, it is unfair to deem the scientist or antiquary – as virtuoso – as a comic type. As far as the Royal Society itself is concerned, it is much more accurate, and polite, to say as Dorothy Stimson does that '[the] Royal Society was in no sense a professional organization in its origins but rather a gentleman's club for the discussion of scientific matters with the whole world as its field of activity.'[115]

But these 'gentlemen,' many of them amateur scientists who were followers of Francis Bacon, had a role to play in the rise of modern science. Paolo Rossi, in *The Dark Abyss of Time*, disagrees with those who undervalue the importance of the 'Baconian' sciences, pointing out that even Thomas Kuhn has revised his thinking to include the value of the Baconian or experimental tradition to the scientific revolution.[116] Rossi's concern is largely with Vico and the earth sciences as they developed in the late seventeenth and early eighteenth centuries. He suggests that objects such as 'fossils' were susceptible to various readings dependent upon the polemical interests of the interpreter, and that while Robert Hooke made fossils the 'medals, urnes, or monuments of nature,' the 'greatest and most lasting monuments of antiquity,' Vico reversed this precedence whereby 'fragments of antiquity' were solely human relics; in the new science human cultures alone had a certain and reliable history.[117]

It is not my intent, in the present study, to launch into a detailed investigation of Rossi's subject matter, but rather to mention that to the limited degree that the regional writers indulged in theorizing, the question of the 'meaning' of fossils serves as good example. 'Meaning' can be interpreted in two ways: the first involves their origin and nature, and the second involves their relevance to other disciplines and to the shaping of man's image of his world. Since Rossi, and Martin J.S. Rudwick in *The Meaning of Fossils*, have done considerable work in both these areas, either for a particular period (Rossi) or for successive ages (Rudwick), only a brief run-down of the situation involving questions of earth history, particularly the 'fossil' issue, is required here.

Renaissance scholars, operating within Aristotelian, Platonic, and hermetic explanations, generally regarded 'fossils' as entities in their own right on the scale of being and, within a series of correspondences, as similar to other unusual objects found in the earth, such as gems. Fossils, then, were classified more as common elements, than as aberrations, of nature. At the same time, naturalists such as Agricola, Ulisse Aldrovandi, Andrea Cesalpino, and Conrad Gesner, were noting the close resemblance of certain fossils, or 'figured stones,' to living creatures. But until the seventeenth century few, if any, people attempted to prove scientifically the hypothesis current among some naturalists that fossils were organic in nature, thus leaving 'fossils' within the 'classification' of the mineral kingdom.[118] Fossils were still generally thought to be stones fashioned by nature in imitation of leaves, fish, insects, and other organic bodies, being *lusus naturae* or *lapides sue generis*, attributable to a 'plastic virtue' in the earth, or to occult influences associated with celestial bodies. What retarded a new concept of the nature of fossils, among other things, was the belief that non-living matter could spontaneously 'generate,' similar to living matter. Metals and minerals could thus grow in the earth out of base material, and stones could grow from 'seeds' that had penetrated the earth's surface: 'The "seed" of a fish, endowed with its specific form, can grow from the material "stone" and generate a fishlike object in the rock, the "matter" of which, unlike a living fish, is, precisely, rock. In other terms, the form of a genuine living organism is combined in fossils with the stonelike matter that is typical of all other "fossils" (or objects-that-lie-beneath-the-ground). The combination is caused by spontaneous generation, or else by the presence of specific "seeds" which have penetrated beneath the earth's surface.'[119]

Various substances were thought to be transformed into stone by a liquid *succus lapidifucus* or *lapidescens* (or 'petrifying juice'), which circulated under the earth's crust, according to Agricola; or by its gaseous form, an 'aurea petrifica' or 'lapidifica.'[120]

The 'seed' or seminal principle, in one form or another, goes back to Aristotle. Interpretations that were more nearly hermetic or neo-Platonic focused more on the 'plastic nature' or 'plastic virtue' theory, which involves the tracing of similarity between certain fossils and some living organisms to 'the secret relationships of similitude that penetrate and give form to every part of the universe.' Plastic virtue governs the growth of organisms, can work within the earth, and 'brings into being those affinities and those "images" that constitute the universe as a unified reality.'[121] Something like the 'virtus lapidifica,' with its power of coagulating, hardening, and reducing matter to various forms, was often taken as evidence of God's handiwork in maintaining the world in harmonius equilibrium.[122] Soon one finds the steady proliferation of hypotheses built upon combinations of mystico-chemical interpretations of Genesis and the Creation, often of a great cosmogonic nature.

By the late seventeenth century, fossils were sometimes, though not always, interpreted as being the remains of organic creatures killed in the catastrophic biblical (or Noachian) flood, dispersed throughout the world, and buried in mud where they became petrified. Their investigation and interpretation 'gave meaning to seventeenth century attempts to find natural philosophical grounding for evidential theology and to impart a sense of directional history to the world view,' and were linked to other contemporary problems in the understanding of nature.[123] Dr Robert Plot, in fact, was one of the last hold-outs in supporting the theory that fossils were generated by a plastic virtue. He was at odds with Robert Hooke's contention that fossils are not 'sportings of Nature ... or the effects of Nature idely [sic] mocking herself,' but remnants of actual living creatures that were least liable to decay, and that were petrified and preserved in rocks.[124] Hooke avoided the flood as being the agent for the existence of fossils, because in his view they could not have been carried to the summits of mountains through a diluvian process, and because this occurrence in fact is supported neither by Scripture nor by profane history.[125] In Hooke's view, the answer ultimately had to lie in a series of catastrophic events, such as earthquakes, eruptions, inundations, and so on, which had altered the earth and life on it since the Creation. In an elaboration of his reasoning, Hooke determined that the Noachian flood had in fact been in essence a great earthquake that was associated with the submergence of land and emergence of new land masses from the sea bottom. Because all such land masses could not have been created simultaneously, there still remained the idea of other earthquakes throughout earth's history.[126] Plot questioned Hooke's postulation of a series of floods or earthquakes, since the historical evidence, 'the Records of time,' held little support for such catastrophic and frequent 'concussions.'[127]

Hooke certainly was not the first to hypothesize on the organic origin of fossils, a development partly attributable to 'advances in animal and plants anatomy, the discovery of the relation between morphology and physiology of the organs and the local observation of geological and geographical changes.'[128] For example, Morello describes the case of the Neapolitan botanist Fabio Colonna, who, in his *De glossopetris dissertatio* (1616) demonstrated 'experimentally' the organic nature of fossilized sharks' teeth, or glossopetrae.[129] In this Colonna was a forerunner of some of the natural historians of the second half of the seventeenth century: in his emphasis on observation, especially of local geographical modifications, which he thought were responsible for the conservation and fossilization of marine animal remains, and on experientialism; in his use of experimentation; and in his critique of Aristotelian and pseudo-Aristotelian theories on the formation of 'figured stones.'[130]

As for Colonna, his studies never led him to the idea of the geological history of the earth; that is, he did not realize that in the past the earth might have been substantially different or have contained different species than in the present. He did not use the localization of fossils in strata to understand the sequence of geological events.[131] Part of the reason why he did not is that he was restricted by a biblical chronology that compressed the history of the earth and humanity into a few thousand years. Hooke, also, operated within this short time span of sacred history, as did many other regional natural historians — even if some expressed doubts about it. John Ray, even if inadvertently, and others started men rethinking the time span of the earth. And, while the regional natural historians were not always explicit about their views on this subject, it should be remembered that they conducted their investigations at a time when the great controversy involving the 'world makers' was swirling around them. In Ray's case, since fossils were localized and in some places occurred in large beds, he rejected the view that the Noachian flood was the active agent in the process, for it would have washed the organisms from the mountains and deposited them evenly. A rising of the mountains from sea level was possible but, according to Ray, would have required more than the four or five thousand years that had elapsed since the Creation. Ray could not believe that the entire earth was once covered by the sea — and again, this was hard to reconcile with Scripture. He had other objections to the notion of fossils as organic remains, but the important point is that others who were familiar with his theories realized that the difficulty lay in having to work within the confines of a short chronological span of earth history.

The essential point is that men started to realize the need, in natural history, to study 'the alterations and transformations that nature undergoes during the course of time.' So Hooke believed that science should investigate how 'shells' 'came to be disposed, placed, or made in those parts where they are, or have been found,' even if 'it is very difficult to read them, and to raise a Chronology out of them.'[132] But before this stage could be reached, the spadework had to be done; the regional natural historians first had to participate in the Baconian project of describing and classifying nature. By the end of the seventeenth century, in many instances, this activity became concurrent with those espoused by Hooke.

Also, in spite of Vico's later pronouncement, natural historians were taking a cue from their civil counterparts, seeking solid evidence or 'documents' for the past history of the earth. 'The criteria of verification that are valid for human history can thus be applied to fossils as well, since they are the "Medals, Inscriptions, or Monuments of Nature's own stamping."'[133] Men could seek traces of the history of nature in civil or even in human history, even if, say, catastrophes were disguised within the context of 'fabulous' stories, which accounts the history of fossils could document.[134] For that reason alone, Hooke and other scientists could not always disregard mythological or fabulous accounts, nor therefore the literature in which they are found.[135] Be it in the realm of human or of natural history, the historical record was incomplete. But where the natural historian had an advantage over the civil was that 'despite the absence of such human records, the "natural antiquary" possessed trustworthy documents provided by nature: fossils. And nature, unlike man, did not falsify evidence, i.e., there are no *lusus naturae*.'[136] Often the regional natural historians covered in this study left the theorizing to Hooke, Ray, and other prominent scientists, although they, too, began to realize that 'the problem of the interpretatio naturae tended to exceed the exclusively spatial and structural dimensions on which it had been based.'[137]

This base, it could be argued, stretched back to the natural historical observations of the ancients, although Baconism greatly added to it. What many people do not realize is the extent to which scientific investigations were being conducted in the intervening period, or, at least, in the Renaissance, that is, before the seventeenth-century revolution in science. Until fairly recently, humanism and science were usually regarded as antithetical, though the odd observer, including Burckhardt himself, did allude to manifestations of Renaissance 'science,' such as the collection of unusual flora and fauna for the amusement of pleasure-seeking princes.[138] What counted for the humanists, supposedly, was not how man could control nature, but how he could control the human world.[139] Lately, with the studies of Barbara Shapiro and others, the pendulum has slowly started to swing in the opposite direction. For example, humanist popularizers are now being given credit for at least 'arousing an interest in subjects that the Galileans were later to transform into sciences among the public at large'; here we have shades of future regional natural historians, vis-à-vis their audience, and also the 'great' figures of the scientific revolution.[140]

Aspects of future Baconism can be detected in the enthusiasm of Renaissance humanists for collecting and communicating new observations in botany, for instance.[141] (A favourite work of theirs was Pliny's *Historia naturalis*, which first appeared in print in Venice in 1469.) The universities held lectures on Galen and Dioscorides, and there were visual demonstrations of plants.[142] Almost anticipating the admonitions of Sir Thomas Browne to his protégé Henry Power in the next (seventeenth) century, the earlier Renaissance student of nature was encouraged to stroll through fields and meadows, examining 'the trees and plants, comparing them with the descriptions in the books of such ancients ... and they brought back whole handfuls to the house. These were in charge of a young page called Rhizotome ['root cutter,' the Greek word for 'herbalist'], who also looked after the mattocks, picks, grubbing-hooks, spades, pruning-knives, and other necessary implements for efficient gardening.'[143] In her article, Karen Meier Reeds makes the important point that unlike the field of anatomy, where Galen held sway, no botanist – at least by the time of the late Renaissance – held Pliny, Dioscorides, or any other ancient in such awe: 'Renaissance botanists asserted their confidence in their own discoveries.'[144] Still, unlike many of the seventeenth-century regional natural historians, they placed considerable emphasis on the study of literary classics, at least where a sound knowledge of *res herbaria* is concerned.

In describing plant life, or for that matter animal life, in their 'histories,' the humanists utilized a system of knowledge that differed from that of the later 'natural histories.'[145] The aim was to compile *all* the information available on the subject, including external characteristics, heraldic uses, magical and medicinal powers, and so on, and to organize it in accordance with the principle of 'similitude' or 'resemblance.' The importance of literary sources to this process is seen in that previously written – and spoken – descriptions of the subject matter were included in such studies because they were thought to form part of its very 'nature.'[146] In the seventeenth century natural historians ignored similitude, concentrating on description of the external structures and upon establishing a correspondingly exact nomenclature. Referring to the opinion of Michel Foucault on the subject, Albury and Oldroyd add that:

This reliance upon visible structure, Foucault argues, depended upon the principle of 'representation,' which replaced 'similitude' as the basis for order in classical natural history. The notion that the 'essences' of bodies were 'faithfully' represented by their external features, and that these features themselves were subject to exact analysis, made possible the establishment of ordered, hierarchical tables of classification based

Introduction: 'Where the Choir Was ...'   33

upon the systematic comparison of visible structures. The works of Linnaeus himself, particularly in botany, best exemplify the tabular order which natural history sought to discover in nature.[147]

In the mineral world, Aldrovandi and Gesner, who also acquired a reputation for their study of plant and animal life, described minerals according to the same system of knowledge as did Renaissance botanists. Agricola, coming from a background of mining and metallurgy, similarly described minerals in terms of similitude, discussed them in terms of their legendary medicinal properties, and the like, and included plenty of literary references.[148] But Agricola's reputation rests largely on his taxonomic scheme, on an underlying classificatory system based on a theoretical account of the genesis and composition of minerals, and on a consideration of certain observable properties – items that were omitted as taxonomic characters by later natural historians.[149]

Unlike Agricola, Boethius de Boodt is presented by Albury and Oldroyd as an example of a Renaissance scholar who constructed taxonomic tables presenting a system, or systems, of mineral classification. However, de Boodt's classificatory schemes also are dissimilar to those of later natural histories in that they 'took as taxonomic characters a wide range of features alien to "natural history," such as aesthetic appeal ("beautiful," "ugly") and place of origin'; furthermore:

> the variety of taxonomic characters used in each of these [de Boodt's] two tables shows the difference between de Boodt's principle of classification and that underlying the 'artificial systems' of the later 'natural historians.' It was indeed possible for the natural historian to construct alternative 'artificial systems' of classification (rather than a single 'natural method' of taxonomy). But artificial systems were always formed on the basis of an exhaustive description of one specific part or feature of the body in question (for example, the Linnaean system of botanical classification concentrated upon the sexual organs of plants), not by taking a wide range of characteristics into account as did de Boodt. We may conclude, then, that the appearance of tables in [de Boodt's] *Gemmarum et lapidum historia* [1609] does not signal the beginings of a natural history of minerals, since the organizing principle of these tables was quite different from that of classical [ie, later] taxonomies, both 'artificial' and 'natural.'[150]

'Renaissance' natural histories, in this situation, were superseded by the type of seventeenth-century natural histories dealt with later in this book, partly because the latter usually deliberately discarded 'information concerning the secret "virtues" of substances, their literary and heraldic significances, and other associated items of Renaissance lore.' The new standards thus dictated 'an increasing "clarity" of exposition, arising from the elimination of "superfluous" data and the fixing of attention upon external features for the purpose of identification.' Again, the turning point is recognized as coming near the mid-seventeenth-century mark, by Foucault, with the issue of Jonston's *Historia naturalis de quadripedibus* (1657), where 'the question of "animal semantics" – the literary and symbolic histories of animals so important to Renaissance scholars – was entirely omitted,' thus 'producing a considerable restriction in the type of information dealt with.'[151] Such developments, plus the gradual expunging of animistic principles from natural historical inquiry and the injection of a good dose of Baconism, resulted in new approaches to the field as a whole. Natural history, at the same time, complemented the mechanical philosophy, 'because both approaches to the natural world merged the living and the nonliving together.'[152] With the mechanical philosophy came the image of a Creator God distanced from constant involvement in the operations of a universe. This made it difficult to know Him in nature through any acts. The primary means of accomplishing this, at least as the century progressed, was through the study of the complexity and harmony of his creation; as Hankins states, 'natural history described this complexity in great detail.'[153] Such study was certainly promoted by the Royal Society, which gave it a social standing and some respectability, helped put its adherents and promoters in contact with one another, and published their findings. But it is hardly necessary to elaborate the point that in some respects, at least, the influence of the society was less than overwhelming; much of the research into particular natural historical subjects was done – often by necessity – by individuals working alone or in small groups at the peripheries of the society's orbit, where easy access to it, geographically and sometimes otherwise, was often lacking.

Members of the society followed, or claimed to follow, Francis Bacon in attacking the 'Monstrous Imagination' of certain ancients, such as Pythagoras, whose philosophy was 'full of superstition,' especially with regard to his belief that the world was 'One Entire, Perfect, Living Creature.' Hence the idea of the 'world soul' or 'world spirit' was under attack. Even Paracelsus, whom Bacon slotted into the category of 'darksome Authors of Magicke,' could not escape his invective.[154] Bacon's example not only was to have its effect on the natural philosopher but also helped lead to the declining prestige of the occult among members of 'polite' English society. Thus, it was not merely the changing world-view – attributed by some to latitudinarianism, which did not allow for astral intervention in human affairs by the Deity – that contributed to the polarization between 'polite' and 'vulgar' culture.[155]

Introduction: 'Where the Choir Was ...' 35

Although there is no lack of studies of Bacon, early-modern British regional studies deserve more recognition as a group, or even as a type, than they have received. Few modern scholars recognize the connections that exist between chorography and the natural histories, seemingly preferring to emphasize the dissimilarities. But just as one has to look to the pre-Newtonian and pre-Royal Society days to mark the rise of science, so, too, can one trace the intellectual origins of British regional study of the second half of the seventeenth century to the earlier chorographic literature. Also, one technique of the chorographers, that of travel and firsthand observation, was eminently fruitful when expanded and applied to natural history.

The entire field straddles a sort of 'twilight zone' between historical, scientific, and geographical scholarship, with relatively few specialists making incursions into a discipline not their own – although recently this situation appears to be improving. It seems that travel diaries such as voyage literature, road books, general geographies, foreign guides, and surveying reports have commanded greater attention. The intellectual background of the Elizabethan or Stuart antiquaries has been discussed by Fussner, Levy, Sharpe, and Kendrick. Even these writers, however, tend to overlook the minor figures, and devote little attention to the even smaller field of regional study.[156] This is not surprising in the light of the fact that overall the scholarship about historiography in Britain during the period under study is not great.[157]

Standard works, such as J.W. Thompson's *A History of Historical Writing*, due to their broad range, treat English historiography of the sixteenth and seventeenth centuries either in general or in disparaging terms.[158] In his work *British Antiquity*, T.D. Kendrick perhaps comes nearest to placing topography accurately in its proper setting among antiquarian studies. He devotes one chapter to Leland (chap 4) and one to prose topographies, especially Camden's (chap 8). Topography, however, is only a secondary theme of the book, and therefore the minor chorographers are given very little attention there. A.L. Rowse, in *The England of Elizabeth*, briefly surveys the work of several important topographers and map-makers, cutting off his study at 1603.[159] *English Scholars, 1660–1730* by David C. Douglas traces the growth and achievements of Anglo-Saxon scholarship in the Restoration and the eighteenth century, yet Douglas basically overlooks the work done in the field of natural history.[160] Articles exist on some of the more famous chorographers (Camden, Leland, Norden), and information on them can often be found in modern-day introductions to their works.

The history of British geography, or aspects of it, has not fared well, either. A major attempt to present a composite history of the geography of Tudor and Stuart times is that of E.G.R. Taylor, in the 1930s, although Margarita

Bowen's *Empiricism and Geographical Thought* also contains sections relating to the early modern period.[161]

With regard to the post–1650 period, only recently has the situation improved. Except for Sir William Dugdale, the men involved in regional study after 1650 are often given brief notice, and that only for their specific involvement in the various fields of science, or for their association with the more famous figures of the time – not for their regional studies. A notable exception to this statement, however, is Michael Hunter's *John Aubrey and the Realm of Learning*, in which Hunter discusses 'Aubrey's Antiquarian Method' and 'Natural History and Antiquities.'[162] M.W. Greenslade and Emery have done valuable research – on Robert Plot and Edward Lhuyd respectively – towards explicating the efforts of two of the writers described in the present study, but much work remains to be done when it comes to assessing accurately the contributions of the 'minor' virtuosi and the connections between them. David Elliston Allen's *The Naturalist in Britain*, meanwhile, very briefly covers some of the figures examined in my study, embracing a broad spectrum of natural history and environmental science, but here geography, topography, and archaeology are squeezed to the sidelines. In like manner the study by F.D. and J.F.M. Hoeniger is but a meagre pamphlet on the growth of natural history in Stuart England, which primarily covers the pre–Royal Society period and makes no mention of the contributions to the field by the regional writers.[163] Not to be overlooked are the excellent works on local history by W.G. Hoskins and the figures associated with the Department of English Local History, which Hoskins founded, at the University of Leicester; their studies, however, seem to focus on more contemporary concerns and present-day methodologies in local history. Often important information can be culled from English regional journals such as *Midland History, Northern History, East Midland Geographer*, and numerous similar publications. Also relevant to regional study are works on cartography and cartographers, or works dealing with the early history of anthropology.[164] Often biographical data can be gathered from the *Dictionary of National Biography*, the *Dictionary of Scientific Biography*, or the works of sixteenth- or seventeenth-century figures who wrote about their contemporaries; Anthony à Wood's *Athenae Oxonienses*, for example, remains an invaluable source.[165]

Concerning the areas of Britain other than England itself, there still exists work to be done. Although Scotland, Wales, and Ireland have longer traditions in regional geography and history than is generally realized, studies that have been accomplished to date are patchy in treatment.[166] Chorography and natural history, in fact, appear to have gained more attention from the

Introduction: 'Where the Choir Was ...' 37

eighteenth- and nineteenth-century county historians and topographers than from many recent historians and geographers. Richard Gough, for example, was an eighteenth-century country gentleman who, after going to Corpus Christi College, Cambridge, devoted his time to antiquarian travels in England. He is remembered today for his vast collection of topographical material – maps, plans, notes, drawings – which forms, in the Bodleian Library, Oxford, a source of valuable information.[167] Other information may be gained from recent works dealing with seventeenth-century scientists, with science in general, or with science in relation to other phenomena – politics, society, humanism, puritanism, and the like. Excellent studies in this regard include the accounts of all Hunter's other investigations, especially with regard to the early activities of the Royal Society, and Charles Webster's *The Great Instauration: Science, Medicine and Reform 1626–1660*, which is crucial to an understanding of the intellectual/scientific milieu of the mid-seventeenth century. Barbara Shapiro's *English Scientific Virtuosi in the 16th and 17th Centuries* and *Probability and Certainty in Seventeenth-Century England* are also invaluable sources for the ties between science and humanism, illustrating how the historical and natural historical works display closely related heterogeneous blends of information about both society and the natural world.

But according to Glyn Daniel there is no general account of the English antiquaries, certainly none that is comprehensive and treats the subject in depth. This is doubly true of the regional writers, who composed a major segment of the antiquarian school of thought in seventeenth-century Britain.[168] Perhaps some scholars of today regard these early regional studies as the dullest of compilations at best, and relegate them to that no man's land between history and science. However, if properly written these can, at least, become 'works of entertainment, of importance, and universality. They may be made the vehicles of much general intelligence, and of such as is interesting to every reader of liberal curiosity. What is local is often national. Books of this kind, in the hands of a sensible and judicious examiner, are the histories of ancient manners, arts, and customs.'[169] That it is possible to link effectively the history of science with the history of geography has been demonstrated by several impressive works.[170] The present study hopes to build upon that edifice, though in a slightly different manner, adding antiquarianism and the history of history to the mill. It also aims to address partially John Fothergill's 1769 unfulfilled intention of writing a 'Review of the Rise and Progress of Natural History in Great Britain.'[171]

CHAPTER TWO

'A Hole Worlde of Thinges Very Memorable'

That I would restore antiquity to Britaine, and Britaine to ... antiquity ... that I would renew ancientrie, enlighten obscuritie, cleare doubts, and recall home Veritie by way of recovery.

William Camden *Britain*

IT WAS FOR THE ABOVE REASON that William Camden and others like him undertook the study of Britain or of its various regions in the sixteenth century. British chorography may be regarded as having been established as a distinct type of topographical-historical-antiquarian literature at this time, but in fact its genesis goes back further than this. This chapter examines the extent and the quality of chorographic literature before the 1600s, thus setting the scene for future developments.

In tracing the origins of regional study one can, of course, go all the way back to the ancients. Many of the early-modern chorographers were familiar with the contributions to the field of not only Ptolemy and Strabo but also Pliny, Herodotus, Pausanias, Mela, and others. In some cases, when describing the physical features of the countryside, the character of the local inhabitants, and so on, the sixteenth-century writers were merely, and directly, imitating the methods of these ancients. The Greeks and the Romans, incorporating a mishmash of information in their works, originally set the tone for chorography, taking it well beyond topography.[1] But Strabo, at least, did manage to set out a system of thought for regional geography.[2] As a historian turned geographer, Strabo was preoccupied with combining the two arts. In some ways chorography was but a version of geography that, however, restricted itself to impressionistically sketching the nature and identity of an individual region – *picturae similitudine observata prosequitur*; while geography dealt with the synoptic view of lands in their entirety. According to one sixteenth-century Continental observer, Joachim Vadian, 'Strabo, Pliny, and others were of much help to poets and historians in their descriptions of localities.'[3] History, geography, topography, politics – all these branches of knowledge became, therefore, part and parcel of chorographic study. As in geography *per se*, specialization was unheard of:

The science of geography, which I [Strabo] propose to investigate, is, I think, quite as much as any other science, a concern of the philosopher ... In the first place, those who in earliest times ventured to treat the subject were, in their way, philosophers – Homer, Anaximander of Miletus, and Anaximander's fellow citizen Hecataeus ... In the second place, wide learning, which alone makes it possible to undertake a work on geography, is possessed solely by the man who has investigated things both human and divine – knowledge of which, they say, constitutes philosophy. And so, too, the utility of geography – and its utility is manifold, not only as regards the activities of statesmen and commanders but also as regards knowledge both of the heavens and of things on land and sea, animals, plants, fruits, and everything else to be seen in various regions – the utility of geography, I say, presupposes in the geographer the same philosopher, the man who busies himself with the investigation of the art of life, that is, of happiness.[4]

Strabo's advocacy of wide learning and Ptolemy's *Geography* (1475, and subsequent editions) provided a basis for British chorography;[5] *Geography* and the *Itinerary* of Antoninus (1512) proved to be especially useful to the chorographer.[6]

As noted in chapter 1, in Britain a native tradition of chorographic-like writing existed before John Leland's time, although it consisted mainly of prose works of varying qualities. This tradition goes back, at the least, to two Christian Celts, Gildas and Nennius. Gildas wrote *De Excidio et conquestu Britanniae (The Ruin and Conquest of Britain)* in the sixth century. Concerning Christian Britain, this work contains accounts taken from local folklore and history, a brief description of the site and extent of Britain, and praise of the island's richness in waterways, castles, and other physical features. Gildas's study was more accurate historically, and oriented more to the Roman tradition, than that of Nennius, the purported author of *Historia Brittonum (The History of the Britons)*, written about 830 AD, probably in south Wales. Nennius's description of Britain in this work displays 'a keen interest in the surrounding countryside, its landmarks and ancient monuments,' and attempts to demonstrate that the Britons had no less glorious a history than the biblical peoples or the Greeks and Romans; of course, the account of marvels already, even at this early date, seems obligatory.[7]

The Venerable Bede (672?–735), famous monk of Jarrow in Northumbria, used Gildas as a source for his pious *Historia ecclesiastica gentis anglorum* (*History of the English People*), completed in 731. In five books, this study examines ecclesiastical history, but contains a geographical portrait of the British Isles, in essence realizing 'the importance of geography in early history, putting events in their geographical setting,' at the same time recognizing the value of the study of place-names and their derivation.[8] Bede begins his work with the flora and fauna, mines, waterways, salt springs, hot baths, and the other natural advantages of the land. The actual situation of Britain is described as follows:

Britain, formerly known as Albion, is an island in the ocean, lying towards the north west at a considerable distance from the coasts of Germany, Gaul and Spain, which together form the greater part of Europe. It extends 800 miles northwards, and is 200 in breadth, except where a number of promontories stretch further, so that the total coastline extends to 3600 miles. To the south lies Belgic Gaul, to whose coast the shortest crossing is from the city known as Rutubi Portus, which the English have corrupted to Reptacaestir. The distance from there across the sea to Gessoriacum, the nearest coast of the Morini, is fifty miles, or, as some have written, 450 furlongs. On the opposite side of Britain, which lies open to the boundless ocean, lie the isles of the Orcades ...[9]

Bede's *Historia abbatum* (*History of the Abbots*), written 716–31, also contains elements of local history, and is indicative of how biographies and hagiographies were linked to local history. As Gransden states: 'the earliest local histories are a series of biographies of the heads of the institution concerned ...' wherein 'the author made what he chose of the holiness of the subjects of the biographies, but emphasized facts relating to the place's history'; hence 'the place gained, in varying degrees, dominance over the biographical framework, achieving a personality of its own.'[10] Gransden presents many such examples of the biographical-local historical form, those stemming from the Anglo-Saxon and continuing into the Anglo-Norman traditions. Goscelin, a monk of St Bertin's, a 'foreigner,' and a writer of the lives of about twenty saints connected with English monasteries, is credited as the 'first man known to have studied English local history on the spot, travelling the land and staying in monasteries to write the lives of their saints,' a 'forerurnner of William of Malmesbury who admired and borrowed from his work.'[11]

Aside from these works, the seeds of later chorographies are found in medieval accounts of Britain that often served as introductions to more general chronicles and histories. For example, the twelfth-century monastic

'A Hole Worlde of Thinges Very Memorable' 41

chronicler Henry of Huntingdon utilized a considerable amount of Bede's description in his *Historia Anglorum* (*History of the English*), written c 1129–54. Here we find graphic, stirring portrayals of Henry's home, Lincoln; the legend of Arthur, taken from Nennius and Geoffrey of Monmouth; other myths; accounts of marvels, Roman roads, and the Anglo-Saxon heptarchy; and lists of shires and bishoprics. In his *Gesta pontificum anglorum* (*History of the Prelates of England*) (1125), William of Malmesbury combined history, topography, and antiquarianism to good effect in presenting a survey of ecclesiastical England. According to R.W. Southern, William's aim was 'a total recall of the past in order to give the community its identity in the present':

> He used the materials of every monastery he could visit, or from which he could get information – their chronicles, charters, legends, ornaments, inscriptions and buildings – to make a survey of the whole kingdom. He travelled widely to gather material for his book ... His brief descriptions of Canterbury, Rochester, Glastonbury, London, Hereford, York, Durham, Crowland, Thorney, the vale of Gloucester, and the fens, are the first accounts of ancient places in this country seen through the eyes of a man with a critical and developed sense of the past.[12]

William Fitzstephen, a Londoner, produced a detailed twelfth-century study of the topography and social customs of London in the prelude to his life of Thomas Becket. Giraldus Cambrensis (Gerald of Wales) wrote seventeen books, all in Latin, several of which were chorographic in nature. These included *Topographia Hibernica* (*Topography of Ireland*) (1187), one of the first treatises to describe the island and its peoples; *Itinerarium Cambriae* (*Itinerary of Wales*), a travel diary of a preaching tour that Giraldus and Archbishop Baldwin undertook in 1188 to gain support in Wales for the Third Crusade; and *Descriptio Cambriae* (*Description of Wales*), an account of the geography, everyday social life, popular customs, antiquities, and even a bit of the natural history of that country, in total combining firsthand observations with folklore.[13] Giraldus depicted the Welsh as being light, agile, fierce, and trained in war, apparently caring for little else than their horses and weapons, and constantly cleaning their teeth with hazel shoots. The population in general is represented as prone to plunder, greed, vice, and so on. None too flattering portrayals such as this make Giraldus's works 'very human withal, too human for many modern Welsh and Irish critics' – not surprisingly![14]

Few non-monastic or non-religious figures produced outstanding chorographies in the mid to late Middle Ages, although there are a few such examples.[15] Chorography occupied the mind of Ranulph Higden (d 1363), a monk at Chester, as can be seen in his great universal history, the *Polychronicon*, whose earliest MS version was completed in 1327. In many ways the *Polychronicon* is a continuum of the medieval descriptive tradition. Higden uses classical and Christian writers as sources, but is more critical than most of these, especially in his approach to their authority. Then there is the unique way he links his first book on geography with the subsequent books on the history of mankind.[16] Although he is familiar with Pliny's natural history, there is little indication that Higden himself took to the field to examine natural phenomena. The scope was there for making greater use of such first-hand observations. Perhaps he was more interested in antiquities, such as the ancient monuments and the Roman trunk roads that he discusses.[17] His critical attitude is reflected in his scepticism of the more extreme claims of Geoffrey of Monmouth with regard to Arthur; Higden usually dismissed accounts not borne out by several reliable witnesses, corroborating one another.[18] The entire first book of the *Polychronicon* consists of a descriptive geography of the entire world, followed by particular descriptions of Ireland – based largely on Giraldus – Scotland, Wales, and England. The term 'chorography,' then, may be applied to his work only in the broad sense of the word; as Gransden says, 'Higden's wish to write a balanced universal history also appears in his virtual suppression of local attachments.'[19]

Higden's *Polychronicon* became a vehicle for acquainting sixteenth-century antiquaries with the earlier topographies. It was translated from Latin into English by John Trevisa in 1387, and later anonymously.[20] In 1480 Caxton published Trevisa's translation of a topographical part of the *Polychronicon* under the title *The discrypcion of Britayne*. This section was reprinted again several times later in a variety of forms, until it became not only well known but also popular among sixteenth-century readers.[21] In one sense, therefore, Higden's first book aided in defining the topographical-historical-antiquarian or chorographic genre. The description ranged freely over the details of physical and political geography, and included such antiquarian concerns as the locations of Roman roads and the names of ancient cities and towns, a history of the language, and accounts of legal organization and terminology. In addition, it contained a commentary on the manners and morals of the writer's contemporaries. And yet, as Arthur Ferguson states, 'Higden used his geographical description, including the familiar speculation on the diverse effect of climate on peoples, as a static setting, a stage of fixed props on which men acted in a sequence of events that bore little continuing relation to their social or geographical context. Geography may have conditioned the original character of a society, but it was not something which, by stirring its inventive genius, might continue to act as a factor in the history of that society.'[22]

'A Hole Worlde of Thinges Very Memorable'

Higden was followed by William of Worcester (1415–82) or William Botoner as he used to call himself. Worcester was physician and secretary to Sir John Fastolf of Caister in Norfolk. After Fastolf's death in 1459 Worcester returned to his native Bristol where history, topography, medicine, architecture, literature, and astrology all occupied his attention. Between 1478 and 1480 Worcester had the opportunity to gratify his antiquarian interests by travelling through Norfolk, Bristol, and the southwest collecting materials firsthand. From these itineraries it appears that he was collecting information for a comprehensive chorographic description of Britain, while keeping a day-to-day account of his journeys. These antiquarian notes, alternating with astrological speculation or private letters – a mixture later popular with John Aubrey – make his *Itinerary* fascinating reading:

Note-book in hand he went forth on his tours, always ready to pick up information from chance acquaintances upon any subject of interest. The pages of his *Itinerary* reveal him almost in the character of a modern interviewer, eager to put down whatever the person he has captured can tell him about places or people. To take some examples: He meets a Dominican friar, one John Burges, at Exeter, and finds that he knows a good deal about the saints of the district. Out comes the note-book, and down go the details: 'Ex informatione Fratris Johannis Burges ...' Other information he gets from a priest at St Mary's Ottery, 'loquendo et potando;' from a ferryman, and from the keeper of a prison at Bristol, and from a workman to a 'plump-maker' in the same place, from whom he had inquired about a tree growing in the streets. A Scotchman tells him all about Scotland and the Isle of Skye, so called – at least so he says – because its mountains are so high. A merchant from the Isle of Man speaks about that island, and also about Ireland.[23]

Worcester was personally acquainted with many of the foremost Continental humanists, but he never seemed capable of attaining their level of scholarship in this work, entertaining though it may be.[24]

The regional studies written in Britain up to the end of the fifteenth century do not form a very unified whole, lacking the sense of purpose that united the chorographies of the sixteenth and early seventeenth centuries. Even so, they do constitute a tradition. Almost all the forms of topographical-historical-antiquarian literature existed by 1500: descriptions of an entire country and of individual regions, cities, and various ruins.[25] In some instances these were annalistic in form, relying mostly on previous narratives that were received uncritically. Their form, from a modern viewpoint, is too chronological, and the topography was often overshadowed by the descriptions of political and religious events – just as some of the seventeenth-century chorographies also tended to emphasize the non-topographical element, that is, genealogy and heraldry. The new sense of purpose is evident in the works of each of the three 'giants' in the field of sixteenth-century chorographic scholarship: John Leland, William Lambarde, and William Camden. These men shared one thing: an almost fanatical love of Tudor England. They explored the country, observing and recording, delighting to produce not so much a history as a 'Speculum Britanniae,' a looking-glass of Britain in which every aspect of the nation of their day should be faithfully reflected. Thus, before too long, the nation-wide surveys were succeeded by more detailed studies of individual counties.

The first important and influential British antiquary was John Leland (1506?–52), 'the father of English topographers,' as H.B. Walters called him.[26] The study and investigation of Britain gained new direction in the 1540s when Leland began work on his never published description of Britain.[27] His efforts extended in a number of directions: topographical description, map-making, and the study of the language, institutions, and remains of the British past. This entailed many preliminary duties. He had to discover and examine the ancient chronicles as well as the archaeological remains that survived from the past. He had to search the libraries and the countryside for information, reading the ancient languages and inventing new 'languages': the interpretation of sphragistics (seals), numismatics (coins), epigraphy (inscriptions), and diplomatics (documents). His notes, moreover, came to constitute 'the ultimate source of a good deal of our knowledge of early English history and ... of the appearance of Tudor England.'[28]

Leland's life story has been ably told by others, and requires only brief reiteration here.[29] In 1533 he was appointed, so the story goes, the official 'King's Antiquary.'[30] Thereafter he was commissioned to:

peruse and diligently to serche al the libraries of monasteries and collegies of this yowre noble reaulme, to the intente that the monumentes of auncient writers as welle of other nations as of this yowr owne province mighte be brought owte of deadely darkenes to lyvely lighte, and to receyve like thankes of the posterite, as they hoped for at such tyme as they emploied their long and greate studies to the publique wealthe; yea and farthermore that the holy Scripture of God might bothe be sincerely taughte and lernid, al maner of superstition and craftely coloured doctrine of a rowte of the Romaine bishopes totally expellid oute of this your moste catholique reaulme.[31]

Leland began his task soon after his appointment and spent nearly ten years travelling and collecting materials for his projected chorography of Britain. In

1546 he presented his plans to Henry VIII in a document edited, or enlarged, by Leland's fellow antiquary, John Bale. This was printed in 1549 under the title *The Laboryouse Journey and Serche of Johan Leylande, for Englandes Antiquitiees*.[32] It is sometimes referred to as Leland's 'New Year's Gift,' and is more commonly known today, thanks to Thomas Hearne and Lucy Toulmin Smith, under the title of *Leland's Itinerary*, or simply the *Itinerary*. After his travels Leland settled in London and began reworking his materials until he became insane.

Leland's *Itinerary* was a statement of the aims of Leland's antiquarian research. These included describing the realm for posterity, setting out the evidence for the nobility of King Henry VIII and his progenitors, and recounting the destruction of Rome's power in Britain. Leland claimed to have compiled a list of Britain's scholars, 'begynnyng at the Druides,' and to have assessed the historians and their 'historiographies,' a job that stirred his desire to see the areas they described. He told Henry: 'I truste that this yowr reaulme shaul so welle be knowen, ons payntid with his natives coloures, that the renoume ther of shaul gyve place to the glory of no other region.'[33] In order to attain these lofty goals, Leland was forced to visit almost every distinctive geographical feature of England. At the end of his research he proposed several specific projects, including a topographical map of the realm, to be made of silver; publication of the record of his travels in a 'Liber de topographia Britanniae'; three books entitled 'De nobilitate Britannica,' revealing to the English nobility their 'lyneal parentage'; and, finally, a history written by himself to be entitled 'Liber de antiquitate Britannica' or 'Ciuilis historia.' The proposed organization of this last work is interesting for its influence on later writers: 'this worke I entende to divide yn to so many bookes as there be shires yn England, and sheres and greate dominions yn Wales. So that I esteme that this volume wille ... conteyne the beginninges ... and memorable actes of the chief tounes and castelles of the province.'[34] Leland's plan, therefore, was to divide the work into several books, each of which would describe the history and topography of a particular shire; the total survey would encompass all of Britain. Accounts of the adjoining isles would also be included. But it was a plan beyond the powers of a single man. Anthony à Wood was to ascribe Leland's insanity to the frustration of so great an undertaking. However, as Levy states: 'that the plan was in its essentials the correct one can be deduced from the fact that almost all the antiquarian research of the sixteenth century followed it; that it was visionary is obvious when we consider the number of men necessary to elaborate it.'[35]

Almost all of Leland's work went unpublished during his lifetime, remaining in manuscript. But the manuscripts themselves were influential and important. After his death his notes were passed on to the antiquaries who followed him. Aside from Bale, John Stow, William Harrison, Camden, Lambarde, Dugdale, and others also transcribed and borrowed from Leland's work.[36] Leland set an important precedent by actually going out in the field to look at what he described rather than merely scouring the chronicles for his data. In this, his method resembled that of Worcester. But again, Leland was remembered by those writers who came later, while Worcester was not. Leland and his antiquary friend, Robert Talbot, used Antoninus's *Itinerary* – a meagre compilation of cities in the Roman Empire, compiled in ancient times, and containing distances between cities – and actual remains, in an attempt to determine the Romano-British sites, thus paving the way for Camden.[37]

The *Itinerary* itself resulted from Leland's voluminous, often disorganized, compilation of rough notes made during his journeys, and covers a vast diversity of topics: 'there is almoste nother cape, nor bay, haven, creke or peere, river or confluence of rivers, breches, waschis, lakes, meres, fenny waters, montaynes, valleis, mores, hethes, forestes, chases, wooddes, cities, burges, castelles, principale manor placis, monasteries, and colleges, but I have seene them; and notid yn so doing a hole worlde of thinges very memorable.'[38] Armed with the king's commission and letters of introduction, which Leland put to good use, he searched dozens of monastic and cathedral repositories, as well as local records and other literary sources and so on for information to supplement his observations made during his excursions; in the process recording local customs, legends, inscriptions, and Roman antiquities (archaeological remains). However, with regard to the latter, it is generally recognized that Leland lacked the capacity to realize fully the potential of archaeological remains as evidence for the reconstruction of the past. Whereas, for instance, John Aubrey was able to see the intrinsic worth of extensive field-work and therefore examined in close detail various hill forts, stone circles, and barrows, for their own sake, Leland shows no clear understanding of how such material might effectively fit into his own studies. He was better at estimating distances between the major ancient cities, and at delineating the total area of a shire. He described the contemporary local scene in detail. He also established a pattern to be followed by future regional writers – including some natural historians – usually describing in a consecutive manner various features found along the course of each local river; noting the houses of the nobility and the etymologies of local place-names as he went along. According to Kendrick, Leland 'was in love with Tudor England, and he loved his glorious present all the more because he so loved his British past.'[39]

William Lambarde's (1536–1601) *A Perambulation of Kent* (1576) owes much to the influence of Leland in its admixture of past and present, but there is a twist in its antiquarianism, if not in its rambling account of topography and history.[40] The focus of the *Perambulation* is to a considerable degree the Kentish gentry, in particular its genealogies and liberties.[41] The work is based on the author's study of Old English law, especially the custom of gavelkind, once employed in Kent.[42] Gavelkind was a socage providing for descent at law to all the sons or heirs to the nearest degree. Lambarde's first work reflects this interest in 'legal antiquarianism.' It consists of a collection of Anglo-Saxon laws with a translation, undertaken at the request of his teacher and friend, the Saxonist Laurence Nowell, and was based on Nowell's research and collection.[43] The *Perambulation* was also based partly on *Dictionarium Angliae topographicum et historicum* (commonly called the *Topographical Dictionary*), a work that also embodied much of Nowell's research.[44] The *Topographical Dictionary*, not printed until 1730, circulated widely in manuscript and, according to Lambarde, was a rough assortment of notes, being 'but a Breviate, for Store,' which 'was meant to be enlarged, as the *Perambulation of Kent* is.'[45] In the second edition of the *Perambulation* Lambarde revealed that:

I had some while since gathered out of divers auncient and late histories of this our Ilande, sundrie notes of such qualitie, as might serve for the description and Storie of the most famous places thorowe out this whole Realme: which collection (because it was digested into Titles by order of Alphabet, and concerning the description of places) I called a *Topographicall Dictionary*: and out of which, I meant in time ... to drawe (as from a certaine store house) fit matter for each particular shire and Countie.[46]

Lambarde was anxious to provide, for others contemplating a county chorography, a reference work – the *Topographical Dictionary* – to 'serve as the source for [in fact] a whole series of county histories.'[47] Here again he is following in Leland's footsteps.

Lambarde's notes, as Southern points out, show his wide range of interests and the varied nature of his friends; the earliest notes are dated 1560, when Lambarde made extracts from Giraldus. These interests later grew to include 'Perambulations of the Forest, fiscal returns, surveys of the courts of Chancery and Exchequer, lists of castles, and landowners in Kent who had obtained an alteration in the terms of their tenure.'[48] The *Dictionary* itself reflects such preoccupations, beginning with a list of counties, each with a chorographical description, Proceeding alphabetically from 'Albion' to 'Wyrisdale.' Lambarde's familiarity with the Anglo-Saxon and British languages is evident throughout this work, for example, in his conjectures on the derivation of place-names. The *Perambulation* warrants our greater attention because its publication released a veritable tide of regional studies, aiding 'any, that should endevour further Progresse therein; *Facile est inventis addere, difficile invenire.*'[49]

It opens with an account and a map of the Anglo-Saxon heptarchy. The shire is discussed next in terms of its physical features and population, the latter in terms of the first settlers of England. The hundreds, lathes, towns, and boroughs, and the taxes of each, as well as the nobility and gentry, are duly listed. There are tables on franchises, knight's fees, forests, parks, cities, markets, castles, bridges, and the like, enough to make one's head spin. A final table lists the ancient and contemporary Kentish writers, taken largely from Bale's *Scriptorum illustrium majoris Britanniae ... catalogus.*[50] The main portion of the *Perambulation* then covers the topography, as well as the history and antiquities of each place in the shire, which is divided into two sections: the dioceses of Canterbury and Rochester. The study ends with 'The Customes of Kent,' in which gavelkind, peculiar to the shire, is discussed.[51] Lambarde's interest in etymology is evident in the fact that every chapter commences with an elaborate title that tries to trace the derivation of the place-name, often providing the equivalents in Latin, British, Anglo-Saxon, or all three. 'Rochester' provides an example: 'Rochester, is called in Latine, Dorobreuum, Durobreuum, Durobrouae, and Durobreuis: in Brittish, Dourbryf, that is to say, a swift streame: in Saxon ... Rofi ciuitas, Rofes citie, in some olde Chartres, Rofi breui.'[52]

Lambarde's critical acumen is revealed in his citation of the literary sources, where he dismisses the suspect, as in the case of one passage from Bede: 'this writer is called Venerabilis: but when I read thus, and a number of such, which make the one halfe of his worke, I say with my self as sometime did the Poet ... What euer thing thou shewest me so, I hate it as a lye.'[53]

He included transcriptions of a number of documents in the text, because of his interest and reliance upon documents not only as sources but as texts instructive in themselves. At one point a lengthy discourse occurs on the changing custom of affixing seals to charters, and at another the usefulness of a knowledge of seals to the dating of documents is established.[54] In this respect, Lambarde anticipated the seventeenth-century work of Dugdale. There are other examples of his linguistic skills, including his brief statements on the ancient manner of forming personal names, or on the relationship between topography and Anglo-Saxon place-names.[55] In criticism of Lambarde's attitude, we can note that his scholarly objectivity was in some instances impaired by his anti-Catholic stance; he took every opportunity to ridicule papistry. But his patriotism remained intact, held together, it seems, by his interest in the Kentish beacons as vital to national defence.

Lambarde lived to the end of Queen Elizabeth's reign, rewarded for his lifetime of devotion to country and sovereign by a celebrated interview with Her Royal Majesty.[56] Of even greater importance in assessing his standing as a scholar is the respect paid him by his fellow antiquaries, including William Camden (1551–1623).[57] John Nichols, meanwhile, writing more than two hundred years ago, assessed Lambarde's work in these words:

In his *Perambulation* we may consider him as opening a new source of learning, as an original author. It was the first book of county antiquities ... The *Perambulation* of Lambarde may justly challenge a comparison with any other county history, for clearness of method, variety and accuracy of information, and for comprehensive brevity. At the same time it has its defects. The Roman antiquities of the county he gives up almost entirely as too remote and obscure, concerning which there certainly are not wanting any useful data; nor has he touched upon the natural history; and he might easily have enlarged ... on the geographic description of the county, as the course of the hills, rivers, and etc. He is most full on the Saxon and English antiquities and historical anecdotes relating to each place.[58]

The first edition of the *Perambulation* contained an appeal by Lambarde for others to amend his work and to complete the description of the rest of the realm. His hope was that 'some one in eache Shyre, would make the enterprise for his owne Countrie, to the ende that by ioyning our pennes and conferring our labours (as it were) *Ex symbolo* wee may at the least by the union of many parts and papers, compact a whole and perfect bodie and Booke of our English antiquities.'[59] In the final analysis it was the publication of Camden's *Britannia* in 1586 that prompted Lambarde to abandon his original plan for a chorography of the entire realm, covering all the counties of England.[60] That Camden made use of Lambarde's keen criticism and detailed suggestions is demonstrated by the fact that he sent the *Britannia* to Lambarde in manuscript for his comments.[61] Both men realized that their efforts were complementary. Whereas the *Britannia* was the true fulfilment of Leland's hopes, providing a panoramic sweep, the *Perambulation* anticipated, in a sense, the *Victoria County* series, with its sharper focus. The different scope of the two works allowed each man to exploit his own subject matter to particular advantage.

Camden's antiquarian researches made him well known even among the Continental scholars, who called him 'the British Strabo.'[62] In fact, he maintained steady correspondence with many of the outstanding European scholars of the day, a group including Casaubon, De Thou, Gruter, Mercator, and Ortelius. In the cases where he did not personally know a scholar, more likely than not he was familiar with his works; he had read Biondo and was familiar with the work of the Germans.[63] It was the chorographic and cartographic activities of Ortelius, and his letter to Camden urging the latter to 'restore antiquity to Britaine, and Britaine to ... antiquity,' that provided the immediate inspiration for Camden.[64] Mercator, the famous geographer, did his share by sending Camden a copy of Ptolemy's 'Tables.'

The *Britannia*, by presenting to the world the Roman antiquities of Britain, was designed to entrench the author's native land among those European nations that claimed the mighty Roman Empire as their origin. The attempt to dispel the myth of the Trojan Brutus aided this cause to a large extent, but the race to establish the glorious past of the nations saw Britain lagging behind much of Europe. For example, Tacitus's *Germania*, brought to light once again by the German humanists, served to stimulate a tide of exact and detailed topographical-historical-antiquarian surveys of Germany and of its various regions. In the forefront of this movement was Konrad Celtis, who in the late fifteenth and early sixteenth centuries called for a collective effort towards the elucidation of Germany's heritage, to be embodied in a great chorographic project that he entitled the *Germania Illustrata*. Thus, utilizing a plan followed later by Camden, Celtis was able to initiate fact-finding correspondence with the other German humanists with whom he was personally acquainted, and with his other distant admirers. This he supplemented with observations made through his travels across Germany. Celtis also inaugurated a program of publishing classical and medieval texts, to be used as source material for his intended project. In similar fashion, Camden prepared himself by studying Greek, Latin, and history, and by keeping abreast of the latest antiquarian scholarship abroad. Celtis's plan was to develop a parallel description of ancient and modern Germany, which he based on Flavio Biondo's scheme of 'illustrating' Italy – in the true Ptolemaic style – in *Italia Illustrata* (1474). Camden likewise described the early Britons, using sources such as Strabo, Tacitus, and the other classical authors, with reference to Britain's development through Anglo-Saxon and medieval times.

Biondo (1388–1463) has already been mentioned in chapter 1. A native of Forli in the Romagna, he is often regarded as the founder of antiquarianism. His *Roma Instaurata* (1466), a topography of imperial Rome, and the *Roma Triumphans* (1459), however, reflect a combination of antiquarian as well as non-antiquarian interests, containing information on religious and political

'A Hole Worlde of Thinges Very Memorable' 51

history, customs and manners, and so on. Since the barbarian tribes had destroyed Roman civilization and rule, and because there was no one left to transmit current events to posterity, Biondo took it upon himself to restore to life, so to speak, the ancient places and peoples of his homeland in the pages of the *Italia Illustrata*, which is the first full-fledged chorographic treatise of the Renaissance. Biondo surveyed Italy region by region, describing the towns, the physical features, the ancient noble houses, and the long-forgotten origins of place-names. For his information he drew upon the entire Latin literature of antiquity, including maps, chronicles, itineraries, and inscriptions on monuments. This sort of activity did not escape the notice of future Continental and British scholars, including William Camden.

Camden, however, was also completely familiar with the native antiquarian tradition. Apart from his knowledge of the work of Leland and Lambarde, Camden knew the topographical-historical achievement of the whole British tradition.[65] 'The native antiquarian movement,' writes Denys Hay, 'was to develop in the sixteenth century, with John Leland, until it combines with continental scholarship in men like Camden to form one of the characteristic features of seventeenth-century scholarship.'[66] Being a close friend of Richard Hakluyt, and coming into direct contact with voyagers such as Cavendish and Drake, Camden was undoubtedly also aware of the current progress in the general field of geography. And so he was stimulated to write of the voyages of discovery in his *Annales ... regnante Elizabetha* (1615).

Unlike the work of Leland and Lambarde, the *Britannia* was the product of a collaborative effort. Apart from his Continental connection, Camden was associated with several circles of scholars some of whom formed themselves into the first Elizabethan Society of Antiquaries.[67] These men, and other correspondents, sent him information from various parts of Britain, information that Camden occasionally incorporated verbatim into his book – at the same time acknowledging his contributors' assistance. Composed in Latin, the *Britannia*, or *Britain*, was first published as a quarto volume in 1586. It rapidly achieved great popularity; within Camden's lifetime five corrected and enlarged editions appeared. The sixth (1607) edition, a large folio volume, represented a major revision, substantially enlarged and containing maps by John Norden and Christopher Saxton. In 1610, Philemon Holland translated the work and made additions of his own; by 1806 seven posthumous editions had been produced by renowned scholars, including Edmund Gibson and Richard Gough.

It is difficult to give an adequate indication of the contents of the *Britannia*. In his work Camden covered history, geography, topography, anthropology, and antiquarianism in general; thus, he included an incredible mass of detail.⁶⁸ He began with the geography of England and then moved on to more 'historical' topics – the nature of England's early inhabitants and the historical development from the Roman through the Norman period. As he dealt with each county separately, noting the particular antiquities of each location, he sometimes allowed himself long digressions on associated historical topics, for example, the degrees of nobility in England. But Camden was primarily an antiquary in the sense that more often than not he neglected locales that had no ancient monuments. He was most interested in those sites and customs that were best known to the ancient authors. For example, in the section 'The Maners and Customes of the Britains,' he leaned heavily on excerpts from Strabo, Tacitus, Caesar, and other classical authors.⁶⁹ He assured the reader that the *Britannia* was chorography, not history. After one of his digressions, he apologized: 'Now remembering myselfe to be a Chorographer I will returne to my owne part, and leave these matters with our Historiographers.'⁷⁰

Unlike Lambarde, Camden was less interested in the Anglo-Saxons than in tracing the Antonine itineraries, identifying Roman names with current sites, and recording Roman antiquities in general. But the *Britannia* does include a review of English law courts that parallels Lambarde's treatment of the Kentish gavelkind. Also, several of Camden's county descriptions repeat the pattern of the *Perambulation*, covering the same topics, with the addition of discussions of climate and produce. Here is Camden's introduction to the section on Kent: 'I am now come to Kent; a Country, which William Lambard, a person eminent for Learning and piety, has describ'd so much to the life in a complete Volume, and who has withal been so happy in his searches; that he has left very little for those that come after him. Yet in pursuance of my intended method, I will survey this among the rest; and lest (as the Comedian says) any one should suspect me of Plagiarism, or Insincerity I here gratefully acknowledge, that his Work is my Foundation.'⁷¹ The overall general plan of the *Perambulation*, in fact, was adopted by Camden. Both it and the *Britannia* begin with a section on the naming of their respective regions. The topics that follow, in Camden's work, correspond in many respects to Lambarde's introductory topics, somewhat rearranged, and, as in the *Perambulation*, the major portion of the book is a place-by-place topographical-historical survey.⁷²

In the long introductory essay Camden discussed the first inhabitants of Britain and then described the various peoples that settled the island – the Britons, Romans, Picts, Scots, Danes, and Normans – concluding with a study of the heptarchy and the ancient law courts. In keeping with his stated concern for historical truth and impartiality, and impeaching 'no man's credit, no not Geoffrey of Monmouth whose history ... is held suspected among the judicious,' he left the veracity of Brutus's being the founder of Britain to the reader's judgment. But then he devastatingly reviewed Geoffrey's history in a reasoned attack that went far towards discrediting the Brute fable.[73] Utilizing evidence from Caesar, Tacitus, and others, he argued that the origin of Britain's inhabitants is to be found in their descent from the Gauls, not the Trojans.[74] His objectivity extended to his story of the Romans in Britain. The fact that Camden's interest, like that of the other chorographers of his day, was mainly antiquarian and his materials almost all literary is most evident in this section. Much of this source material was taken from the ancient Roman chorographies, usually presented verbatim, without any attempt made to shape the sources into a consecutive historical narrative. Camden did, however, use a variety of additional sources such as monumental inscriptions, old law codes, and coins – reproductions of which accompany the text. He emphasized the coins in particular, since from them 'there ariseth very much light to the illustration of ancient histories.'[75] But, unlike Leland, he actually had little interest in purely archaeological evidence. Such evidence usually appears in his writing as an afterthought to the history of a site. The description of Folkestone provides one example of this: 'a flourishing place in times past, as may appear by the peeces of Romane coin and Britaine brickes daily there found.'[76] In some respects this also parallels Lambarde's stance.

In general, Camden styled himself a second Lambarde, conceiving the long essay on the individual counties as a 'perambulation thorow the Provinces or Shires of Britaine,' at the same time constructing it along the models established by Strabo, Ptolemy, and the other ancient geographers.[77] Camden, again like Lambarde, personally surveyed the British countryside while preparing the *Britannia*. This helped make his work one that did not deal solely with the ancient past, but that also described the contemporary British scene. Each county's topography, towns, rivers, economy, early inhabitants, and prominent families were described, and a map of the particular area accompanied the text. The larger divisions, those delineating Britain, Scotland, Ireland, and the offshore islands were preceded by a brief general history. Etymological discussions, topographical description, and history were all combined in the following typical sample, on Buckinghamshire:

Chiltern got that name according to the very nature of the soile of Chalkie marle, which the ancient English men termed Cylt or Chilt. For, all of it mounteth aloft with whitish hilles, standing upon a mixt earth of clay and chalke clad with groves and woods, wherein is much Beech, and it was altogether unpassable in times past by reason to trees, untill that Leofstane Abbot of Saint Albans did cut them downe, because they yeelded a place of refuge for theeves. In it, where the Tamis glideth at the foote of those hilles with a winding course, standeth Marlow ...[78]

Camden received a considerable portion of this kind of information from contemporary chorographers other than Lambarde, a group that included Sampson Erdeswicke, Dee, Owen, Stow, and Carew. Carew in particular was an old friend. His *Survey of Cornwall* and Camden's section on Cornwall may both be considered almost as works of joint authorship.[79] Correspondents from farther afield sent Camden information on some of the remotest parts of the country. For example, Reginald Bainbrigg, a schoolmaster, made several excursions along the Roman wall and surrounding countryside, gathering inscriptions and other information: 'The wall crosseth Eden at Carelile and goeth to Stanwiggs, wher ther stands a verie ancyent churche, but ruinous as commonlie, all the churches on the bourders are: frome thence it goeth to Blaytarne not far from Scalbie castle wher I found this inscription in faire letters, this stone was laitelie digged up and put in a howse newlie buylded ...'[80] Bainbrigg was to painstakingly transcribe copies of dozens of inscriptions, which he forwarded to Camden.

The primary motive behind Camden's work was patriotism. Denys Hay, for one, believes that Camden, by means of the various editions of the *Britannia* that had appeared by the time James proclaimed himself king of Great Britain in 1604, 'did more to unite Britain in the long run than did King James.'[81] Other modern scholars see this book as a work of major significance for historical studies. Stuart Piggott views it as a milestone in the history of British antiquarian thought, describing the changing nature of antiquarianism in the early seventeenth century in these words:

The appearance of the English versions of the *Britannia* show in themselves a changing antiquarian public in this country. The original Latin work was addressed to the world of European scholarship ... But by Jacobean times a new class of reader had grown up in England, anxious to read antiquarian literature in English: a taste which the *Britannia* itself had gone far to create. We have moved out of that Latin speaking fraternity of learning which, up to the time of Elizabeth, had carried on the tradition of the scholars' lingua franca, and are in the new, self-confident, national state in which, with the increase of literacy, an interest in local history was no longer confined to the learned professions, but was as likely to be found in the merchant or the country squire.[82]

The chorographers who followed were not, for the most part, as scholarly or erudite as Camden. None of them could boast of being as well versed in Anglo-Saxon studies as Lambarde, or in Roman antiquities as Camden. Only a minority had any significant connection with Continental scholarship. In a sense, therefore, what followed the *Britannia* is largely considered by modern scholars as anticlimactic; so they tend to overlook the importance of the chorographic activity of the first half of the seventeenth century. In few, if any, instances was Camden's book closely followed. Many of the seventeenth-century chorographers were more interested in their contemporary surroundings than in the distant Roman or Anglo-Saxon past. But once Camden provided the 'overview' by his description of all of Britain, each chorographer could now concentrate on describing his own part of the British Isles, as Lambarde had done, thereby not only helping fulfil Camden's original plan but also continuing the native chorographic tradition.[83] From the early studies of the ancients, to the biographical-local history of the monks, to the formulation of antiquarianism in the Renaissance, chorography had gained momentum in Britain, where it was firmly established by the end of the sixteenth century.

R.W. Southern made the following comment about the 'researchers' of the period after 1560, whom he called the 'secular successors of the post-Conquest monks':

They [both groups] were engaged in the same task of bridging a gap between past and present which made them uneasy and diminished their stature in society. But ... the post-Conquest monks were sure that they had a great past, but they were uncertain of their present and future. Their post-Reformation secular successors were relatively sure of the present but uncertain of their past. The monks felt the danger of losing their lands; the new landowners felt the danger of holding their lands without the ancient respectability which would give dignity and stability to their position. The monastic antiquaries searched the records to give detail and lucidity to their inherited conviction of greatness; the secular antiquaries searched to discover what it was they had inherited. Hence their interest in family history, and the consequent importance of the College of Heralds in the historical researches of the Tudor and Stuart periods. Hence also their interest in institutions and in the descent of landed property, and the consequent importance of lawyers in the historical movement. Heralds and lawyers were the men who handled ancient documents as part of their daily work, and they became the interpreters of these documents to their generation, just as the English monks in the post-Conquest monasteries had been the interpreters in their time ... The Tudor researchers had a better fortune. The methods they revived and the materials they unearthed have continued to attract workers from that time to this[84]

As historical and antiquarian researchers, at least, the chorographers fitted into this mould quite nicely, and set the stage for future developments in the seventeenth century, when the regional writers came to the fore in bringing about a sort of 'scientific enlightenment'; or at least – through natural philosophy, whose application to the other branches of learning and the arts became obvious – in providing the groundwork for the 'real' Enlightenment of a later date.[85]

CHAPTER THREE

Speculum Britanniae

This our Britannia, for the fertility and fruitfulnes thereof, matcheth the best ... And above all other blessings it hath greatest cause to reioyce in the free use of the true knowledge of Christ, wherein it triumpheth above all other kingdomes or Countries of the world ... Our Englande may be truely called Olbion a happie Countrie.

John Norden *Middlesex*

BY THE LATE SIXTEENTH CENTURY more and more men were using chorography as a means of studying their native regions of Britain. Their main motive – glorifying Britain – was being plainly exhibited for all to see. By the time John Norden wrote the above words (1593) in his regional study of Middlesex, which was the first instalment of his projected survey of many of the individual shires of Britain, the issue of William Camden's *Britannia* had already established chorography next to history among the esteemed scholarly pursuits.

The first important chorography to follow the *Britannia* came from the hands of Norden. It so happens that this man also proved to be the most prolific writer of chorographic literature to soon follow Camden. His work, with its origins in late Elizabethan times, connects the *Britannia* with the mainstream of historical-topographical literature in the first half of the seventeenth century. The large amount of material available for a study of his writings, and the fact that Norden on his own carefully surveyed more than a dozen counties in their entirety, warrants the allocation of nearly an entire chapter to his work.

Norden is primarily remembered today as a professional surveyor and a writer of devotional, or religious books. Because his efforts in these other fields are interrelated with his chorographic undertakings, and since they aid in demonstrating that typical Elizabethan quality – versatility – found in many antiquaries, these wlll also be examined here briefly. This man's story is characterized by frustration, a frustration due mainly to an inability to secure sound financial backing for his most ambitious chorographic project. Norden was determined to write a series of county chorographies illustrated by small maps, calling the entire project 'Speculum Britanniae.'[1] Unlike Lambarde, Norden did not envisage a corporate effort, and was determined to undertake the entire work on his own.[2] However, like Lambarde, who had a similar plan but completed only a study of Kent, Norden failed in his scheme and published only his studies of Middlesex (1593) and Hertfordshire (1598).

Little is known of the particulars of Norden's life. Born in 1548, the son of a Somerset yeoman, in 1564 he entered Hart Hall, Oxford, taking his BA in 1568 and MA in 1573. After graduation he found employment as a private secretary to Lady Ann Knyvet in Wiltshire.[3] In 1581 he travelled to London, seemingly to establish a career in that city. This trip in itself is significant; it is possible that Norden's original interest in topography was aroused during the journey by his chance encounter with Don Antonio, the pretender to the throne of Portugal, who had sought refuge in England. Norden recorded this meeting in these words: 'At the arrivall of Don Antonie, the supposed K[ing] of Portugall in the weste partes of this Realme for refuge. It so fell that I traveyled certayne dais iourneis, in companie of him, and his followers who, as they were for the moste part verie learned, so seemed they desirous to enquire and learne the etymon and significations of the names of towns, Riuers, howses, Bridges, and what so ever thinges of note, by which we traveyled.'[4] Norden may have undertaken the role of a guide for these refugees from abroad, at least for a few days. This role would not have been incompatible for a man thoroughly familiar with this part of the country, having been born and raised here. He had already trained himself as a land steward and estate surveyor, and in matters concerning religion he was 'discontented with the organization and still more with the ministers of the new Reformed Church.'[5] During this period he apparently acquainted himself with Camden's *Britannia*, Christopher Saxton's *Atlas*, and with several other chorographical works.[6] At the same time he was busy collecting rubbings of old inscriptions and gathering other antiquarian material.

There used to be some doubt as to whether John Norden the chorographer and John Norden the writer of devotional books were one and the same. Anthony à Wood believed that there was only one writer by that name. However, Charles H. Coote, contributor of the Norden biography to the *Dictionary of National Biography*, noting that two separate works were published in 1597 under Norden's name, *The Mirror of Honor* and *Preparatiue to His Speculum Britanniae*, assumed, strangely, that one man could not have written two such dissimilar works in the same year. He reasoned, therefore, that there must have been two John Nordens. This viewpoint, with no solid contemporary evidence to warrant such a distinction, soon came under attack. It was eventually laid to rest by the investigations of A.W. Pollard, the results of which were published in his article establishing the unity of this supposed dual existence.[7] An examination of the prefaces and dedications in several of Norden's devotional works supports Pollard's position. For example, in the dedication to *An Eye to Heaven in Earth*, Norden refers to the 'ordinary imployments,' those involving surveying, and states that he is not a 'Divine': 'If it should be demaunded ... why I would venture to undertake a matter of this subject: so farre (in opinion) differing from my ordinarie imployments: It may please you to conceive that it is not altogether contrarie to my publike professions, as being a *Christian*, though no professed *Divine*.'[8] Helping to eliminate any remaining doubt is the dedication to his last religious work, *A Good Companion for a Christian*, where his son states: 'My deceased Father very often survaied the Kings Lands, but now by me he humbly tenders himselfe to be survaied by you.'[9]

It was after his arrival in London in 1584 that Norden began composing his devotional books and pamphlets.[10] The decision to do so was the result of two factors. First, his devotional works were intended to provide a way of earning money in order to support the study of law, which he undertook while in London. He dedicated *A Sinfull Mans Solace* to Sir Edmund Anderson, the lord chief justice of the common pleas, in an attempt to find an influential patron to support his various activities:

Although (Right honorable, my very good Lord) tender yeeres and slender capacitie doe hold back my willing minde from executing and performing that shew of zeal towards you, that my duetifull affection could gladly affoorde, I here, humbly crave your patient acceptaunce of this my poor Mite, wherin resteth a Million of duetifull endevours, if knowledge were according my addressed willingnesse of minde, to make my duetie truly knowen unto you: but that I account my selfe (and not unworthilie) among the poorest, with the simplest and of the yongest, of those of my profession in that place: and yet as a chicken under the wings of your Lordships favour, I rest in hope of defence, fro the scarring practises of such, as (deservedly or undeservedly) make profers and manifest meanes to bereave the credite of all and some of those that are or should be, as I (for functions sake) desire to bee: namely, as one able to execute, and with truth to prosecute our vocation, the office of Attorneis; which office, as it hath and ought to be of credite, so ought it to be executed with honesty and trueth (as it is not unknowen to too many of us) is growen into contempt of many. In respect (I thinke) of the ignoraunce that the rudest sort are blinded with, that they see not their owne ease and benefite, slaundering the honest minded for the misbehaviour of the unsatiable, and unhonest dealer's sakes....

Although (perchance) it may be obiected against me, that I am of an other profession, unlerned, without experience, a greene head, and of no iudgement: I yeeld me guilty of all: onely, my profession, as a Christian, I stand to defend: though not a professed Divine, but a poor Pupil, that is willing to be instructed of the more learned and godly divines: presuming herein, to make trial, whither it wil winne me condemnation or good liking: praise or dispraise: favour or ill-will.[11]

The second factor was directly related to the first. Apparently unsuccessful in his pursuit of a career in the law – there is no other reference in his works to the law, and his name is absent from any of the registers of the Inns of Court – and unable to secure patronage, Norden sank into a state of melancholy that, paradoxically, appears to have strengthened his faith in an eventual heavenly reward, one transcending his material privations in this world.[12] This, along with his already established religious zeal, spurred him on to write devotional books. Although some of these achieved considerable popular appeal, his career as a religious author proved to be singularly unsuccessful.

It is not clear why Norden then chose to direct more attention to chorography and to surveying as a means of livelihood. Lynam suggests that in the true Elizabethan antiquarian spirit, his travels about the countryside as a surveyor and lawyer had reinforced his appetite for chorography.[13] It is more likely, however, that it was Lord Treasurer Burghley, to whom Norden often appealed for help, who suggested that he attempt this undertaking.[14] Burghley resided in Northamptonshire; this led Norden to complete in manuscript, by 1591, a description and map of that county, one that included a dedication to the lord treasurer.[15] It is here that Norden acknowledges Burghley's assistance and expresses his own desire to describe all the other shires similarly. Apparently, he had originally conceived of this venture as basically a map-making project in the tradition of Saxton's *Atlas*. However, the chorographic prose descriptions that were to accompany his maps gave his work that added dimension lacking in Saxton's *Atlas*.[16]

Upon receiving a manuscript copy of the survey of Northamptonshire, Burghley helped Norden procure the privy council's endorsement of his project. Encouraged by this action, Norden proceeded to survey Middlesex, publishing a description of this shire in 1593. The *Dictionary of National Biography* states that this work was the result of an order in council given by the queen, but it is more probable that Elizabeth issued the order after she had seen a manuscript copy of this survey. Shortly afterwards, in July 1594,

Burghley issued from Greenwich another order recommending to favourable public notice 'The bearer, John Norden, who has already imprinted certain shires to his great commendation, and who intends to proceed with the rest as time and ability permit.'[17] Norden's dedication of *Middlesex* to the queen had helped him gain her attention and approval for his project. Financial aid, however, towards the publication of this particular work came from a different source, namely Sir William Waad, clerk to the privy council and a patron of geographical enterprise.[18]

The firsthand observations recorded in *Middlesex* testify to Norden's familiarity with his subject. The first few pages are intended to serve as an introduction to the whole 'Speculum Britanniae.' Here we find an etymological discussion concerning the origins of the name 'Britain,' followed by an abbreviated historical review of the early inhabitants of the island, beginning with the Angles and ending with William the Conqueror. Norden reveals his detailed knowledge of the sources of British history at various points, acknowledging his debt to Ptolemy, Gildas, Polydore Vergil, Camden, and others, either in the text or by citing these sources in the marginalia. At one point he even goes out of his way to cite Lambarde's expertise in Saxon matters, stating that 'Lamberde' is 'most expert in the Saxon toung.'[19]

The treatment of *Middlesex* itself opens with a rationale for beginning the 'Speculum Britanniae' with a study of this particular shire:

Hauing thus briefly touched the generall, I purpose to proceede to the particular descriptions of this our Britania: wherein (imitating the artificial Painter, who beginneth alwaies at the head, the principall part of the bodie:) I thought it not unfit to begin my *Speculum Britaniae* with Myddlesex, which aboue all other Shyres is graced, with that chiefe and head Citie London: which as an adamant draweth unto it all the other parts of the land, and aboue the rest is most usuallie ferquented with hir Maiesties most regall presence.[20]

This is followed by a history of the ancient tribe of the Trinobantes, and by an outline of the 'lymites and principall bounds of Myddlesex' which centres on a description of the major rivers that flow through the region. The fertility of the soil and the produce are mentioned next: 'the soyle is excellent, fat and fertile and full of profite: it yeeldeth corne and graine, not onelie in aboundance, but most excellent good wheate.'[21] The discussion here, however, is not of a technical nature. Norden, not unlike the other contemporary chorographers, did not base his observations on what we would consider as scientific testing or data. The closest he comes to this is to distinguish, for example, between clay and other types of soil; that is, he relies on readily obtainable information based on visual examination, the accounts of landowners, and so on.

The next section is concerned with the ecclesiastical and civil government of the shire, listing the now almost standard information; hundreds, market towns, 'houses of law,' battles, parks, named hills, and 'olde and aunciant highwaies now unaccustomed.' The remainder of the book is devoted to an alphabetical 'index' in a form similar to Lambarde's *Topographical Dictionary*. It identifies the cities, towns, villages, and 'houses of name.' Quite often the only information given for one of these entries is the map location of the place named. Other entries contain etymologies, historical events, and antiquities such as inscriptions found on monuments. In several instances these are accompanied by Norden's sketches of the coats of arms of prominent families. In this regard Norden was an innovator in chorography. He thought it a useful practice to record the inscriptions and coats of arms found on funeral monuments, a practice that had previously been confined to the manuscript collections of the heralds. He is careful to exclude, in some cases, material already found in Camden's *Britannia*. The section on London, nearly ten pages long, is perhaps the most interesting one. According to Norden, London is

the most famous Citie in all Brytaine, which Erasmus upon the Proverbe Rhodii Sacrificium, saith, is deducted of Lindus a citie of the Ile of Rhodes, Stephanus calleth it Lyndonium ... Leland taketh it to be Trenouant, new Towne, for that in the british toong signifieth a towne: M. Camden seemeth, in some sort, to yeelde that it should be called London of the British word Lhwn, which signifieth a woode, or ... Lhong [ships] ... in regarde that our Thamis yeeldeth such apt accesse for ships euen to the citie.[22]

As for the founding of London, Norden's opinion seems to follow Camden's skepticism of the Brute legend: 'There is a great varietie among writers, who first founded this Citie: Some will haue Brute the Troian to be first builder of it, but Brute, and his historie, is meerely reiected of manie in our daies: It was recdefied by Lud, in the yeere of the worlds creation 5131 who built the wals about it, and erected Ludgate, who also changed the name of Trenounant into Luddestowne, now London.'[23] Then, sounding like the epitome of a modern-day representative of the local chamber of commerce, Norden describes London as 'A Citie of great Marchandize, populous, rich, and beautifull.'[24] In like manner he later portrayed the native inhabitants as being 'faithfull, louing and thankfull.'[25]

Norden focused his history of the city on events such as its fiery destruction by the Danes and its eventual reconstruction – both events occurring during the reign of King Alfred. Upon establishing the date of the selection of the city's first mayor (1209), Norden slips smoothly into an account of contemporary sixteenth-century London politics, which features yet another list, this one presenting the city's wards and parishes. Prominent landmarks and important buildings such as the Royal Exchange, St Paul's, the gates of the city, Black Friars Castle, and Leaden Hall are described from a relatively intimate personal viewpoint, interspersed with the sort of detailed observations expected of a man who is a surveyor by profession: 'On the east part of the Citie, is a most famous and strong Castle, called the Tower of London, the maine tower whereof, some suppose to be builded by Iulius Caesar. It is strong and ample, well walled and trenched about, beautified with sundrie buildings, semblable to a little towne.'[26] While London is omitted in the manuscript version of *Middlesex,* conversely, Norden left out of the published work several sections found in his manuscript. The manuscript deals, for example, with the means of livelihood of the population, while the printed work does not:

Such as live in the inn countrye, as in the body or hart of the shire, as also in the borders of the same, for the most part are men of husbandrye, and they wholy dedicate themselves to the manuringe of their lande. And theis comonlye are so furnished with kyne that the wife or twice or thrice a weeke conveyeth to London mylke, butter, cheese, apples, peares, frumentye, hens, chyckens egges, baken, and a thousand other country drugges, which good huswifes can frame and find to gette a pennye. And this yeldeth then a lardge comfort and releefe ...[27]

Often Norden extended the type of preaching found in his devotional tracts to the manuscript, as in his description of Pancras, a place 'forsaken of all ... usually haunted of roages, vagabondes, harlettes, and theeves,' admonishing the reader to 'Walke not ther too late.'[28] In yet another place he exhibits a tendency common to most antiquaries, an interest in unusual or strange discoveries: 'Not farr from this place was founde the bone of a man of an admirable magnitude of late yeares, by a man laboringe in a gravel pitt, as it is reported, the vew thereof I haue desired, but it is broken and spoyled (as they saye).'[29]

Elizabeth's interest in Norden's project offered some hope to him, so that by 1594 he had completed a chorographical description of Essex. He sent manuscript copies of this work to Burghley and to another potential patron, the earl of Essex. These people however, made no offer of financial subsidies. The inevitable result was that this study remained unpublished until 1840.[30]

Essex was a much less ambitious endeavour on Norden's part than *Middlesex*. Except for the general section on England as a whole, this work, like all Norden's subsequent chorographies, follows the plan of *Middlesex*: general remarks on the soil, produce, and climate, tables of fairs and rivers, a map of the county, and an alphabetical index of places. An interesting sidelight to this list is Norden's comparison of parts of England to the biblical lands, attributable to his religious fervour. In *Essex*, for example, the shire is called the 'englishe Goshen, the fattest of the Lande: comparable to Palestina, that flowed with milke and hunnye.'[31]

This religious enthusiasm, in conjunction with his need for money, compelled Norden to produce several more didactic works over the next few years. In 1596 there appeared *A Progresse of Pietie* and *A Christian Familiar Comfort*, and in 1597 *The Mirror of Honor*. In the latter work he reflected on the type of relationship that should exist between a patron and a writer, revealing the bitterness of his own experiences:

It soundeth neerest to true nobilitie, to give plaine demonstration of his purpose, towards endevoring followers, either to encourage or discourage them, that they consume not more yeres in mourning for time lost in vaine hope, then in recounting the comforts received by best endevours. Fayre words may bee compared to a pleasing sunne, which warmeth onely, but clotheth not, nor feedeth the bodie: and as the sunne shineth upon all, so fayre words are free to all, from all. But relieving deeds are the substance, wordes but the shadow, performance is the fire, and promise but the smoake.[32]

The Mirror of Honor, it should be noted, was directed at the English soldiers who had returned from war in the Netherlands, in Ireland, and against Spain, and was intended to reform their morals.

In 1594 Norden perambulated Surrey, Sussex, and Hampshire, perhaps also visiting the Channel Islands and the Isle of Wight. The textual descriptions of these areas were not published, because of his continuing inability to enlist a patron. Some of the maps, however, were engraved and printed on their own. In desperate need of money, Norden sent, in 1595, a composite manuscript volume to the queen entitled *A Chorographicall discription of the Severall Shires and Islands of Middlesex, Essex, Surrey, Sussex, Hamshire, Weighte, Garnesey, and Jarsey, performed by the traveyle and veiwe of John Norden, 1595*.[33] The descriptions of Middlesex and Essex were abridgements of those previously written. Maps of the three islands were only accompanied by 'A brief commemoration.' The description of Sussex, Hampshire, and Surrey may have been abridged from a fuller text, but except for Surrey, evidence for this is lacking.[34]

Speculum Britanniae 65

Obtaining little reward for these efforts from his monarch, in 1596 Norden issued a *Preparatiue to His Speculum Britanniae*. This was a pamphlet designed to counter criticisms of his earlier chorographies by various learned gentlemen. In it he reminds his critics of his great task: 'I the most unworthye, being imployed (after the most painful and prais worthie labours of M. Christopher Saxton) in the rediscription of England.'[35] He also was not afraid to remind the lord treasurer, in the dedication to Burghley, of the money he had expected long ago: 'Although (Right Honourable) I haue beene forced to struggle with want, the vnpleasant companion of Industrious desires, and haue long sustained foyle, inforced neglect of my purposed business, and sorrow of my working spirit. It may yet now at lengthe please your Honour to effect what you haue begun ...'[36] The nature of this pamphlet indicates that Norden's accuracy had been attacked on several counts, and that doubts were expressed about the way he drew his maps. Norden responded to these in the following fashion: 'Some thinke it a necessarie thing to distinquish as well the Unites of euery parish, as of euery hundred ... As touching the conceite of some that would haue the distinction of the limits of euery parishe, I holde it not so needeful as impossible, and I thinke the most of iudgement will affirme the same.'[37] In general, this work laid down a series of rules for interpreting town names, thus aiding both the historian and the geographer, and also, by stressing the necessity of examining the site, pointed out the importance of history to the geographer.

Burghley again responded by issuing a letter the following year to all justices of the peace, 'requiring them to aid the bearer, John Norden, gent., who has very diligently and skilfully, travailed to the more perfect description of the several shires of the realm.'[38] This was, however, as Pollard states: 'a precarious method of endowing topographical research.'[39] Norden made clear in a handwritten appeal included in the queen's presentation copy of the next instalment of the 'Speculum Britanniae,' *Hartfordshire*, which he published at his own expense in 1598, that he had spent more than one thousand marks in five years of work on his chorographic activities. He went on to sign himself pathetically: 'Quid ego miser vltra, Your Maiesties most loyall distressed subject, J. Norden.'[40] Although, as this message to Elizabeth stated, Burghley had three times reminded the queen of her obligations to her surveyor, she was still unwilling to give Norden the satisfaction he sought.

The Description of Hartfordshire appeared with a Latin dedication to Edward Seymour, the earl of Hertford. The manuscript originally contained an English dedication to Burghley; perhaps Norden thought Burghley could persuade the queen to act on his behalf.[41] But Burghley died in 1598, before the work was published, making the subsequent change in the dedication necessary. This proved to be Norden's only chorography besides *Middlesex* to be published during his lifetime.[42] The contents and their arrangement follow a similar pattern to that of the surveys of the other shires. The boundaries are defined, the general topography of the area and the different types of soil are discussed, while the air, for the most part, is found to be 'very salutarie.' Listed next are the shrieval and ecclesiastical divisions of the county, the market towns, the fairs, the beacons, and so on. This, in turn, is followed by the standard alphabetical index, or table, which contains the etymological derivations of place-names and other items of historical and topographical interest. This index fills over half of the description. Most of the places named there are found on the map, marked by the symbols in the map's key for market towns, villages, castles, and religious places.[43] The remainder of the work consists of a short alphabetical table of the houses of the nobility and the gentry, with grid references.

In general, Norden's methodology varied little from his past investigations and therefore the results were not particularly extraordinary. A latter-day investigator of Hertfordshire, Sir Henry Chauncey, has labelled Norden as 'conceited' for reprimanding Camden over at least one issue, or point.[44] Other than this, perhaps the only notable feature is Norden's interest in antiquities, as revealed in his description of the ancient Roman city of Verulamium:

The brasse monies whereof I haue much but seeme farre more, Import the antique names, and pictures, not only of sundry Emperors, but of some of their Empresses also ... through the fury of the Saxons and Danes it [Verulamium] was sackt and subuerted. ... [At length] one Eadmere [abbot of the Monasterie of St Albans] ... found sundry Idolls, and aulters not a fewe, superstitiously adorned for the honour of these unknown gods of the Pagans, some of these Idols were of pure golde, some of other metall, and withall he founde great store of householde stuffe, and other thinges witnessing the glory both of the citie and the cittizens of the same. Besides sundry pottes of gould, brasse earth, glasse and other metall, some frawght with the ashes of the dead, some with the coyne of the aunciont Britons and Romane Emperours. And in a stone were found certayne Brytish bookes, whereof one imported the historie of Albans martyrdome.[45]

The publication of *Hartfordshire* did little to relieve the author's financial burden. So again, in 1599, Norden made an appeal, this time to the earl of Essex in *A Prayer for the Prosperous Proceedings of the Earle of Essex in Ireland*, in which Essex is compared to Moses, David, and Gideon. Norden, a fanatical anti-Catholic, evidently looked to Essex – regarded by the Puritan party as the champion against the succession to the Crown of the Infanta of Spain – for support. Essex's rebellion in February, 1601, ended any hope that the nobleman's patronage could have held for Norden. When Norden applied to Burghley's son, Sir Robert Cecil, a notorious enemy of Essex, for a renewal of his warrant to survey, it was refused. He brought even greater disfavour upon himself by his attempt to persuade the queen of the existence of a second John Norden, purported author of the appeal to Essex.[46]

Norden finished several more parts of his huge project, none of which were published in his lifetime. A survey of Cornwall and a revised version of the survey of Northamptonshire were both completed about 1610; there are references also to a manuscript description of Kent, which may have accompanied a version of his map of that county published in the 1607 edition of Camden; and surveys of Norfolk and Suffolk have been attributed to him.[47] But the total scheme was abandoned, Norden perhaps eventually realizing that it was too monumental an undertaking for one person. Of these works, his study of Cornwall is the most elaborate and the longest. Norden had finally been able to secure permanent appointment as duchy [of Cornwall] surveyor in 1605, and it is possible that the new monarch, James I, commissioned *Cornwall* from him. In the manuscript that he presented to James, Norden describes the work as 'a member of a greater body, your Majesty being pleased to further the perfection of the reste of the lineaments. Might it stand with your Majesty's good opinion and favour' he adds 'to enable me to proceed in the residue of your Majesty's Kingdom (being by the former travaile and by tedious attendance for my promised recompense, merely undone) such shall be my loyal care and faithful deligence, as nothing shall be omitted.'[48]

The work begins with a general history of Cornwall and a general description of its situation, followed by a particular description of each hundred. Norden starts at Land's End and arranges his material alphabetically. In this study he is much less sceptical of the Brute legend than he is elsewhere, by then apparently sharing Lambarde's view that there might be some truth in Geoffrey: 'But howsoeuer it stande with the opinion of some to disable this historie, and to goe about to proue it, and the whole process thereof to be but fayned, ther be manie probable inducementes to make me more credulous of the veritie therof: The generall acceptance of the historie, so manie generations: Brutes pedegree: ... And sundrye other actes, and stronge probabilities of the truth of this historie haue bene lefte vnto this age from the begining. If all be fayned, what proof can ther be, of the truth of any historye?'[49] He is even credulous of certain other legends, for example, the story that 'a great part of this Promontorie [Cornwall] is swallowed vp of the deuowring sea, namelye, the Countrye of Lioness and other Lande.' Yet his credulity was based on the fact not only that he liked the story but also that 'the Sea, ayded by tyme and tempests, is euery where a powerfull adversarie to the Land, and in moste places the Lande weake to resiste.' In other words, it rested on common sense and on personal observation.[50] This is a key point. Although a large portion of the chorography is virtually a paraphrase of Richard Carew's *Survey of Cornwall*, Norden included a considerable amount of his own original observation. For example, reminiscent of Giraldus, he writes from personal experience of the 'baser sorte of people'; 'manie of them are of harshe, harde, and of no suche civile disposition, verie litigious, muche inclined to lawe-quarrels for small causes.' Or, at another point in the text, he describes a peculiar stone that he stumbled upon in the course of his travels; and at yet another juncture he reflects upon his witnessing, at Trewardrayth, the taking in of a catch of pilchard, 'this silly small fishe.'[51]

At times Norden included material that Carew had not noticed, such as Trethevy Quoit, one of the finest neolithic tombs in Cornwall.[52] Although he did not investigate the site as thoroughly or as scientifically as did John Aubrey or Edward Lhuyd in the latter part of the century, his description is nevertheless detailed, and is complemented by a handsome illustration. The entire book, in fact, contains thirteen drawings of various 'curiosities' of Cornwall. Since becoming duchy surveyor Norden was obliged to illustrate his surveys with architectural drawings, bird's-eye views of the landscape, and the like. As for the maps that accompanied the text, there were originally ten manuscript maps – a general map of Cornwall and a separate map for each hundred.[53]

The alphabetical table, as usual, occupied most of the book. A typical entry reads as follows:

Careck rode, f.18. A Rode in Falmouth hauen, where the deepenes is suche as a Carecke, the greatest ship of Burden, may ryde; whererof it taketh name Careck rode.

or, from the same section:

Pensignance, e.16. A howse and Mannor of Richard Carew of Antony, esquire; but his moste abode is at Antonye.[54]

In the last twenty years of his life Norden was finally able to achieve wide recognition as a surveyor in both public and private employment. Early in James's reign he moved to Hendon in Middlesex, to continue his surveying.

Relatively few details are known about his family, except that in the years 1616 and 1617 he was assisted in his work by his son.

Between 1603 and 1614 Norden's surveying prevented him from putting out any devotional tracts. In 1600 he had published *Vicissitudo Rerum* and in 1601 *A Store-House of Varieties*, which was a second edition of *Vicissitudo Rerum* with a variant title-page. In 1603 he issued *A Pensive Soules Delight*, a poem celebrating Elizabeth's leniency to her enemies. But it was eleven years before Norden published another religious work.

These works reveal a typically Elizabethan versatility – the pursuit of more than one occupation. *Vicissitudo Rerum*, in particular, also reflects the author's melancholy. In it, he concerns himself with the common theme of man's inconstant state, with the problem of unending turmoil in the universe. The work specifically presents, in verse, a description of the planetary changes and their effects upon the affairs of men. The will of God, who controls all things, is seen as being behind all the visible signs of mutability in the universe. Man is simply the victim of chance and corruption:

> Th' inconstant state of *man* inconstant, mooues
> My constant Muse to moure and pause a while,
> Sad and in silence, as my state approoues,
> Beset with sorrowes, comforts in exile,
> Fed with imperfect promise (wounding smile.)
> Reft of releefe, the worlds *change* I sing.
> This first approou'd, a second part I bring.[55]

The signs of change were there for all to see. One only needed to consider the foreign and domestic affairs of the nation. That the foreign scene caused Norden more than a little apprehension is evident in the fact that his *Christian Familiar Comfort* was intended to alleviate the fear and dismay perpetuated among his fellow countrymen by the threat – real or perceived – of Spanish intervention in England's affairs. Domestic turmoil was not lacking either; Elizabeth could not live much longer, but the question of succession had not been settled yet. It is not surprising, therefore, as Marjorie Nicolson states, that 'Elizabeth's death had seemed indeed the end of the world – the end of a world that England was not to know again.'[56]

The recent scientific discoveries merely added to this sense of powerlessness. The heliocentric hypothesis of Copernicus seemed only to instil a greater pessimism. The old concept of geocentricism was no longer in accordance with the newly observed facts about the movement of the heavens. Few Englishmen were able to accept, at this time, the new theory as more than idle speculation. Even the learned Sir Francis Bacon was an outspoken and active opponent of the concept. Who could blame Norden for accepting it as yet another sign of the decay of the universe?[57] Bacon, however, did accept the overall 'Idea of Progress' linked inextricably to eternal change. Norden, on the other hand, grounded in Elizabethan science, in which he had read widely, was unable to do so.

Norden's recognition of the Ptolemaic concept of the universe is revealed in stanzas 6 to 8 of the *Vicissitudo Rerum*. Norden adds one more 'sphere' to the nine of Ptolemy. This one represented the phenomenon of variable precession – the oscillation of the equinoxes as they move along the celestial equator:

> Ten spheres in one, Astronomers do hold:
> The tenth revolving in his fixed tide,
> Twenty foure houres, and then his circle rold,
> Again revolves, powers infinite her guide,
> From East to West, still on the dexter side:
> And by her course most swift and impetuous,
> The rest she moveth, most miraculous.[58]

It was not until his old age that Norden could finally commit himself fully to his main love – instructing others in piety through his religious works. In the dedication of *An Eye to Heaven in Earth* to Sir Henry Hobart, the head of the commission on revenues who was responsible for discharging the crown's financial obligations to Norden, Norden wrote: 'I can no better expend the idle interims of the rest of my libertie (my dutie and care, to answer your expectations in my service, duly respected and performed) then to seeke my owne satisfaction, in that which immediately concerneth not your ordinarie imployments: But my future accompt of another service, enjoyned me by another Master.'[59] In the last six years of his life Norden was able to complete several more works of moral instruction as well as another work on topography.

Norden's work as a surveyor is considerably more extensive than that of Saxton. In 1600 he was acting as surveyor of the crown woods in Berkshire, Surrey, Devonshire, and elsewhere.[60] Also important as an example of his efforts as an estate surveyor is his work on the estates of Sir Michael Stanhope on the coast of Suffolk (1600 and 1601), drawn on twenty-eight sheets. These display all the usual elements of the surveys of the period – waterways, houses, churches, woodlands, meadows, field boundaries, and the names of owners. Norden paid special attention to the coastline in order to show and account for an area of beach, part of the manorial property. His exceptional interest in erosive processes, such as beach erosion, is displayed both here and in some of his chorographical works, for example, in *Cornwall*.[61]

After having obtained the surveyorship of Cornwall, Norden surveyed the Windsor area for the king.[62] For this he received from James a gift of £200 in acknowledgement of a vellum folio manuscript. This work is notable for its beautifully coloured maps, including a 'Plan or Bird's-Eye view of Windsor Castle from the north.' Of even greater interest is *The Surveyors Dialogue*, an important textbook for his profession, published in 1607. This last work is primarily concerned with demonstrating the importance of surveying. It is dedicated to Cecil, with the commendation that the surveying of land will increase the revenue thereof. It consists of five 'books' couched in the form of a dialogue between a surveyor and several other people. Norden constantly appeals to the past, and to scripture, for support in his plea for the necessity of surveying. Referring to Joshua 18, his surveyor says: 'Joshua commaunded the children of Israel, that euery Tribe should choose out three men, that he might send them thorow the land of Canaan, to view, survey, and to describe it.'[63] This book, aside from ending with advice that will help ensure his continued employment – by recommending that an estate survey should be renewed every seven to ten years – lays down the ideal relationship between a landlord and a tenant with respect to tenures. The second book continues along the same lines; the third contains a consideration of graphic surveying in a dialogue between a 'Bayly' and a surveyor. The surveyor describes in detail the instruments available to him, such as the plane table, the theodolite, and the chain. Norden's main concern in his county studies was not necessarily with correct plotting; hence he was little concerned with errors introduced through the use of the plane table, for example. The remainder of the text deals with the method of keeping a court of survey, how to enter and enroll deeds, and related points of law. Book 4 centres on the usefulness of the compass and of tables of computation.

There is also a discussion of various ways of improving the land. Curiously, at times Norden sounds much like a modern-day conservationist. In a discussion of woodland that is being destroyed for the sake of sowing corn, Norden states: 'But it is to be feared, that posterities will find want, where now they thinke is too much [woodland].' He continues: 'Things that wee haue too common, are not regarded: but being deprived of them, they are oft times fought for in vaine.' This is followed by a biting commentary on the wasting of woods in Sussex.[64] The advice he offers with regard to the improvement of land is sound and based on common sense and in this he is closer to the agriculturalists of the Restoration than to his contemporaries.[65] That his opinions on soil improvement were highly thought of may be concluded from the publication of a fourth edition as late as 1738. Generally, *The Surveyors Dialogue* is an interesting work in this respect, but as far as surveying *per se* is concerned it was a statement of things as they were, contributing little to the progress of this science.

In the following years Norden continued to visit the various regions of England. Between 1609 and 1616 he surveyed in detail many of the manors and estates of the duchy of Cornwall. His travels throughout the countryside and his appeals to authority concerning diverse matters are contained in the *Calendar of State Papers*.[66]

Norden's chorographies and professional surveys have proved to be of considerable interest to local historians because of the wealth of information they contain on the legal, economic, and social history of early-modern England. In more than one instance he examined the manorial customs in addition to topographically surveying the land.[67] Norden, it is thought, was publicly employed for the last time in making a survey of the manor of Sheriff Hutton in Yorkshire in the summer of 1624.[68] At the end of his career he published *England, An Intended Guyde for English Travailers*, a thin, small quarto book containing a series of distance tables for England, Wales, and the English counties, based on the triangular scheme invented by him. This, a rare road book of the early seventeenth century, is yet another production of his that established a model for others to emulate in the future.

John Norden's story is a common one in the sense that many in the England of his day struggled against uncertainty just as he had. Norden is remembered today as a cartographer as well as a surveyor and religious writer. His popularity and expertise is indicated by the fact that William Camden chose for inclusion in his *Britannia* (1607 edition) a number of Norden's maps of the various shires. This was in part due to Norden's many cartographic innovations, three of which were very important: he invented a useful table that gave distances between towns; he included on his maps markings that showed the main roads in each of the shires; and he invented new representative symbols that he added to his maps.

As a chorographer, Norden has been justifiably criticized on several counts. His knowledge of history was adequate, but it may be significant that he was not a member of the Society of Antiquaries, probably because he was not socially acceptable, his gentility too slight and too recent. Yet his acquaintance and in some cases close friendship with some of the major antiquaries of his day proved to be immensely beneficial to his undertakings.

(Even this has indirectly inspired criticism of his work, as when Gough suggested that Norden's *Cornwall* drew too heavily on Carew's work on that county.)[69] At the same time, although he was not nearly as knowledgeable in languages as was Lambarde or Camden, he was much less inclined than most of the other antiquaries to indulge in wild etymologies, and in fact he did some valuable work in this field – especially evident in his chorographies of Northamptonshire and Sussex.

The accuracy of Norden's work is rarely assailable. Other evidence for the subjects that he dealt with usually tends to verify his own conclusions. He also had a keen eye for detail. This is evident in his description of the manor of Mincinbury in the parish of Barley, Hertfordshire: 'Ib. Manor of Mincinbury. Demesne Lands there. The lord holds in his own hands the site of the aforesaid manor, the buildings of which are almost completely in ruins except for a barn which was formerly couered with straw. The barn is spacious and capacious. The aforesaid house is situated among woods, dark and muddy and adorned with neither orchard or garden.'[70]

Levy has commented on Norden's 'strong visual sense, a sense that was titillated by the vicissitudes of time as shown in ruins.' This ability was required of a good chorographer. 'Unlike almost everyone else ... he went beyond describing a ruined castle merely because it was there: Norden almost automatically saw it peopled and thriving, as it must once have been.'[71]

It is a pity that this man, who contributed so much to several fields of human endeavour, was frustrated again and again in his efforts to secure a patron. But as Christopher Bateman states: 'It has been just matter of complaint in all ages that those who have most eminently distinquished themselves in behalf of mankind have generally met with unworthy treatment, having been forced to an unequal combat with neglect and poverty while living, and even when the envy against their merit has ceased, the memorial of their virtue has been buried in the same grave.'[72]

By Norden's day interest in chorography was in part a side effect of the excitement generated by voyage and travel literature.[73] Even outside of Europe, chorography grew.[74] In the area of cartography, England obtained personnel and methods from Holland.[75] Many of the early British maps were printed there, and eventually joint partnership in publishing houses was common between these two areas.

There was, as Taylor points out, still an 'absence of any systematic literature of regional geography' within Britain itself, although there were certain intertwining links between individuals or among groups of writers.[76] The Elizabethan Society of Antiquaries only occasionally devoted attention to work of a directly chorographic nature. The contacts that existed among the individual members of the society involved in this field were mainly of an informal nature and therefore not always recorded. Meanwhile, the progress of academic geography in Britain during the seventeenth and eighteenth centuries was 'largely concerned with the University of Oxford which during that period led the way in geographical study and accomplishment, and produced one work of outstanding merit.'[77] That work, *Geography Delineated Forth in Two Bookes*, is by Nathanael Carpenter, supposedly the first Briton to write on theoretical geography as distinguished from mathematical treatises on navigation or narratives of travels.[78] But Carpenter's work, although 'a very important general geography,' had little or no discernible effect on the work of the early regional writers.[79]

Even so, chorographers such as Norden, it may be noted, helped stimulate others to seek new ways of constructing accurate maps and instruments, and so cartography and surveying profited from chorography. The first English book claiming to deal with surveying matters was published in 1523, the *Boke of Surveyeng*.[80] But it was the transfer of lands accompanying the dissolution of the monasteries that generated a great need for accurate land measurement. Auspicious conditions for the growth of surveying were also provided by the revival in mathematical activity. Interest in mathematics was given new life partly by the necessity of developing precision instruments for overseas navigation.[81] Therefore, it is not surprising that scientific methods should be applied to the measurement of land. By the time of Norden's death, then, several works dealing with the art of surveying had appeared.[82] In most instances, though, the practical surveyor was not a man of letters. Literary education was easily available only to a few; quite often improvements made in the methodologies of surveying went unrecorded until brought to the notice of a surveyor who also happened to be a writer – Norden being one example. Most chorographers left the more scientific measurements to the professional surveyors, content to use words, not instruments, in charting their own territories. Surveying, therefore, although given an impetus by the chorographer, developed largely in a separate, more mathematical direction.[83]

Most chorographies written in the early seventeenth century emphasized the historical, antiquarian, and topographical elements. Their authors' reading of Camden, Lambarde, and the civil historians largely accounted for the historical part. But several other factors combined to involve the chorographers in topography. The obvious one was a natural interest in the landscape, in some ways akin to that held by the eighteenth-century romanticists. Yet this interest was not nearly as scientific in nature as that of the later natural historians, who undertook the first modern studies of geomorphological features. To the chorographers it was the *surface* of the earth, its 'visual impact' that really mattered, and not its underlying features or composition. Conversely, it may be argued that the natural historians tended to neglect the topography in favour of ample discourses upon the plants, archaeology, strange accidents, and physiological abnormalities of the regions they described; but their interest in and investigations of earth history – geology, palaeontology, stratigraphy, etc – cannot be so summarily denied. In other words, they placed importance on the complete physical composition of the land, not merely its visual aspect. Few if any of the British regional writers of the early part of the seventeenth century, therefore, even approached a really serious 'scientific' study of the earth's natural history, and saw no reason to do so. 'People' interested them more than 'rocks.' However, a few non-chorographers did try their hand at such studies as natural history. Here again, the classical writers had led the way, and later, during the Renaissance, Leonardo da Vinci; Georg Bauer Agricola, the 'German father of mineralogy'; and Bernard Palissy had all developed 'proto-scientific' theories about one aspect or another of landscape evolution.[84] As we have seen, William Bourne and Thomas Hill, who are not considered to be regional writers, were among the earliest British writers to take an interest in the subject.

Up to the seventeenth century scientific explorations were still inhibited by some fundamental beliefs inherited from medieval times, which placed severe restrictions upon the consideration of natural history. Some of these beliefs persisted into the second half of the century, but their greatest impact was felt before the advent of the new ideas associated with the rise of the Royal Society. For example, while the belief that God's omnipotence was everywhere revealed in nature was common throughout the entire century, in pre-Restoration England both the Anglican conformists and the Puritans held that sin was a paralysing, all-consuming evil that had spread throughout nature, causing universal decay. The seventeenth-century theorists changed the direction of their attention from the decay of man to focus on the degeneration of the sublunar reaches, especially the earth itself. Comets, earthquakes, the infertility of the soil, and so on, as compared to conditions in biblical times, were viewed as manifestations of this degeneration, and of the wrath of God. The utopian existence of mankind, for most thinkers on the subject, had already occurred in the antediluvian world, that before the Noachian flood. An overwhelming amount of time and energy was now devoted to trying to reconcile a growing knowledge of nature with the Mosaic record as contained in the first few chapters of Genesis.

Such bibliolatrous outlooks on nature also centred to a large degree on the problem of the earth's age. The Mosaic tradition generally accepted in the seventeenth century was that the earth was less than six thousand years old and its end was imminent. This by itself tended to stave off speculation about what was considered to be an accompanying feature of the earth's degeneration, namely the process of denudation – which is in itself a major topic of natural history – because if the end of the world was at hand there was no need to consider denudation and its resultant destruction of the continents. Later, once this present danger had passed in the minds of men, denudation and other geomorphological topics such as earthquakes were again considered safe ground for the theorist, who upheld the idea that such processes were slow ones, of little danger to anyone. However, before this view became current, the idea that nature had only six thousand-odd years to complete her degenerative duty made an evolutionary view of natural history impossible.

It was only by the second half of the seventeenth century that scientists began slowly to realize that the age of the earth – as seen in nature – was in fact greater than that allowed for by the Mosaic tradition; only then did they begin to reinterpret the story of Creation less literally, or at least slightly less literally. New questions about earth-chronology were being asked. The study of fossils, for example, was thought possibly to hold new answers.[85] Other natural phenomena were also seen as indicative of the great age of the earth.[86] When Edward Lhuyd, in speculating as to the mountaintop origin of numerous boulders found on the floor of a valley in northern Wales, discovered that within living memory only two or three were witnessed to have rolled down, he told John Ray that at this rate many thousands of years would have been required to account for the dislocation of all of the boulders: '... in the ordinary Course of Nature we shall be compelled to allow the rest many thousands of Years more than the Age of the World.'[87]

It is evident, then, that even before the growth of science in Britain the serious study of natural phenomena was stultified by a variety of factors. These tended to encourage further some strange notions about the subject. Scripture – Psalm 104 – was used to support the theory that the general level of the oceans was higher than that of the continents. One theory, commonly referred to as 'microcosm-macrocosm,' held up the idea that man (microcosm) was a replica of the universe (macrocosm), so that human bones were regarded as the counterpart of the earth's rocks, man's flesh likened to the soil, his pulse to the tides, etc. Because none of the regional writers of the early part of the century was much of a theorist when it came to the discussion of such topics, each tended to confine his attention to readily perceptible surface features, such as the topography. This meant that if one had to examine any features of the earth at all, one could avoid an extensive discussion of its intrinsic scientific properties – of which the chorographer was basically ignorant in any case – concentrating instead on human settlement with its accompanying transformation of the landscape. This had the added benefit of conveying a pleasant picture of the countryside, one unencumbered by distracting side trips into hazardous territory.[88] It is therefore not surprising that this increasing emphasis on the study of man was extended to the point that genealogy largely replaced topography as one of the major components of chorographic study. Only with the growth of natural historical studies in the second half of the century was attention devoted to the examination of the earth *per se*, that is, to the exact science of land-forms.[89]

In the early decades of the century it was widely believed that the earth's topography was essentially of primeval origin, that land-forms were literally moulded by the divine hand on the third day of the Creation. There were some others who maintained that the Noachian flood transformed the contours of the earth into their approximate present shape and form. But by the second half of the century, when one discusses the natural historians, it should be kept in the back of one's mind that they wrote at a time when morphological theories of the origin of topographical features were becoming increasingly popular – with the flood theory close behind – when earthquakes were regarded as the major force in earth sculpture.[90] Despite their lack of interest in land-form creation, most chorographers of the early part of the seventeenth century described the actual topography of their region.[91]

The liveliest, most exhilarating chorography of the time was Richard Carew's *Survey of Cornwall*, published in 1602.[92] Carew was a member of one of the leading families of Cornwall, a justice of the peace, a high sheriff, and, in 1584 and 1597, a member of Parliament. He was also a member of the early Society of Antiquaries. He and Camden were friends; both studied at Oxford. Camden may have been responsible for persuading Carew to write a full account of Cornwall, exceeding in substance the one in his own *Britannia*.[93] Other notable Englishmen later praised Carew's efforts, including Ben Jonson, who in his *Execration* (1639) classed Carew with Cotton and John Selden. Camden, in turn, incorporated much of Carew's material into various issues of the *Britannia*.[94] Sir Henry Spelman was also indebted to the squire from Antony for the valuable criticism of his work on the rights of the churches, the *Epistle on Tithes*; for his assistance Spelman dedicated this book to Carew.

Because F.E. Halliday's introduction to the *Survey* says about all that needs saying about this work, there is no need to cover it in detail here. The *Survey* followed Lambarde's plan of first providing general information according to topics, then a hundred by hundred description. Following

Lambarde's example, the genealogical information is scarce, while the industries and the topographical features of the area are described in detail. And the worthies and the author's friends and acquaintances are depicted with a quaintness peculiar to Cornwall, as more than one observer has noted.

In his study Carew expressed his secret sympathy for the story of Brute, accepting the legend that Cornwall was bestowed by Brutus on Corineus after his exploit in wrestling with Gogmagog on Plymouth Hoe.[95] Staunch West Country man that he was, Carew was unwilling to desert a former fellow Cornishman by birth, King Arthur. There is also a painstaking portrayal of husbandry, an account of typical Cornish delight in traditional sports and pleasures, such as archery, 'hurling,' and wrestling, an historical report of the Cornish government, civil and ecclesiastical; some topographical detail; and so on. Throughout this work Carew's prose has a rhythmic, almost colloquial quality that is fired by a lively use of alliteration. The description of the condition of Cornish houses, for example, is worth quoting: 'Of all manner vermin, Cornish houses are most pestered with rats, a brood very hurtful for devouring of meat, clothes, and writings by day; and alike cumbersome through their crying and rattling, while they dance their gallop galliards in the roof at night.'[96] Such words have made this work 'one of the minor classics of English prose,' according to its editor.[97]

After Carew and George Owen, no one wrote a full-scale county chorography for fifteen years. Yet the work of one man, which covered all of Britain, must be mentioned here. John Speed (1552–1629) was a member of the Merchant Taylors' Company.[98] Antiquarian and historical pursuits were secondary to Speed's interest in cartography, but he was a member of the Society of Antiquaries, and there became acquainted with Camden, Cotton, and Spelman. Unlike Norden, he enjoyed the constant patronage of a man who stood high in the court of King James, Sir Fulke Greville.[99] Freed from financial concerns, Speed devoted himself whole-heartedly to his cartographical and related concerns, free 'to express the inclination of my mind.'[100]

Like the other chorographers in the tradition stretching back to the famous John Leland, Speed was concerned with the face of Britain in both the past and the present. In 1598 he presented the queen with maps of several of the counties, and between 1605 and 1610 he published a series of fifty-four county maps that had been drawn originally by Saxton and Norden. Sometime during these years Speed determined to join their maps to his own chorographic labours and thereby fulfil the design of Leland in a single work. By 1611 he was ready. In that year he published the *Theatre of the Empire of Great Britaine*, a large folio volume that included the maps along with a descriptive chorographic text and an index on the back of each of these. In conjunction with this work Speed issued a history of Britain that, although primarily a reworking of older histories, was widely admired by his contemporaries.[101] It was decoratively illustrated, expensive, and impressive. Speed's own critical abilities were not profound. In the *Theatre* he explained that his borrowings from previous writers was for a worthy cause: '... wherein I have held it no sacrilege to rob others of their richest jewels to adorn this my most beautiful Nurse, whose wombe was my conception, whose breasts were my nourishment ... '[102] Camden provided the main source. Nevertheless, Speed was industrious in making extensive use of discoveries and materials of the English antiquaries. He supplemented his work with his own observations – to which the geographical notes in the descriptive text bear witness – and he claimed to have personally perambulated every province of England. Notwithstanding the work of Saxton and Camden, the *Theatre* proved to be the earliest English attempt at atlas production on a large scale.[103]

The chorographic text that accompanies the maps in the *Theatre* presents some new topographical-historical information. The descriptions all follow the plan in the first section of Lambarde's *Perambulation*, but without a detailed review of the locations. A survey in the *Theatre* treats about a dozen subjects; name, situation, natives, air, soil, past and present civil and ecclesiastical government, various 'wonders,' and the hundreds. All the information for each shire is virtually compressed into one folio page.

Speed chose his authorities carefully, utilizing the *Domesday Book*, Nennius, Bede, Leland, Lambarde, Stow, Camden, Carew, and Cotton. He described Cotton as 'another Philadelphus in preserving old Monuments and ancient Records: whose Cabinets were unlocked, and Library continually set open to my free accesse.'[104] On more than one occasion he wrote to Cotton, sending the latter 'coynes' to be added to his collection, or merely in search of information: 'Yf you will send a Note of all Monasteryes in the Realm, as also the Book of Henry the fourth, I shalbe much beholding to your Worship. Thus you see how bold I am, but it is in love of that Kingdom which your self seeks still to adorne.'[105] It was only natural that Speed should have depended heavily in his massive compilation on the material supplied to him by his fellow antiquaries and by the local gentry of repute. He was not prepared to concoct stories, but relied on the information given him by recognized authorities. Desiring his work to be as accurate as possible, he specifically asked for reliable persons to supply him with the names of every hundred, parish, and notable place within each township.[106]

The account of Speed's native Cheshire is primarily based on his own survey, and provides us with a typical description found in the *Theatre*. The borders and the 'forme' (ie shape) of Cheshire were discussed first, followed by an exaggerated account of the climate and soil: '[In] aire and soile it equals the best, and farre exceeds her neighbours.' The account of the ancient inhabitants contained references to the works of Ptolemy, Tacitus, and other classical authors. The shire was found to be the 'Seed plot of Gentilitie,' producer of many ancient and worthy families; the men were referred to as 'The chiefe of men,' while in beauty, grace, and feature, the women were portrayed as inferior to none. The principal commodities were listed next. The town of Chester was given prominence in the narrative; the city castle and the minster were described, followed by a brief history of the city, and a statement on the contemporary political divisions. Finally, Speed's dismissal of a certain 'wonder' is of curiosity value:

If I should ... inforce for truth the Prophecie which Leland in a Poeticall fury fore spake of Beeston Castle, highly mounted vpon a steepe hill: I should forget myself and wonted opinion, that can hardly beleeue any such vaine predictions, though they be told from the mouthes of credit ... [for example] Leyland for Beeston, whe [who] thus writeth:

> 'The day will come when it againe shall
> mount his head aloft,
> If I a Prophet may be heard from Seers
> that say so oft.'[107]

The *Theatre* is noticed today mainly for its cartographic value. Apparently, Speed had access to many of Norden's manuscripts and printed works, and used them liberally as sources for his own county maps.[108] However, his originality in preparing some of the 'Chards' (maps), especially the town plans, is undeniable. his contribution is explained in the *Theatre*: 'Some [plans] have been performed by others, without Scale annexed, the rest by mine owne trauels, and unto them for distinction sake, the Scale of Paces ... five foote to a pace I have set ...'[109] The utilitarian aspect of such plans had been clearly enunciated by Agas in *A Preparative to Platting* (1596): the surveyor should 'so lay out the streets, waies and allies, as may serve for a iust measure for paving thereof, distance betweene place and place, and other such things of use.'[110] The value of large-scale maps and plans, however, is underrated by historians and geographers alike. Drawn up by sixteenth- and seventeenth-century chorographers and cartographers, these often provide important historical or statistical details; depict the site and structure of buildings, many of which no longer stand; or form a reliable record of the changes in a town's street plan. In some cases, for example, they can elucidate the street plan of a medieval town. In Speed's case, he often supplied the sites of buildings that were later destroyed, as in his plans of Hereford and Gloucester.

Speed's work is distinctive for yet another reason. In basing his *History* on Camden, 'that brightest lampe to all Antiquities,' Speed dismissed Brutus as a 'vulgar received opinion.' In his consideration of the Trojan legend he exhaustively searched out the earliest sources and listed the arguments on both sides, citing authorities from Nennius down to Humphrey Llwyd. Speed's conclusion on the subject reveals that his doubt about the Brute legend, if anything, was even more forthright than that of Camden.[111] But the uniqueness of Speed's work is found in a related chapter, that concerning the manners and customs of the ancient Britons. Here, the ancient Britons are portrayed as a barbarous, polygamous lot who dye themselves, wear long hair, and practice 'diabolical superstition' as religion. Speed went on to draw a comparison between these primitives and contemporary American Indians, and included two pairs of drawings showing the Britons as he thought they appeared in ancient times.[112] Thus he attempted to depict historical figures literally, only as the evidence of authoritative sources indicated. Speed's intelligent pictorial use of coins, and the superiority of his account of Anglo-Saxon England – to Camden's due to his use of family trees of the principal royal houses – has also been noted by Kendrick.[113]

Like Norden, Speed lamented the 'instability, both of Mans life, and Glory: (a point fitting for great Princes ever to thinke on).'[114] He was quick to moralize, but this may have been due in part to his involvement in theological matters; he was an expert in ancient genealogies. His vehement Protestant bias, like that of Norden, Lambarde, and some of the other antiquaries, is revealed throughout his works. In 1610 Speed received a royal privilege that required that for ten years his *Genealogies recorded in the Sacred Scriptures* (nd) be inserted in every copy of the authorized version of King James Bible, thus increasing the price of the Bible. In this respect, again like Norden, Speed sought financial gain outside of his main field.

By the time Speed published his works, chorography had become a national pastime in a sense, diligently researched by some, avidly read by others. As new authors came on the scene, its nature was constantly evolving, although the great changes came only after mid-century.

CHAPTER FOUR

Removing the 'Eclipse from the Sunne'

> I have adventured (in some sort) to restore her [Leicestershire] to her worth and dignity, being animated hereto by the examples of many grave and worthy men.
> William Burton *A Particular Description of Leicester Shire*

IN WRITING THE ABOVE, the chorographer of Leicestershire William Burton was acknowledging his debt to Camden and Speed, and other notable chorographers. This is not surprising since, as the seventeenth century progressed, Camden's *Britannia* and Speed's *Theatre* dominated local history to the extent that no other county surveys actually reached print in the twenty years from the issue of Richard Carew's *Cornwall* in 1602 until 1622. In fact, if one overlooks the various quasi-local histories printed after 1622, this terminal date – with respect to England, at least – may be extended into the mid 1650s. This relatively long period of publishing inactivity, however, does not mean that regional work itself ground to an abrupt halt; antiquaries throughout the nation continued to assemble chorographic and similar types of material, but the emphasis was now directed at producing studies that were local, not national, in scope, and thus these may have been considered less worthy of publication at this time. There was a growing tendency to shift the locus of attention in chorographic studies, from topography and history to genealogy and heraldry. In most cases, the shift meant chorographers deliberately wrote for a limited audience, for example, for a particular land-owner who wished to have the genealogy of his family researched and set out in a manuscript, to be deposited in the family muniment room for the benefit of his descendants. In other instances the chorographies were directed at a larger, county-wide audience – but not a national one – simply because in the larger market-place the competition with the works of Camden and Speed would have been too great. Because these latter two figures had already covered the history and topography of much of Britain, many chorographers apparently now believed that the inclusion of genealogy and heraldry in their work offered the only hope of contributing further to the general goal of uncovering the riches of Britain's physical and human resources.

Since it was much easier to investigate in detail the backgrounds of a few prominent local families rather than family backgrounds throughout all of Britain or England, the former became the usual course of action. Often the patronage of local families, whose histories were reported in such studies, left the authors with little financial incentive to publish. It became common, therefore, for several manuscript copies of a study to be distributed merely hand-to-hand locally. In any case, the few authors who did attempt to get their work into print soon discovered that the majority of publishers were unwilling to accept it, not only because of the fear of competing with the national works, but also because the general public was inclined to believe that chorographies that included genealogy and heraldry were uninteresting, and therefore unworthy of purchase. Such circumstances as the ones described above apply to most – if not all – of the chorographies reported in this chapter, even in those cases where history and topography still had their place alongside heraldry and genealogy. Furthermore, because all of the chorographies covered in this chapter reached their final form in the interval between about 1618 and the outbreak of the English civil war, chronologically they form a cohesive unit for study.

The authors of these works tended to place more emphasis on standard literary sources than had their predecessors, primarily because of the nature of their investigations. These required considerably less visual inspection of the countryside and a greater amount of research into family registers. This fact by itself enables them to be counted among the first to include lengthy genealogical tables, lists of civic officials, and the like into antiquarian study. Also, except perhaps for Thomas Westcote's survey of Devonshire, these chorographic works generally exhibit a pronounced lack of literary flair, which only contributed to the difficulty of finding a publisher. Because most of the authors were well-educated men, trained at the Inns of Court or at similar institutions, this lack of literary panache can be largely attributed to the dry nature of their topic – one based largely on genealogy. This was the age of great scholars such as Edward Coke, John Selden, and others, when the legal profession in England was devising innovative ways and means of examining political and legal issues. So, it is not surprising that a new generation – which was familiar with law, and included men who in most cases enjoyed the opportunity of retiring to country estates where their appetite for chorography was whetted by the beauty of the surrounding countryside – was soon engaged in examining the locales in a slightly different manner from that of its predecessors. These men applied a formal or practical training in genealogical research to their chorographic studies, which enabled them, as William Burton metaphorically indicated, to cast light on the antiquities of the land in a new manner.

We may begin with Robert Reyce (1555–1638), whose chorographic activity was among the first of this period to exhibit a slight digression of interest from that of the previous workers in the field. Reyce spent most of his life at Preston, a village in the south of Suffolk. He belonged to a close-knit circle of prosperous local gentry of Puritan persuasion. His wife, Mary Appleton, was the eldest daughter of a family of wealthy clothiers. Many of Reyce's associates and correspondents were men of antiquarian interests. These include John Winthrop of Groton, one of the Massachusetts Puritans and the first governor of the colony. Reyce also knew Augustine Vincent; Windsor Herald, aide to John Weever in the compilation of Weever's *Ancient Funerall Monuments*, a work of considerable interest to antiquaries; and James Strangman, Essex antiquary, a 'somewhat shadowy figure,' but a member of the Society of Antiquaries, who was generous in helping other scholars.[1] Reyce kept up a correspondence with Sir Simonds D'Ewes of Stowlangtoft, an antiquary to whom he was distantly related, and who was associated with Spelman, Selden, and Cotton. That Reyce was highly thought of by contemporaries is plain in the notice of his work in several contemporary books, including John Guillim's *A Display of Heraldrie*: '... a worthy Gentleman, whose great charge and care in collecting and preserving the Antiquities of that county Suffolk merits a large Encomium.'[2] That he was kind and charitable is evident in his concern for the welfare of the poor, a fact confirmed by the establishment of a charity at Preston that bore his name. No information exists about Reyce's earlier schooling. He may have attended a small grammar school at Boxford near Preston, but he tells us only that Suffolk was the 'Country, vnto the which next vnder God, I doe owe that little that I have, for my birth, education, and habitation.'[3] His name is missing from the records of the English universities and Inns of Court, but like many Elizabethan Puritans he studied in Geneva at Theodore Beza's academy. His familiarity with Latin and possibly Greek is evident in his work; his knowledge of theological issues may be conjectured from his correspondence with John Winthrop. His last surviving letter, to D'Ewes, was written when he was eighty-two.[4]

Reyce wrote the *Breviary of Suffolk*, his major chorographic effort, for his friend and patron Sir Robert Crane, knight of the shire and high sheriff of Suffolk, and sent him the manuscript with the stipulation that it be kept in the

Crane family and not published. The *Breviary* is the most substantial of the three works on Suffolk attributed to Reyce. One, a manuscript anthology of Latin records relating to East Anglia, survives only in a copy in another hand.[5] Another consists of brief accounts of past royalty, nobility, and gentry of Suffolk, accompanied by descriptions of their arms.[6] Reyce probably began work on the *Breviary* itself shortly before the death of Queen Elizabeth in 1603, although surviving copies were given their final form between 1618–19, and 1627–31. Several items of internal evidence point to the 1602 date for the original of the work. The fact that Reyce was working on it by this date is demonstrated, for example, by a remark about 'this last yeares price 1602 of hopps.'[7]

At first glance Reyce's work seems to follow the classical pattern that William Lambarde had pioneered in Kent a quarter of a century earlier. The organization of material and the choice of headings is common to both works. The title appears to have been adopted from *The Breviary of Britayne* (1573), Thomas Twyne's translation of Humphrey Llwyd's *Angliae Regni Florentissimi Noua Descriptio* (1573). One of Reyce's immediate inspirations was likely Carew's *Survey of Cornwall*, which was published at about the same time Reyce began his own chorography. Both of these works agree not only in the general ordering of content and the titles of the majority of the sections, but often also in the phraseology.[8] The *Breviary* survives only in two recensions. The core of the work – a general description of Suffolk – remains relatively constant in both of these. One was made for Sir Robert Crane around 1618–19, a transcript of which is in the Harleian Collection (MS 3873); a further transcript of this was used by Hervey for his edition. The second recension survives in the Ipswich Central Library. It was made at a later date, between 1627 and 1631, for an unknown person.

Reyce begins the *Breviary*, 'the best Jewell that I haue,' with a series of now familiar topics. First come brief notices of name, climate, size, borders, rivers and lakes. Being in the 'Fatt of the land' (that is, wealthy), this shire, says Reyce, is first to be spoiled in time of domestic insurrection or strife. In addition, its close proximity to the centre of governmental affairs renders it susceptible, in Reyce's mind, to an abnormally heavy financial burden. Sorely lacking is a complete topographical description of the county, with the exception of the following general statement: 'This country delighting in a continuall evenes and plainnes is void of any great hills, high mountaines, or steep rocks, notwithstanding the which it is nott always so low, or flatt, butt that in every place, it is severed and divided with little hills easy for ascent, and pleasant Ryvers watering the low valleys, with a most beautifull prospect which ministreth vnto the inhabitants a full choyce of healthfull and pleasant situations for their seemly houses.'⁹ The description of the soil type is not superior to that found in similar studies, and as for minerals, we are told only that in ancient times there may have been a gold mine near Banketon. From a discussion of stone quarries, Reyce abruptly switches to the 'things of life,' which he takes to include wild flowers, timber, and other like products. Buildings and other dwelling-places are next on the list, but Reyce's description of these is superficial and considerably inferior to that of others.

Much of the *Breviary* is concerned with genealogy and heraldry. Reyce utilized original documents as sources of information, especially when he was unable to trace a certain name 'either in history or others discent, or in any records of Knights fees, aydes or tenures.'[10] He also relied on various printed works, such as Holinshed's *Chronicle* and Hakluyt's *Voyages*. The antiquarian nature of the *Breviary* is clearly revealed in this section. Reyce obtained the rest of his information from blazons of arms and epitaphs in churches. On the whole, he shows a scepticism for family tradition, as in his comments on the genealogy of the Waldegraves: 'I confesse I have seen this discent following which I dare nott express it for truth, although some of the family doe much doate in it.'[11] Many of the regional studies undertaken from this point on seem to resemble Reyce's work as much as they do the model established by Camden. Many of these, like the *Breviary*, remained unpublished for many years. It may be that people were more interested in travel accounts, descriptions of voyages and exploration, and in popular rather than local histories.[12] William Webb's *A Description of the City and County Palatine of Chester*, for example, was written in the early seventeenth century, but not published until 1656.[13] Webb, under-sheriff to Sir Richard Lee in Cheshire, rode over the entire shire personally visiting the places he describes. His account is therefore trustworthy, being at the same time quaint. Webb concentrated on 'painting' a picture of the lordly houses and county seats in graphic detail.

Webb commends the work of John Norden, while regretting that Norden was unable to complete his avowed aim of chorographically describing all of the shires. Likewise, John Speed is noticed for his work. But Webb was 'transported with I know not what longing desire, that some particular Descriptions of other parts and Countyes of the same Kingdome, not yet by any man published might be taken in hand.'[14] He decided, therefore, to partly fill the gap by producing a study of his native city of Chester, a study purportedly based on Stow's *Survay of London*.

Clearly, Webb's main concern rests in the holding of land and officeholding. In this respect *Chester* falls somewhere between Stow's *Survay* and Reyce's *Breviary*. In some ways Webb's work fits neatly into the general tradition of chorographic literature. The situation and shape of Cheshire are covered first, and the reader is then swiftly plunged into descriptions of each of the seven hundreds of that county. There we find a list of dwelling-places and their occupants that resembles in many ways a modern county directory:

'... upon which Brook or River, from Coghal towards Chester, lies next the Lop of Wirvin, the Lands of Iohn Hurleston Esquire; to which also joyns a Demayn of his called Piton Farm; and the next Neighbor to Wervin; upon the said Brook, is Moston, not long since purchased and beautified with a delicate house of Brick, by Mr. Iohn Morgel ...'[15]

The account of the town of Chester dominates the entire work. First, Webb attempts to uncover details of its foundation and the origin of its name, citing the many sources that he has consulted. These include Higden's *Polychronicon* and Camden's *Britannia*. Chester, he speculates, is derived from an abbreviation of the Roman *Legescestria*. Glossing over the early history of the town, Webb briefly describes its walls, towers, and gates, in an attempt to emulate Stow. Streets are located in relation to one another, but are not examined house by house as in Stow's *Survay*. Next the civil and ecclesiastical governments past and present are discussed, and excerpts from *Domesday Book* for Cheshire are included. In general, compared with the chorographies of the flamboyant Richard Carew, *Chester* lacks a certain literary flair. We obtain, for instance, little notion of the nature of the local customs; was 'hurling' one of the sports enjoyed by the people? And what was the physical appearance and dress of the inhabitants like? Many questions such as these remain unanswered, perhaps partly because Webb was one of that group of writers who had no pressing desire to publish their findings.

By the time Webb was at work on his chorography, about 1620, antiquarian activities were in fashion and were being described as the knowledge proper to a gentleman. It is also clear that curiosity about the history and the geography of one's countryside was often stimulated by brief residence at one of the universities, particularly Oxford, or at the Inns of Court. At the Inns of Court lawyers were already sifting through the annals of antiquity, preparing the constitutional case against absolute government. Others, who undertook the management of landed estates on behalf of big and monied patrons, became involved in real-property law and title search. Such research in fact, spurred many on to wider, chorographical descriptions of their localities. There was thus a practical side of chorographic research as well, for such knowledge was of use to the real-property barristers and to administrators of local government.

An example of how these influences bred an interest in antiquities may be found in William Burton (1575–1645), Squire of Lindley on the border of Warwickshire and elder brother of the more famous Robert, author of *The Anatomy of Melancholy*. Burton's *Description of Leicester Shire* (1622) was the first chorography of that county and also proved to be the only chorographic survey of an English county to reach print between 1602 and 1656.[16] Burton was responsible for placing the papers of John Leland in the Bodleian Library for safekeeping, and his improvements on Saxton's map of Leicestershire were incorporated in John Speed's *Theatre*. Among Burton's friends and acquaintances can be numbered Robert Cotton, William Somner the antiquary, Augustine Vincent the genealogist, the poet Michael Drayton, and William Dugdale. Some of these friendships were developed while Burton was at Brasenose College, Oxford, where he obtained his BA in 1594, and at the Inner Temple. Like Carew, Burton knew several languages, and in 1597 he published with Thomas Creede a translation of *The History of Cleitophon and Leucrippa* from the Greek of Achilles Tatius. In his manuscript 'Antiquitates de Lindley' he states that he combined the study of law with literature, and in 1596 he wrote an unpublished Latin comedy, 'De Amoribus Perinthii et Tyanthes.' Burton was called to the bar in May 1603, but because of failing health he soon retired to his estate near the village of Falde in Staffordshire. He devoted himself seriously to his survey of Leicestershire.

Leicester Shire is dedicated to the Earl of Buckingham. In the dedication Burton stated his intention of removing the 'Eclipse from the sunne, without Art ... [of] Astronomical dimension, to giue light to the Countie of Leicester, whose beauty hath long beene shadowed and obscurred ...' In the preface he indicated that although he was preoccupied in the profession of law, he was inclined to the study of antiquity: '... there is no Study or learning so fit or necessary for a Lawyer, as the Studie of Antiquities, and Species thereof.' It is there, also, that Burton described being animated by the performance of Camden and the other antiquaries mentioned earlier.

For the most part, *Leicester Shire* is a digressive book. It is not a systematic survey of the county despite a brief introductory description of its topography – a topic that literally permeates some of the earlier chorographies that Burton claimed to follow. Here we find the standard pieces of information: the site and situation of the county, according to Mercator, and its shape and extent; an account of the air, soil, and rivers; and a list of the castles, religious houses, market towns, and parks. All this is painstakingly compressed into about seven pages of text, whereas the earlier chorographers usually devoted large tracts to these subjects. Once the topography is disposed of, except for short topographical descriptions that later accompany the entries of places, the remainder of the book is given to an alphabetical index. Burton's main interests are the people of Leicestershire, the noble families and their arms, the churches, the value of the church livings, and especially the holders of the advowson. These items are contained in the alphabetical listings of the towns, manors, and hundreds in which they are located. This is in accordance with the author's goal of setting out titles of lands, manors, and tenements so 'that the continuance of them in a Name or Bloud, might be discouered, and the ancient Owner (so farre as could be found) might bee knowne.'[17] Burton's book contains many genealogical tables of well-known families and their coats of arms. It includes such items as a list of the sheriffs of the county from the time of Henry II onwards, and tables of the abbeys and other religious houses. Burton was especially fond of quoting the inscriptions found on tombs in the parish churches. A standard type of entry, that for the village of Poultney, reads:

Poultney, in the Hundred of Guthlakeston, not far from Misterton. This place gave name to the ancient family of Poultney, of whom I have spoken before in Misterton. In the 23. yeere of Edward the third, Sir John de Poultney was Lord heereof. This was sometimes a village, as appeareth by an olde Roll, made in the time of King Edward the first, setting downe the Townes in every Hundred; but now it is utterly decayed, not one house remaining. It is in the parish of Misterton, and had a Chappell, which now is also ruinated and gone.[18]

Numerous digressions are scattered throughout the work. In describing, for example, Bardon-Park's quarry hill as being not as imposing as Athos, Olympus, or Tenerife, Burton listed the authors who wrote about these other hills. Furthermore, he felt obligated to give his own hypotheses about the origin of hills, along with samples from other opinions on the same subject. At diverse points in the text we also find a history of leprosy, a discussion of the incorruptibility of bodies and reasons for embalming, and a story of the origin of knight's service, with a definition of scutage. Any useful knowledge inherent in these digressions was for the contemporary reader probably not as important as their curiosity value and the fact that they tended to inject a certain amount of levity into the otherwise relatively dry and tedious commentary.

Apparently, Burton began compiling his survey as early as 1597, without any 'intendment that it should ever come to public view, but for my own private use, which, after it had slept a long time, was on a sudden raised out of the dust,' and then, 'by force of an higher power drawn to the press, having scarce an allowance of time for the furbishing and putting on a mantle.'[19] The 'higher power' was Buckingham.[19] Aware of its defects, Burton intended to publish a new edition and so he spent many years collecting new material and making corrections to his original publication. The outbreak of civil war, however, prevented any further issue.[20]

Burton was not the first man to make collections for a study of Leicestershire; he was anticipated by both John Rous and by Henry Ferrers of Baddlesley Clinton, Warwickshire. Ferrers, another obscure figure in the annals of chorography, perambulated both of these counties, beginning his trek many years before Burton commenced his own survey. Ferrers's genealogical and heraldic collections were liberally used by Sampson Erdeswicke, and William Dugdale, both of whom acknowledged their indebtedness to him.[21] Burton was also encouraged to undertake a survey of Warwickshire and began to put together all the notes and material on this county that he had collected in the course of his researches, being 'willing to furnish him therewith that shall undertake the illustration of the countye.'[22] As it turned out, William Dugdale, a generation later, was the first to publish a study of Warwickshire, having been inspired originally by Burton's efforts. Burton, it seems, was at the centre of a Midland antiquarian circle that included Dugdale.

Kendrick describes the quality of Burton's antiquarian studies as 'scientific,' but mainly in the sense that Burton attempted to be scrupulously accurate and thorough in recording what he personally found in every church in Leicestershire.[23] Most of the chorographies continued to be styled, at least in theory, in the Camdenian mode. The new direction, however, was definitely towards placing more emphasis on genealogy and heraldry than was formerly the case with Norden, Carew, and other earlier chorographers, and, in fact, interests such as Burton's in heraldry, value of livings, and so on were out of the deed-box! Quoting from Hoskins and Finberg, Oruch points out that the tale of topographical prose from 1622 to 1640 was one of unfulfilled ambitions, and is illustrated in the attempts at a description of Devonshire, though three men laboured individually on chorographically surveying this county: 'There were between 360 and 400 indisputable gentry in Devon ... the descent of whose property is the main concern of the contemporary county historians – Pole, Westcote, and Risdon.'[24] Not one of the chorographic works of these men was published in the seventeenth century. And yet, together they form a distinct group, not only because of their common subject – Devonshire – but also because of the productive interchange of ideas that circulated among their authors.

Tristram Risdon (1580?–1640) illustrates the interest of Puritan gentry in chorography. He is described in the *Dictionary of National Biography* as 'apparently a Puritan, somewhat inclined to preach and moralize, but his observations are nowhere obtrusive.' Born at Winscot St Giles, near Torrington, Devonshire, Risdon was admitted into either Exeter College or Broadgates Hall (now Pembroke College) – the existing evidence for his education is unclear – near the end of the reign of Queen Elizabeth. He left Oxford without taking any degree, and retired into his own county upon having obtained the patrimony of the estate at Winscot. This life of relative leisure afforded Risdon the time to indulge in collecting material towards the compilation of a survey of his native county. With help from his intimate friends, the aforementioned Sir William Pole and Thomas Westcote, he was able to produce such a work. Supposedly only manuscript copies existed until 1714, and his survey was not properly printed until 1811.[25] It was begun in 1609, according to Anthony à Wood, and finished in 1630. There is evidence, however, that Risdon kept adding material to his survey well into the 1630s, and that his intention from the start was to have it published.[26] This was prevented both by his death and because his executors either neglected to publish the work posthumously or else were hindered in this task by the outbreak of civil war.[27]

Risdon's chorography is not entirely devoid of topographical detail. He locates the towns, hundreds, rivers, and roads, and generally notices topographical features throughout the book. The first few pages of *Devon* concentrate on the ancient history and geography of the region, citing Ptolemy and Camden as authorities. In discussing the origin of the name 'Cornwall,' Risdon, like Carew before him, is unable to discard totally the Brute legend:

Some I know would have this country when it was all one province, to be called Corinea, of Corineus, cousin unto Brute, a special man of account under him, whom he rewarded with this region at his arrival; which relation, others do think, carrieth no other truth than an ancient tradition; yet, forasmuch as it is left unto us from our ancestors, it were against humanity to reject the same, and to derogate credit from that which hath so long time been received, and found so many learned patrons.[28]

It may be that Risdon was more than just a little influenced by Carew, a man whom he regards as having described Cornwall 'so eloquently and learnedly.'

We next read of the extent of the shire, the climate, and the courage of its inhabitants – which apparently has, over the ages, abated due to a decrease in the amount of 'manly exercises'. In Risdon's time the roads were exceedingly bad throughout Britain, and they were probably in worse shape in Devon than elsewhere, being, as Risdon tells us: 'rough, and unpleasant to strangers travelling those ways, which are cumbersome and uneven, amongst rocks and stones, painful for man and horse ...'[29] There follows an account of the principal soils in the county, which comprise four main types. The southern part of Devon is esteemed by the author for its fruitfulness, and is considered the 'Garden of Devonshire' in contrast to the lean and barren north and west.

In other respects, too, Risdon's work exhibits his strong interest in developments in agricultural production, such as the use of lime as a manure. His comments on this practice are enlightening: 'Of late, a new invention has sprung up, and been practised, by burning lyme, and incorporating it for a season with earth, and then spread upon the arable land, hath produced a plentiful increase of all sorts of grain amongst us, where formerly such never grew in any living man's memory.'[30] This section on agriculture and husbandry concludes with an account of the commodities and livestock found in Devon, and contains a quaint comment on the value of 'cyder,' typical of the many digressions in the book: 'a drink very useful for those that navigate long voyages: whereof one tun serveth them instead of three tuns of beer, and is found more wholesome drink in hot climates.'[31] Risdon has much more to say of the inhabitants, but not before briefly commenting on the county's minerals, rivers, bridges, and havens – like Carew, he mentions the herring and pilchard industry. The usefulness of a 'miraculous' loadstone for sailors is mentioned, but the exact location of the loadstone in Devon, and its properties, were not reported on in a scientific manner until a much later date.[32]

Next, Risdon ventures to describe the various classes: gentry (including noblemen), merchants, yeomenry, artificers, 'mechanikes' (tradesmen), and labourers. The pastime of hurling is said to be on the decline, for the 'common sort' of people are no longer able to afford to observe all the holidays during which such pastimes are enjoyed. The labourer's existence is described as one of misery; this is particularly so for the 'spadiard,' the daily labourer employed in the tin works, whose 'apparel is coarse, his diet slender, his lodging hard, his drink water, and for lack of a cup, he commonly drinketh out of his spade or shovel ... His life most commonly is in pits [literally] ... and in great danger, because the earth above his head is in Sundry places crossed over with timber to keep the same from falling.'[33] Risdon has very little to say specifically about the role of women within the framework of Devon society of the time, an omission that his reviewer finds unpardonable: 'after having said so much on the Bravery of our Men, not to take Notice of the other Sex; as if their Artillery did not do as much Execution on its proper Objects, as the Swords and Guns of the former when engaged with an Enemy.'[34]

Except for a short discussion of the ecclesiastical and civil administration of the county, the remainder of *Devon* concerned, as Risdon states, with:

> ... the particular places, with their ancient and most eminent families; or any other memorable matter, that hath come within the compass of my knowledge, worthy the leaving to posterity; wherein many ways may be used: as by taking the tythings, or having the hundreds for my guide; by the archdeaconries as they are limited, or by the divisions, as they are favoured, or by the course of the rivers. But propounding herein an order to myself, I purpose my beginning in the east part of the county, and with the sun, to make my gradation into the south, holding course about by the river Tamer, to visit such places as are offered to be seen upon her banks. Lastly, to take notice of such remarkable things as the north parts afford.[35]

This is precisely what Risdon proceeds to do, much in the manner of Reyce or Burton, but with the addition of slightly more topographical content. There is certainly more than enough genealogical information, but at least we are not inundated by inscriptions. Interesting bits of information are passed on to us: descriptions of statues of antique beauty, such as those found at Thorncombe; brief accounts of historical events, such as the defeat of the Danes at Axminster; mention of prominent contemporary Devon sons, including Nathaniel Carpenter, the geographer from Uplime; and lines of poetry, whose authorship is usually not disclosed — for example, these concerning the Tamar River, which divides Cornwall from Devon and the rest of Britain: 'On this side Tamer the English sees, / And thence the Britons eke it eyes.'[36]

In compiling his chorography Risdon relied to a greater degree on written sources – Camden, Carew, Pole, Westcote, and Richard Hooker, the author of a sixteenth-century survey of Exeter –than on his own personal observations of the contemporary scene.[37] Risdon also left in manuscript a 'Notebook' containing further genealogic and heraldic collections on Devon, which was edited in 1897 from the original manuscript by James Dallas and Henry G. Porter.[38] Written mostly between 1608 and 1628, it was therefore contemporaneous with *Devon*. It is basically a mass of heraldic information, which includes the arms of many families not found elsewhere. More space is devoted there to an account of the feudal baronies of Devon than to any other topic. It also contains lists of the sheriffs, justices of the peace, mayors, and knights of the shire. Much of the information found there is taken from the original deeds, which passed through Risdon's hands at one time or another.[39]

In *Devon*, Risdon's entry on Calcombe House acknowledges the value of the labour of Sir William Pole (1561–1635) towards his own work:

He [Pole] was the most accomplished treasurer of the antiquities of this county; and, had he been pleased to have been the author of this work, the worth of this county, the natives thereof, and his own sufficiency, would have been better known. Such a gift had he of rare memory, that he would have recited upon a sudden the descents of most eminent families; from whose lamp I have received light in these my labours.[40]

Pole's story is not unlike that of many of the other chorographers. Having obtained an education at the Inner Temple, he was placed on the Commission of the Peace for Devon, served as high sheriff of Devon in 1602–3, and represented Boissiney, Cornwall in the Parliament of 1586. He was knighted by King James I at Whitehall in 1606. Pole spent most of his life at Colcombe, which estate he had inherited from his father. It is not difficult to imagine him spending many an evening conversing with Risdon and Westcote, trading with his friends pieces of information of an antiquarian nature. After all, Pole was 'learned also, not only in the Laws, but in other polite matters. He was very laborious in the study of Antiquities, especially those of his own County, and a Lover of that venerable Employment.'[41] At his death in 1635 Pole left only his unfinished papers for posterity, never having completed in its entirety an individual survey of the county. Judging by a letter he wrote to a man named Reynell, who sought genealogic information from him, in which Pole stated his 'purpose (God willing) to set out something for the Antiquities of Devonshire,' Pole had already begun his self-appointed task of researching the antiquities of that county as early as 1604.[42]

Of special interest here are the two manuscript folio volumes entitled 'The Description of Devonshire,' which were printed in 1791 by John William de la Pole. Apart from these, Sir William Pole left several other volumes of manuscripts, including one containing deeds, charters, and grants, compiled in 1616; a thin volume containing coats of arms; and a volume of deeds and grants to Tor Abbey.[43] Exactly what sort of information is contained among these papers? Certainly it is not topographical, for there is little enough of this in Pole's work. For the most part it is the genealogic and heraldic material that Risdon valued so highly. The 'Description' contains an account of the ancient baronies of Devon or, more specifically, of their holders from the time of William I; lists of the knights of Devon and of the more prominent statesmen and military men of the county; and a catalogue of the high sheriffs of Devon. Pole also included an account of the hundreds – beginning, like Carew and Risdon, in the east – focusing on the holders of the manors within each parish, for example, 'Uphay, before sett downe standinge in the side of an hill, upon the West of Axmister, and over ye river Ax, standinge in an advanced ground, tooke his name of his scite, and gave his name unto his dwellers; for it appears yt many of that name successively enjoyed the same, untill it was transferred.'⁴⁴ Pole used the standard literary sources: Hooker, Holinshed, *Domesday Book*, old deeds and charters, inscriptions, and the like. The only topographical information of any consequence is that for Exeter. The site, shape, and extent of the city are outlined, the principal buildings, gates, and bridges are mentioned, and the general street plan is described.⁴⁵

The third member of this informal Devon circle, Thomas Westcote (1567–1636), is the author of *A View of Devonshire in 1630*.⁴⁶ Westcote's work is based upon the same plan as that of Risdon, but is less comprehensive. The respect these men held for each other is displayed in Westcote's entry for Winscott:

My worthy friend, Mr. Tristram Risdon, we are imboldened to visit you in our travayle, to have only a collation of your collections, observations, witty and pretty conceits, antique names of places and families, and therewithall your company awhile, the better to illustrate and make known the worth of your country and the natives thereof; and to give these gentlemen fuller satisfaction and content to such questions as out of their curiosity they shall demand: you may not play the nice musician with us, who requested, would scarce tune his instrument, but voluntarily will crack all his strings. Here is good company, and yours added, we shall need no more; but shall pass our journey pleasantly, with celerity.⁴⁷

The details of Westcote's life are sketchy. Born at Shobrooke in Devon, in his youth he was at one time or another a soldier, a traveller, and a courtier. It seems that he was not the only adventurer in the family; his brother George, a captain in the army, died at the age of twenty-seven in the disastrous expedition against Lisbon in favour of Don Antonio, the Portuguese pretender. Thomas lacked the formal legal training experienced by many of the other antiquaries, but was apparently self-educated to a large extent, 'having by ordinary reading, observation, search, and discourse, collected long since some few particulars of the antiquities and other notes and observings of this County.'⁴⁸ In middle age he retired to a private country life, probably residing at West Raddon with his eldest brother, Robert.

Westcote was anxious to undertake a chorography of Devon similar to that accomplished for Cornwall by Carew. Aside from Risdon and Pole, the 'primum mobile,' as Westcote puts it, behind his discourse was Edward Bourchier, the earl of Bath. Bath, 'cheerfully animated' and 'seriously required' him to compile this work.⁴⁹ It appears, however, that in the end Westcote was too modest to have it published. At one point he refers to his effort as a 'bundle of waste paper' (446).

Even if it has its share of egregious errors, *Devonshire* provides some entertaining and amusing reading. Westcote's aim is to intermix a 'pleasant tale with a serious discourse,' or, 'an unwritten tradition with a chronicled history, old ancient armories and epitaphs, well near buried in oblivion ... some etymologies seeming and perchance strange and far fetched.' His hope is that these matters will give recreation 'to a wearied body and mind (that reads for recreation), with more delight and content for variety, than dislike the severe critic for simplicity, vulgarity, or doubt of verity.'[50] In its overall plan *Devonshire* does in fact resemble some of the other contemporaneous chorographies in its arrangement of topics. Book 1, for example, treats general topics, and books 2 through 5 survey the shire systematically by major divisions and hundreds.

Westcote tackles his task in a manner quite familiar to us by now. The first item on the agenda is a disquisition on the origin and etymological derivation of the name *Devon*, which leans heavily on Camden's influential *Britannia*. Westcote is especially anxious to impress upon the reader that the origin of names is most uncertain. He then delineates Devon's geographical boundaries while sketching, in broad strokes, her market towns, forests, parks, and rivers. Preliminaries out of the way, he commences a lively account of the inhabitants. The forefathers were hardy – 'whereof the quantity of a bean would satisfy nature in such sort' – but 'delighting in the sweetness of foreign dainties,' they acquired a weaker constitution.[51] This has apparently improved because the current natives are physically strong, 'bold, martial, haughty of heart, prodigal of life, constant in affections, courteous to strangers, yet greedy of glory and honour.' Westcote is inclined to quote long and irrelevant passages from the classics, and here he applies a verse from the poet Pindar, on Lacedemonia, to his own country:

> Their grave advice is found in aged brains;
> Their gallant youths are lusty lads indeed,
> Which can both sing and dance in courtly trains,
> And daunt their foes with many a doughty deed.[52]

He also relates a story about the heroics of his countrymen, which he found in Speed.[53] The section on the inhabitants is concluded by a discussion of various classes, one in which such obvious terms as 'nobility' and 'husbandry' are nevertheless given lengthy definitions.

Westcote then embarks upon a methodical discussion of the agriculture, minerals, shipping, and commodities of the shire, one that catches the reader's attention and fixes it on the unique properties of the region. But the reader is referred to Carew's *Cornwall* for a complete account of the manner of discovering and treating certain metals. Westcote's pride in his native 'country' is revealed in the account of Devon's mariners: '... the whole wide world brings forth no better, whether you will impress them for valour to adventure, or knowledge to perform any action; painfulness to undergo, or patience to endure, any extremity, adversity, or want whatsoever: all which in one I may boldly aver, and yet not be taxed for over-valuing of them or their worth.'[54] He extends this pride, like many of the chorographers, to cover the entire nation, describing the sovereign's navy as 'the sinews of our strength.'[55] What follows is a record of the political, ecclesiastical, and military divisions of the county, one that by now seems so natural a part of the narrative in chorographic description that it is easy to forget its documentary value. Westcote borrows heavily from the works of his fellow chorographers; obviously, like them, he felt little compunction about appropriating work by other scholars, if desirable, with or without citation of sources. Originality was not necessarily a main aim among the antiquaries. What mattered was that such borrowing was, in reality, little more than an exchange of information towards the attainment of a common goal.

The core material is contained in the last four books. These are particularly concerned with the ownership of land and with tracing the descent of property. Westcote considers the inclusion of this 'particular view' of the hundreds, towns, parishes, etc, necessary so that 'nothing may be defective of what is spoken of in the survey or description of other counties.' Even here, however, the author's extreme modesty shows through. He considers himself too defective in good judgment, learning, and time to write a chorography worthy of his audience. Thus, he is able to present them with only a 'bare relation, broken and independent fragments.'[56] And yet we find ourselves transported across the countryside in a pleasing manner, accompanying Westcote in his literary perambulation from one hundred to the next: 'But let us Spend no idle time; and for our easier and better proceedings let us again return to Exmoor ... After this pause we will with an easy pace ascend the mount of Hore-oke-ridge, not far from whence we shall find the spring of the riveret Linne; which in his course will soon lead us into the north division: for I desire you should always swim with the stream, and neither stem wind nor tide.'[57] Along the way we encounter gossip or fables that Westcote has knowledge of, such as the tale of the 'nymphs of Torridge Spring,' or a relation of a 'professor of physick' (doctor) whom the author knew personally.[58] Also, there is slightly more topographical material here than in the surveys of his two Devon compatriot antiquaries, of the type relevant to a study of descriptive geography:

Barle yields nothing to Exe in quantity, and seems as if she would strive for superiority, as having the first bridge of stone, as otherwise unpassable, and that in the midst of the forest; near which is a large deep pool which they name Symon's Bath, as a place where one Symon used to bathe himself, and is said to have been ... another Robin Hood, and standing in outlawry kept this forest; and in the moors of Somerset there is a burrow or fort called, by the inhabitants, Symon's Barrow.[59]

Besides his chorograhic survey Westcote was the author of the 'Pedigrees of most of our Devonshire Families,' a compilation containing much additional genealogic information. This work is also included in the 1845 edition of his survey.

Two of the last county surveys to be completed before the outbreak of civil war were those illustrating the counties bordering Westcote's Devon, namely Dorset and Somerset.[60] Both works were executed by Thomas Gerard (1582–1634), who was primarily interested in genealogy and the ownership of land.

Gerard was a member of a family long settled in Dorset. It is possible that he personally knew Camden and members of his circle, for he had studied at Oxford (Gloucester Hall), matriculating in June 1610. Eight years later he married Anne, the daughter of a Robert Coker of Mappowder, Dorset, and settled at Trent, Somerset. Gerard was familiar with both shires. His two surveys appear to be contemporaneous, written over an extended period of time during the first quarter of the seventeenth century and completed about 1633. Both works are arranged on the same plan; that is, they name a river, then proceed to enumerate all the towns along its banks:

For the order of this View or Survey of Dorset, I shall neede noe better, or more warrantable President, than the learned and judicious Camden, and therefore with him I will beginne at the surest and certainest Bound, the Sea; which from the first Westerne Limite I will follow Eastward, untill it forsaketh this Countie; and by the way observe what Rivers runne into it, whose Streames shall be my Guides, even from their Springs and Fountaines, untill they take up their Lodgeing in the Ocean; for on these Rivers and their Branches are generallie Seated those Places of note, which I shall in this my journey observe, I willinglie imparte unto you.[61]

In both works particular emphasis is placed upon the importance of exactitude in noting heraldry, and both contain the same type of unusual epithets and quaint expressions.[62] Gerard's chorographies exhibit a more professional quality than does Westcote's *Devonshire*. However, contrary to the opinion of E.H. Bates, editor of Gerard's *Somerset*, who believes that the 'topographical descriptions are not confined to dry details, but reproduce the scene as Gerard saw it,' the reader of Gerard's surveys finds little satisfaction in his rather curt treatment of the picturesque landscape.[63] Certainly there are some vivid descriptions based on personal observation: the floods seen from Langport and Somerton, the pleasant walks through orchards and gardens, and so on. But these are few. There is little attempt at local colour here, and only occasionally is historical material of a non-genealogic nature included.

Both surveys begin by describing the farthest western limit of each shire, progressively reporting the scene on a west-to-east schema. A typical entry outlines the derivation of the name of a place, its 'antient' and current owners (or residents), its settlement history where documentary evidence is available, and on occasion, the topography as well. The latter is given prominence in the account of Weymouth, one of the seaside towns that engaged in a steady commerce with France and Newfoundland:

Weymouth as nowe it is but little, consisting chiefelie of one Streete, which for a good space lieth open to the Sea; and on the back of it riseth an Hill of such steepenesse ... from whence you have a faire Prospect of the Towne and Haven lieing under: And from the other side you may see Weeke, the Mother Church of Weymouth Melcombe on the other side, though the River much surpasseth the other for Conveniencie of Scite; for this standing on a Flatte affordeth roome for Buildings, with a Market Place and convenient Streetes, and also Yardes for their Wares.[64]

Gerard, on the whole, is averse to paying attention to the kind of gossip that Westcote delighted in recollecting. The closest he comes to this is in his rare presentation of pieces of trivia particular to a place, as when he tells of a Dutchman who discovered the alabaster mines at Minehead.[65] Had he been more concerned with such incidental details we might today have a more vivid picture of what he personally experienced.

Gerard made use of a variety of sources, including the standard epitaphic inscriptions, parish registers, old deeds and cartularies, and the like. His history of a manor in lay hands usually begins with a plea of King John's reign, or an inquisition post mortem of the early thirteenth century. The muniment rooms of his many friends among the gentry were also open to Gerard's researches, and he may have known personally several of the other chorographers. He refers to Carew's survey in the text of *Somerset* (179), to Pole's collections (127), and on several occasions to William Burton, the chorographer of Leicestershire. Gerard may also have had access to the manuscripts left by Leland, which had passed into Burton's possession.[66] Gerard also refers frequently to Camden's *Britannia*, and he consulted other contemporary works, such as Weever's Ancient *Funerall Monuments*.[67]

Weever's book proved useful in the work of many antiquarian researchers, even though it covers only the dioceses of Canterbury, London, Rochester, and Norwich. Weever, a friend of many virtuosi, had access to Sir Robert Cotton's library and to the Office of the Herald. But *Funerall Monuments* is not at all a chorographic study. It contains inscriptions from the various monuments found in parochial churches, cathedrals, tombs, sepulchres, etc., information of a type useful to all learned men. They no doubt shared Weever's sentiments concerning the preservation of these relics of the past: 'And also knowing withall how barbarously within these His Maiesties Dominions, they are (to the shame of our time) broken downe, and utterly almost all ruinated, [their] brasen Inscription erazed, torne away, and pilfered, by which inhumane, deformidable act, the honourable memory of many vertuous and noble persons deceased, is extinguished, and the true understanding of diuers Families in these Realmes ... is so darkened, as the true course of their inheritance is thereby partly interrupted.'[68] And yet Weever seldom recorded archaeological finds, nor did he exhibit an appreciation of their intrinsic value. In this respect he was no different from his contemporaries, including Gerard.

The only other survey of an entire county executed before the outbreak of civil war that incorporated a substantial amount of genealogic or heraldic information was Sir John Dodridge's *The History of the Ancient and Modern Estate of The Principality of Wales, Dutchy of Cornwall, and Earldome of Chester* (1630), a work that commanded a large following among government officials perhaps, but not a large general audience. Dodridge (1555–1628), a native son of Devon, graduated BA from Exeter College, Oxford (1576), became a member of the Society of Antiquaries, and held various political and judicial posts.[69] His *History* is very different from nearly every one of the chorographies examined thus far. Thus, chorographers such as Carew, according to McKisack, 'would have had little to learn from Doddridge, apart from a few points of law and finance.' McKisack rightly states that the *History* adds little to our own knowledge of the local history of the region it describes.[70] In contrast to what one generally thinks of when considering the limits of a chorographic study, this work contains virtually no topographical information. Also, except for the work of Giraldus, Matthew Paris, David Powel, and Humphrey Llwyd, Dodridge reveals little knowledge of many of the usual literary sources. The form of the book is that of a historical treatise on the succession and manner of the goverment of Wales and the succession and revenues of Cornwall and Chester. Dedicated to King James, the *History* is shorter than most other chorographies. It is divided into three major sections, one for each of the areas under consideration. Dodridge introduces each section with a brief general description, but even these introductions are historically, not topographically, oriented. For example:

The uttermost part of this island towards the West, stretching it selfe by a long extent into the Ocean, is called the County of Cornwall; lying over against the Duchie of Britaine in France. The people inhabiting the same, are called Cornishmen, and are also reputed a remnant of the Britaines, the ancient Inhabitants of this land: they have a particular language, called Cornish, (although now much worne out of use) differing but little from the Welsh ... This territorie was anciently reputed a Dukedome, but a little before, and also after the Norman Conquest, it was an Earledome, and so continued untill the eleventh yeere of King Edward the Third, at which time it was of new constituted a Dutchie, and the first Dutchie that was erected in England after the said conquest.[71]

The section on Wales begins with an examination of the principality previous to its conquest by Edward I. We are informed of such diverse matters as the origin of the 'Baroyes of Marchers,' the political and judicial administration, the yearly revenues during the time of the Black Prince, the present revenues, and the fall of Richard III: 'But for that the prosperity of the wicked is but as the florishing of a greene tree, which whiles a man passes by is blasted dead at the roots, and his place knoweth it no more.'[72] The other two sections proceed along the same lines, describing the revenues, listing manors, and so on. In his account of Cornwall, however, Dodridge does take time to extemporize on the different kinds of 'tynners and tynne,' and on the coinage, topics already covered by Carew. Here Dodridge was more concerned with the classical writers than with contemporary literature. In his treatise on mining, for example, he quotes from Diodorus Siculus, a writer on mining from the Augustan age.[73] Nevertheless, the remainder of Dodridge's material is derived from official or legal sources such as charters, statutes, chancery warrants, and patent rolls. Dodridge's quasi-chorographic study was typical of the trend that saw few – if any – of the remaining surveys of the first half of the seventeenth century adhere to the 'standard' chorographic models established earlier, whether they focused on history-topography or on genealogy-heraldry-topography – ignoring for the moment the role of antiquarianism.[74]

CHAPTER FIVE

'Rapidly Thinning Wisps and Patches'

... the neatness and method of his work enabled him [William Dugdale] to present the results of his research in a form vastly superior to that of any previous history of a similar nature. For this reason later investigators have found the work as satisfying as did Dugdale's contemporaries ... Gough gave it as his considered opinion that Dugdale must 'stand at the head of all our county historians.'
David C. Douglas 'William Dugdale'[1]

ALTHOUGH THE CIVIL WAR had placed obstacles to chorographic writing, in England, at least, in other respects headway in scholarship continued that also aided in a spectacular – if brief – revival of chorography. By the mid-seventeenth century advances had been made, for example, in developing library catalogues, sigillography, numismatics, and so on. The first chorographer to take full advantage of these advances, and to contribute significantly to them, was Sir William Dugdale (1605–86). In so doing, Dugdale attempted, in the 1650s, to revive singlehandedly a tradition of chorographic writing upon which, as we have seen, regional study had been based. This was no mean feat, as the genre had already degenerated into variant forms that in most instances only slightly resembled the earliest chorographies.

In its focus on history and antiquities, Dugdale's *The Antiquities of Warwickshire* (1656) is clearly of the established chorographic tradition. However, as David Douglas states in the quotation that opens this chapter, in many respects Dugdale's methods and his completed work differ from those of an earlier generation of chorographers, or antiquaries.

There are, then, two sides to his *Warwickshire*. The work did, in fact, establish new standards of accuracy and method in the documentary study of many fields of learning. Dugdale, along with other English researchers such as Hickes, Gale, and Rymer, was instrumental in encouraging the systematic and critical presentation of source material through the use of original documentary evidence, putting considerably less emphasis than did most of the earlier antiquaries on the mere copying of inscriptions or charters.[2] Dugdale's reliance on informal public opinion in the county of which he wrote exceeded that of his predecessors. For these reasons, *Warwickshire* went beyond anything that had hitherto appeared, 'remarkable for its accuracy and its constant references to original authorities.'[3] Yet, as T.D. Kendrick notes, 'the fogs of medieval antiquarian thought' still lingered, if only 'rapidly thinning wisps and patches.'[4] Natural history – the element so crucial to nearly all the influential regional studies that followed – did not even so much as overlap the antiquarian topography in Dugdale's book. In this respect Dugdale did not herald the interest in natural phenomena that became so popular later in the seventeenth century. Furthermore, there are few indications of modern archaeological or scientific investigation in *Warwickshire*. Dugdale commented on stone axe heads – weapons used by the Britons before the art of making arms of brass or iron was known[5] – and he also took notice of various tumuli and barrows, but for the most part his sources were literary.[6] Personal observation of the landscape and detailed evaluation of individual antiquities seemed to be almost ancillary activities when compared to his use of other sources.

Although *Warwickshire* has never been closely examined in its entirety – a fact that is surprising – there is plenty of information about its author.[7] Born at Shustoke in Warwickshire into a family of yeoman background, William Dugdale exhibits a career pattern similar to that of many of his scholarly contemporaries. This is true even though he never attended university; after studying at the Free School at Coventry (1615–20), Dugdale studied law at home under his father's direction. (His father, John Dugdale, enjoyed a prolonged residence at St John's College, Oxford, as a bursar and steward.) William Dugdale married early, at the age of seventeen, in order to please his father, who was by then old and infirm. Shortly afterwards he bought Blyth Hall near Coleshill, Warwickshire, and settled into the comfortable life of a country gentleman.[8] Under his father's tutelage Dugdale was encouraged to investigate, like most country gentlemen, the history of his estate and the pedigree of his family. Instruction in works such as Sir Thomas Littleton's *Tenures in Englysshe* (1525?) whetted his appetite for heraldry and genealogy, and for antiquarian research in general. In this respect Dugdale's motives typify the subtle change that was taking place in the reasons why men were drawn to the study of antiquities. In the sixteenth century they had above all a zeal to voice the praises of Tudor England and to portray the wealth of the land. However, under the shadow of the English civil war, as Douglas points out, the titles to their estates interested men more, and so questions of origin became of immediate concern.[9]

Dugdale struck up friendships outside the Midlands circle with which he is now closely associated, and personally befriended many of the leading scholars of his day. In 1630 he was taken to see Sir Christopher Hatton – 'a person highly affected to Antiquities' – in his lodgings near Temple Bar, who welcomed him 'with all expressions of kindness, with readiness to further him in these his Studyes.'[10] Both men took delight in their friendship and Hatton used his wide influence to obtain for Dugdale, in 1638, an appointment as pursuivant extraordinary with the title of Blanch Lyon. The following year Dugdale became Rouge Croix pursuivant in the College of Heralds, where he lived for the next several years while engaged in antiquarian pursuits. After Strafford's execution Hatton foresaw civil unrest, and so encouraged Dugdale to undertake a remarkable project. Dugdale, accompanied by William Sedgwick, a heraldic painter, proceeded in 1641 to ride around the country – starting with London and then from shire to shire – copying the inscriptions in the religious houses and drawing the various monuments and coats of arms, 'to the end that the memory of them, in case of that ruine then imminent, might be preserved for future and better times.'[11]

The 'fanatique rage of the late times,' as John Aubrey was to observe later, did indeed destroy many an old monument.[12] But once the civil war began, Dugdale was in a better position, as a member of the College of Heralds, for antiquarian observation. An ardent royalist, he travelled around England with warrants from Charles I demanding the submission of garrisons. Being stationed at Oxford also enabled him to examine original material at the Bodleian and other libraries, whereas most of the other scholars – including Aubrey – retired from that town upon the outbreak of conflict.[13] Dugdale's wartime experiences and his ultra-conservatism inspired him later to publish *A Short View of the Late Troubles in England* (1681). Unfortunately, as Royce MacGillivray states, in this study Dugdale was unable to grasp 'the complexity of personality, of historical processes, and of the machinery of society ... he ruthlessly systematizes the events of the war to accord with his black-and-white picture of a guiltless king struggling in vain against a band of evil politico-religious conspirators.'[14]

Nevertheless, Dugdale's antiquarian skills were already being finely honed. Aside from the education he received from his father, his first introduction to the study of history and antiquity was through the influence of a kinsman, Samuel Roper, a barrister of Lincoln's Inn. In time Dugdale fell into a pleasant circle of scholarly friends, beginning with William Burton, author of the *Description of Leicester Shire,* and Sir Simon Archer of Tamworth, who had collected materials on the early history of Warwickshire. It was after Dugdale went to London in 1635 that he made the acquaintance of Sir Henry Spelman, Roger Dodsworth, and several other eminent researchers. Spelman prevailed upon Dugdale to join Dodsworth, a Yorkshire antiquary, in the latter's study of the foundation of the monasteries. This research came to fruition as the *Monasticon Anglicanum*, the work on which Dugdale's fame today is largely based. It contains histories of the abbeys as dipicted in their charters, names of the successive abbots and their deeds, descriptions of the buildings and inventories of their farms and granges, with lists of the properties seized during the Reformation; thereby it 'made known for the first time a whole range of documents whose true significance had hitherto been unappreciated, and by so doing it illustrated almost every phase of English social and economic history in the Middle Ages ... The *Monasticon* taught English scholars the importance of charters for history, and it published these in such numbers that a comparative study of them became for the first time possible.'[15]

Later writers charged Dugdale with plagiarism for apparently appropriating Dodsworth's collections without acknowledgment, and for publishing the *Monasticon* – in its third edition – under his own name only. The *Monasticon* was, in fact, a cooperative effort on the part of both men, with copious assistance from other scholars, and Douglas – though he admits that the work was primarily Dodsworth's – expressed his doubt that Dodsworth ever would have succeeded in publishing it himself (Dodsworth died before the *Monasticon* was ready for the printer).[16]

Of more concern to us is *The Antiquities of Warwickshire*, which also owed much to other scholars. Dugdale knew William Burton had already envisaged a chorographic study of the county.[17] Burton's collections for Warwickshire were added to those of Sir Simon Archer, who, like the gentlemen of note of the county, 'being desirous ... to preserve the Honour of their Families by some ... worke as Mr. Burton had done by those in Leicestershire,' in turn communicated these to Dugdale, hoping Dugdale might make use of them towards a study of this county.[18] Archer (1581–1662), in fact, maintained a steady correspondence with Dugdale.[19] A relatively well-off Warwickshire gentleman, Archer, like his correspondent, was motivated by the practical desire to work out his family pedigree and to trace the history of his own estates. *Warwickshire* might early on have been intended as his own work. In any case, he collected a great number of original documents, mostly deeds and charters, which he eventually handed over to Dugdale. It is worth noting that both men knew Thomas Habington, the chorographer of Worcestershire, and that after Habington's death Archer urged Dugdale to persuade Habington's son to agree to the publication of the survey of that county.[20]

Burton's acquaintance with Dugdale involved, naturally, the exchange of antiquarian information. For example, Burton informed Dugdale of an unpublished survey of Staffordshire.[21] Likewise, in 1636 Burton appealed, through Dugdale, for Archer to undertake a survey of Warwickshire; and so the project was at last firmly launched. Originally, Dugdale was to have been the 'junior' partner in this endeavour, but eventually he came to assume full responsibility for it. This was probably because of Archer's ebbing enthusiasm and because Archer's financial resources were beginning to dwindle as the project got underway. Dugdale, however, was able to obtain financial backing, though only on the proviso that he assume sole responsibility of authorship. This arrangement was presumably satisfactory to all the parties involved. Archer's direct involvement was relegated for the most part to one of contributing material on the arms and monuments in the churches.[22] Much of *Warwickshire* was completed by the time civil war broke out, but Dugdale kept working on it, even if at a slower pace once the hostilities began, until the time of its publication in 1656.

Archer and Hatton both rendered Dugdale an invaluable service by procuring for him access to the muniment rooms of their friends. Also, Dugdale regularly wrote to Archer in the hope of obtaining local information, such as that concerning genealogy, from him. In most cases the cooperation of the landowners was readily obtained. Many of the newly established families may have considered that even a fleeting notice of their pedigree or property in Dugdale's book substantiated their claim to gentility. There were times, however, when such assistance for whatever reason was withheld; on one occasion Archer found it necessary to inform Dugdale that 'Men doe promise me much, but I finde them slowe of p'formance.'[23] Hatton went further than merely introducing Dugdale to his own powerful friends and associates; he backed Dugdale's enterprise financially and obtained for him access to the public records then housed in the Tower of London, and to Sir Robert Cotton's famous collection. Dugdale's gratitude to his patron for his generous assistance was expressed publicly by the dedication of *Warwickshire* to Hatton, where he is described as a 'principall Mecoenas of learning, and especially of Antiquities."[24]

Despite its great length and numerous evidences of prodigious scholarship, *Warwickshire* at least resembles the earlier tradition of chorographic writing in its admixture of topography and genealogy. In typical Camdenian fashion Dugdale divides the shire into its hundreds and proceeds along the rivers and streams to describe each parish in turn:

'Rapidly Thinning Wisps and Patches' 107

For the order and methods of this present work, I have followed the Rivers (as the most sure and lasting marks) where they lye proper for my course; and sometimes have taken my aime from those great and well-known Roman ways, viz. Watlingstreet and Fosse; which thwarting each other upon the borders of this Countie, extend themselves many miles, through it, or as a boundarie thereto. And whereas the Hundreds are so few, and the Rivers, with their branches very many, I have taken each Hundred by it self ... discoursing in order of the Towns, as they lye adjacent thereto, or neer those petty streams which run into it; beginning always with that wherein the Church is seated, and then proceeding with the severall small Hamlets or places of note, whether depopulated or otherwise, contained within the same Parish; setting forth a succession of their antient possessors; by which the rise, growth, continuance, and decay of many Families, with their most memorable actions, are manifested.[25]

Dugdale, far from anticipating the natural histories that formed the next generation of regional studies, compares his work to that of the previous generation of chorographers, specifically referring by name to Lambarde, Erdeswicke, Carew, Burton, and others.[26] His 'survey,' in fact, was an attempt at 'illustrating' the antiquities of the county in much the same way that the ancient chorographers desired to 'paint' – verbally – their lands. Dugdale viewed his diligent searches 'into the vast Treasuries of publique Records' as a continuation of the kind of investigation done by Polybius, Suetonious, Livy, Tacitus, and others 'who made speciall Use of the publique Records of Rome.'[27]

Dugdale begins his study by commending the gentry of Warwickshire for promoting his 'publique work,' that is, by granting him access to their archival repositories. At the same time he recognizes the great honour they deserve for the pious respect they pay their ancestors by 'representing to the world a view of their Tombes, and in some sort preserving those Monuments from that fate which Time, if not contingent Mischief, might expose them to.'[28] Next, in the way of a general introduction to the whole study, Dugdale, relying on *Domesday Book*, informs the reader of the names of the hundreds and towns that existed in the county at the time of the Norman settlement. Although unable to find a definitive answer in the ancient accounts, he nevertheless speculates as to whether these names had already existed by the time of King Alfred. He concludes that there probably had occurred some alteration over the course of time; there were ten hundreds at the time of the Normans but only four when Dugdale wrote, and none of those four are found by the author to retain any of the old names. The four were Barlichway, Hemlingford, Kineton, and Knightlow. They are represented on the general map of the shire which appears near the front of the book, and later appear separately in individual maps.

Now we arrive at the main portion of the book. Commencing with Knightlow Hundred Dugdale gives an account of the former possessors of each place, the ecclesiastical institutions, old manners and customs, wakes, and general local information. Sometimes his account stretches back to Anglo-Saxon times but, as Dugdale states, the spoliation wrought by the wars of petty Saxon monarchs, and the lack of sufficient 'Light of Storie' for guidance, left him with few 'memorials' to guide him in his investigation of pre-Norman times.[29] Furthermore, in most cases he is unable to adequately give the history of each parish, pleading his 'own disabilitie to perform it.'[30]

Throughout *Warwickshire* topography takes a back seat to the genealogies, the blazons of arms, and the epitaphs and inscriptions. In this, Dugdale follows the pattern established by earlier chorographers, though on an enlarged scale. The following example, the opening paragraph outlining the city of Warwick, contains about as much topographical information as one is liable to find for any one entry in the text:

The first place of note that presents it self to my view, on the banks of this fair stream [River Avon], is Warwick, standing on the north side thereof; which, as it is and hath been the chiefest town of these parts, and whereof the whole County, upon the first division of this Realm into Shires took its name, so may it justly glory in its situation beyond any other, standing upon a rocky ascent from every side, and in a dry and fertile soil, having the benefit of rich and pleasant Meadows on the South part, with the lofty Groves, and spacious thickets of the Wood-land on the North; wherefore, were there nothing else to argue its great antiquity, these commodities, which so surround it, might sufficiently satisfy us, that the Brittans made an early plantation here to participate of them.[31]

Usually there is considerably less topographical description than this for a particular place, while natural history is almost totally disregarded; Baconian science had not, by this date, found favour among the majority of regional writers.

Social, economic, and linguistic historians have been able to reap the fruits of Dugdale's researches on at least two related fronts. First, in many instances Dugdale identifies settlements that no longer exist and, where appropriate, he comments on the nature of the depopulation that has occurred between 'ancient times' and his own day.[32] Second, Dugdale went further than almost every scholar before him in attempting to discover the derivation of place-names. In etymologizing the names of towns and places, however, he was not, in his own words 'over-bold, because most of them had their originall denomination from the Britans, or Saxons; and that Time hath much varied the antient name.'³³ Yet he did have the advantage of being able to utilize the results of other recent studies in this field, including those of Spelman and Somner.³⁴

A closer look at one small section of *Warwickshire*, that describing Coventry, will clarify our conception of the kinds of topics that interested Dugdale. In the narrative he arrives at Coventry by 'following the stream of Shirburn' until it leads him there. The town is represented as still a city of eminent note, even if lacking the glory and riches that it had previously possessed. After this brief general introduction, Dugdale delves into the origin of the name and foundation of the town. Again, however, the 'Light of Stories' is too diffuse to accurately guide him through those olden times. Once more he is forced into sheer speculation, wondering whether the term 'coven' was occasioned by some covenant of religious persons – since there is evidence of an ancient nunnery here – or whether, as others think, the word derives from a local brook, whose true name might have been 'Cune.' Unable to find concrete evidence, Dugdale decides not to argue either one of these positions. Dugdale's detailed examination of the foundations, endowments, and histories of religious houses and chantries forms the core of the section on Coventry. His interest in these topics was stimulated no doubt by his Church of England background and by his collaboration with Dodsworth in researching and writing the *Monasticon*, a treasury of this type of information. The account of Coventry continues with a historical examination of the events leading to the destruction of the local nunnery by Canute in 1016, and of the subsequent establishment of a monastery by the earl of Mercia near the same site. Dugdale finds the record of these events in various sources, including the Rous collection in Cotton's library. From this point on the narrative remains centred on the religious houses, and on the genealogies of prominent families throughout the town's history. Occasionally other bits of historical information are interspersed in the text, for example: 'But in K.H.4. time, I find nothing memorable, excepting that the K. held a Parliament here in ann. 1404. 6 of his reign.' There is the odd mention of this or that fair, or of the state of the town's trade.³⁵ Architectural features and conspicuous landmarks are also found worthy of observation. A prominent cross, built in 1541–4, is noticed as 'one of the chief things wherein this City most glories; which for workmanship and beauty is inferiour to none in England.'³⁶ Finally, several pages are devoted to an account of the various chantries, replete with notes on the monumental inscriptions found on the walls and elsewhere.

The text of *Warwickshire* is devoid of superfluous verbiage and is meticulously documented by references to Dugdale's authorities. Naturally, in such a massive compilation, errors were unavoidable; more than one contemporary reviewer was able to point out slips in the critical discernment of the evidence.[37] But usually Dugdale's work allowed little justification for such criticism. His scrupulousness is revealed in some sound valedictory advice he gave to Archer, warning his associate of the pitfalls of depending on 'any mens collections or transcripts, without comparinge them with the originalls,' since even the most judicious may have 'here and there gleaned according to their fancye, and left behinde them as materiall things as they have taken.'[38]

It is therefore principally Dugdale's mastery of technique that was to influence future regional writers. Published before the new approach to regional study was expounded by members of the Royal Society, *Warwickshire* did not deal with the natural history of a place. That Dugdale was at least interested in pursuits such as those that we would today refer to as 'archaeological,' however, is undeniable. For example, he aroused a colleague's interest in 'a notable discoverie' that he found at Tamworth, Warwickshire, 'by the digging for Marle to manure Mr. Archer's land.' For a depth of more than four feet, a width of eighteen feet, and a length of fifty-five yards, the earth there was found to be very black. It was believed by some that at this site at least two thousand bodies were buried in a mass grave. Finding 'a speare head of Iron, much eaten with rust; and ... potshards, some of large magnitude,' Dugdale was ready to believe, like Camden, that Tamworth had been a royal village under the Mercians.[39] Dugdale was also interested in 'a faire [Roman] stone taken up about two miles west from Newcastle upon Tine' by a Mr Shafto, a lawyer of Gray's Inn, who wished to present this object to Oxford University.[40] But overall, Dugdale's amateur interest in such antiquities was like that of most chorographers; they had little interest in purely archaeological evidence, which therefore plays very little part in their published work.

Warwickshire won immediate praise from Dugdale's fellow antiquaries. William Somner thought it 'so copious and well stored for the matter; so curious and well contrived for the forme ... (in one word) a Master-piece.'[41] It turned Anthony à Wood's life into 'a perfect Elysium' by satisfying his 'insatiable desire for knowledg.' At the Restoration Dugdale went on to become Norroy Herald and later, in May 1677, Garter King-at-Arms. During this period he published several other scholarly works, including *The History of St Paul's Cathedral* (1658), *The History of Imbanking and Drayning of diverse Fenns* (1662), *Origines Juridiciales* (1666), and *The Baronage of England* (1675–6).

Three other chorographies, inspired by Dugdale's example, were published in quick succession in the late 1650s, none of which attained the learning displayed by many of their predecessors. Kent received an inordinate amount of attention; two works dealing with that county were issued in the space of one year, 1659. Richard Kilburne (1605–78), a Kentish justice of the peace and steward of the manors of Brede and Bodiam, Sussex, published *A Topographie; or Survey of the County of Kent*, which contained some valuable information about Kilburne's own parish, Hawkhurst, but was primarily a meagre gazetteer.⁴² It contained little description of natural phenomena, but did not lack the usual lists or tables naming the political divisions of the county.

John Philipot (1589?–1645), Somerset Herald (1624–45), made considerable collections for a county history of Kent, and the *Villare Cantianum: or Kent Surveyed and Illustrated* is generally credited to him.⁴³ This work is closer to Dugdale's *Warwickshire* than that executed by Kilburne; Philipot also acknowledged his debt to Somner's work.⁴⁴ The aim of Philipot's study is to

take the County of Kent under Survey or Prospect, and represent to the publique view, those several Antiquities which in my search I found to be wrapt up either in common Records, or shut up in the private Muniments, Escripts, and Registers of particular Families; from whom I have endevour'd to pluck off the veil that they might for the future stand as an Alphabet to point out those Families that are yet in being, that are totally exinguish'd or that lye entomb'd in other Names and Extractions, which by Marriage have swallowed up the Heir generall.⁴⁵

Obviously, the publication of such a work indicates that natural history had not yet made a substantive impression on some quarters of regional study. Neither is there much description of a topographical nature. *Kent Surveyed*, on the other hand, has been noticed for its merit as an early history of property.⁴⁶ The account of the holding of manors still played a conspicuous role there, as in evident in the following passage: 'Hawking in the Hundred of Folk stone contains two little Mannors within its Verge, which must not be passed over in Silence. The first is Bilchester, which belonged to the Knights Templers, but upon their suppression, in the second year of Edward the Second, it escheated to the crown.'⁴⁷

The same year Philipot's work was published, Edward Leigh (1602–71) had published *England Described*, a chorographic work that covered the entire country. It was entirely derivative and of small historical value, anticipating the sorry state into which chorography was to fall. Much of Leigh's information was derived from the earlier chorographies, including Camden's *Britannia*: 'I have made much use of Camden, and if I could have added to his Chorography, some new and memorable things of each county, which he had not observed, I should have thought it might be usefull.'[48]

The initial brief revival of enthusiasm for chorography – generated by the publication of *Warwickshire* – gradually subsided, and the dearth of chorographic writing becomes increasingly pronounced as the seventeenth century wears on, especially when compared to the growing number of natural histories that were by then being produced. Only one man, Robert Thoroton (1623–78), followed closely in Dugdale's footsteps. Although Thoroton's major antiquarian production, *The Antiquities of Nottinghamshire*, is often elevated to the same lofty pedestal granted Dugdale's *Warwickshire*, and the two works are similar in many respects, the simple fact remains that Thoroton never quite attained the same level of scholarship or originality. Due to the pronounced stylistic and topical similarity between the two works, there is no need to go into graphic detail of the contents of Thoroton's study.[49] Thoroton's biographer, John T. Godfrey, puts the matter in a nutshell: 'This work [*Nottinghamshire*] is almost entirely genealogical and heraldic in character, and its compilation appears to have occupied Thoroton about ten years [1667–77]. He did not, apparently, visit every village and hamlet in the county, but derived much of his information from a mass of valuable manuscript placed at his disposal by his numerous friends among the county gentry.'[50]

More may be said, however, of Thoroton's personal life. Descended from a family that had long held considerable peoperty in the county of which he wrote, Thoroton became sizar of Christ's College, Cambridge, in 1639, graduating BA (1643) and MA (1646). Having received from the university a licence to practise medicine he combined his practice with the occupations of a country gentleman. Unable to keep people alive for any time he consequently chose 'to practise upon the Dead' by ascertaining, through the contemplation of deceased Nottinghamshire worthies, what was to be gained from 'the shadow of their names.'[51] Although a royalist, Thoroton took little part in the civil war. After the Restoration he became a justice of the peace and a commissioner of royal aid and subsidy.

Nottinghamshire was dedicated to Gilbert Sheldon, Archbishop of Canterbury, and to Dugdale, both personal friends of the author. In the dedication Thoroton tells us how, during the course of Dugdale's visitation of the county, he and Dugdale were staying at the home of a mutual friend, Mr Gervase Pigot, when Dugdale suggested that Thoroton attempt a description of the region. At the same time Dugdale offered his assistance towards the proposed venture.[52] In his compilation, as Godfrey indicates, Thoroton utilized as his sources family registers, estate conveyances, epitaphs, inscriptions on old monuments, old deeds and rolls, etc. Being a man of great wealth, he was able to employ paid assistants in his work. Although Thoroton spared no effort to assign large portions of his book to the documentation of his own family history, the work itself was composed along traditional chorographic lines, describing the county on a hundred by hundred basis.[53] But by the time Thoroton's work was issued, fewer and fewer parts of Britain were left 'unpainted' in the chorographic mode. Coincidentally, Robert Plot's *Oxfordshire* was released in the same year, marking the new course being plotted – no pun intended – in regional study. That course had been laid out by Sir Francis Bacon and by his followers. Even Dugdale's book, it may be argued, despite its erudition, set the tone for change: Camdenian purists of the time, no doubt, were less willing to continue supporting a brand of chorography that had degenerated into the mere illustrating of pedigrees and heraldry, to the exclusion of topography and other antiquarian concerns.[54]

CHAPTER SIX

'Men Wake As from Deep Sleep'

> It is idle to expect any great advancement in science from the superinducing and engrafting of new things upon old. We must begin anew from the very foundations, unless we would revolve for ever in a circle with mean and contemptible progress.
> Francis Bacon *Novum Organum*

IN THE PREFACE to his recent book, Jerry Weinberger openly states his intention of restoring the eighteenth-century view that Sir Francis Bacon should be ranked the greatest of all the 'Moderns,' an act that in itself signifies that the ancients vs moderns question is not dead yet.[1] But leaving aside such issues for the moment, as well as the broader aspects of Bacon's overall philosophy, the intent here is to briefly examine some of his writings that had the most impact on the course of regional study, especially his views on civil and natural history. Much of the effect he had on the regional writers may have been indirect or imprecise, sometimes more ideological than scientific or technical, and often not even acknowledged – or when it was, the original intent of Bacon's comments was distorted. But what is clear is that for the individual regional writer Bacon (1561–1626) loomed largest. A negative effect of this might be that Bacon's empiricism, which led him to overcalculate the amount of data required towards the compilation of natural history, sometimes encouraged the indiscriminate collection of a multitude of it. Nor was he as utilitarian as some believed him to be, though this belief discouraged some from pursuing science for the sake of learning *per se*, or made them wince at theorizing. This is not necessarily to go so far as, say, R.F. Jones, in promoting the view that utilitarianism was *the* overriding concern of the Royal Society or that Baconism overwhelmingly dominated the age as a whole, suggestions that by and large have already been put to rest. Descartes, for one, is currently enjoying some popularity as a promoter of the doctrine of utility, notably as applied to the work of physicians and artisans, especially in part 6 of his *Discourse on Method*, which was widely read in England. In other words, one needs to look beyond the 'bark' of the Baconian apologists of the Royal Society to see that it was not necessarily equaled by the 'bite.'[2]

Having recognized this, however, we need also to remember it was Bacon who was most vocal in arguing that in both human and natural history the utility of any work was determined mainly by the extent to which it depended on measurement and induction rather than on metaphysics and deduction. Also, in both branches of history, the appeal must be made to 'efficient' or secondary rather than final causes. Thus, secular or natural causes were to be the new measuring stick applied to both civil history and natural history, although for Bacon this did not involve denying the existence or omnipotence of God, who ultimately controls both nature and men – even if that control is not always readily perceptible to humans. History and science in this mode of thought could become separated from theology as well as scholasticism.

Bacon certainly was not the first to insist upon the close observation of nature, and as historians of science have pointed out, in his day an empirical philosophy of science was not novel, but was present in the tradition of classical science. However, as Thomas Kuhn has noted, Bacon and his followers established a 'different sort' of empirical science, which, for a time, coexisted with its predecessor.[3] Their concern was not with 'thought experiments'; they rarely sought to demonstrate what was already known, but desired

> to see how nature would behave under previously unobserved, often previously non-existent, circumstances. Their typical products were the vast natural or experimental histories in which were amassed the miscellaneous data that many of them thought prerequisite to the construction of scientific theory. Closely examined, these histories often prove less random in choice and arrangement of experiments than their authors supposed ... The corpuscularism which underlies much seventeenth-century experimentation seldom demanded the performance or suggested the detailed outcome of any individual experiment. Under these circumstances, experiment was highly valued and theory often decried. The interaction which did occur between them was usually unconscious.[4]

Kuhn listed the other novel features of the Baconian movement. These include emphasis on experiments that constrained nature –for viewing under man-made conditions – and employment of scientific instruments and chemical apparatus previously non-existent or restricted in use. In this scenario Baconism, if contributing little to the development of the classical sciences, nevertheless 'did give rise to a large number of new scientific fields, often with their roots in prior crafts.'[5] At the same time, if in the seventeenth century scientific pursuit 'produced few transformations more striking than the repeated discovery of previously unknown experimental effects,' a similar assessment can be made of the collection of previously unknown or uncollected natural historical data. In neither of these cases did 'a body of consistent theory capable of producing refined predictions' arise, but such activity marked a beginning to the 'Experimental Philosophy' – as distinguished from the classical or 'mathematical' sciences – with its gradually increasing systematization and precision.[6] Unlike most others, Newton participated in both traditions, Baconian and classical. He collected and used data, but often to elaborate theory, a stage not always reached by many regional natural historians.

While Bacon's so-called inductive method has drawn much attention, for him the reconstruction of the sciences had to be founded on a base of natural history of a novel kind. In turn, natural history is also at the base of Bacon's hierarchical pyramid of knowledge – with metaphysics, as generalized physics, at the apex – and is inestimably important to the whole:

let such a history be once provided and well set forth, and let there be added to it such auxiliary and light-giving experiments as in the very course of interpretation will present themselves or will have to be found out, and the investigation of nature and of all sciences will be the work of a few years. This, therefore, must be done or the business must be given up. For in this way, and in this way only, can the foundations of a true and active philosophy be established; and then will men wake as from deep sleep, and at once perceive what a difference there is between the dogmas and figments of the wit and a true and active philosophy, and what it is in questions of nature to consult nature herself.[7]

The term *history* itself, for Bacon, has a broad connotation, similar in meaning to the Greek *historia* or 'factual inquiry into' than merely denoting past events. He first breaks down the word into two broad general categories: 'civil' history and 'natural' history. The former consists of three types: ecclesiastical; civil – in the narrow sense, or political; and history of learning and the arts, or literary; these may then be further subdivided. Civil history, then, generally consisted of inquiry into human affairs.[8]

For Bacon, civil history may also be arranged into 'antiquities' or 'remnants' of the past, 'memorials' – rough drafts of history – and 'perfect history' – portion of time, person worthy of mention, or distinguished deed.[9] For Bacon antiquities consisted of 'history defaced, or remnants of history which have casually escaped the shipwreck of time,' and included coins, monuments, proverbs, genealogies, etymologies of words, traditions, and so on, out of which 'acute and industrious persons' 'may recover some of the things lost to time.'[10] 'Memorials' – or 'preparatory history' – provided the raw source material, facts and information, for use by writers of perfect history. Memorials included commentaries, registers, edicts and decrees, etc. The fact that, according to Bacon, civil history could also be divided into pure civil history and mixed civil history, especially with regard to the latter, tended to cloud the distinction between 'civil' and 'natural' for some regional writers and/or encouraged some to jump freely from one to the other. F.H. Anderson cites one example of a 'mixed' civil history, namely cosmography: 'a mixture of natural history in respect to regions with their sites and products, of civil history in respect to the habitations, government, and manners of the people, and of mathematical calculations in respect to climate and to the configurations of the heavens beneath which the regions of the earth lie.'[11] Bacon did not necessarily view the two types as naturally exclusive, so why should anyone else? As far as civil history is concerned, while recognizing the documentary value of official records, Bacon did not emphasize careful, specialized research in source material or exact documentation, as did Dugdale later. This is reflected in his *History of Henry VII*, which lacked a developed technique of criticism and is characterized as an example of the 'literary' historical method, where 'the role of the historian "proper" was to portray the course of events together with "the causes and pretexts, the commencements and occasions, the counsels and orations, and other passages of action."'[12] Bacon was clearly more interested in history proper, or perfect history, as distinct from a brand of civil history composed of memorials or antiquities. Ironically, his call for data collection and accurate assessment did not go unheeded by many civil historians. Stuart Clark spoke of the 'major contribution to the new way of thinking about the uses of the past' made by Bacon:

in the *De Augmentis Scientiarum* all '*Philosophia*,' human as well as natural, was secularized at a blow by the total segregation of the worlds of sensory perception and divine revelation, of reason and faith. Rational knowledge, whether of nature or man, was henceforth to be made up entirely of exact sciences of the material world conceived of amorally and in physical terms. The truth of these sciences was to be guaranteed by a radical empiricism and by the logic common to them all, the inductive-deductive method of the *Novum Organum*. Speaking of the science of man (*Doctrina de Homine*) Bacon said, 'it is but a portion of natural philosophy in the continent of nature.'[13]

Clark goes on to say that 'in reducing all sciences to natural sciences, Bacon in effect transformed all history into natural history,' and that he reduced civil history 'to an empirical, neutral, methodologically passive record ... "Literary" history was thus separated from, and made logically prior to, the theoretical structures which could be built on it.' The result was that literary history 'was to perform the identical empirical function with regard to the sciences of man as natural history performed with regard to the science of natural phenomena.'[14]

The natural historians of the second half of the seventeenth century were caught up in the Baconian project of making 'faithful Records of all the Works of Nature ... which can come within their reach.'[15] Where the regional natural histories are concerned – where the organization of ideas is not necessarily advanced and the material more complex – the basic acquisition of exact data seemed to be more worthy a task than premature attempts at conceptualization. So, Bacon's call for experimentation and information gathering established the pattern for future developments. Whereas the chorographers had written about the physical properties of places alongside their social, political, economic, legal, or religious history, by this period many interested men arrived at the conclusion that the only proper way to describe such places was by concentrating on the study of their natural phenomena, almost to the point where little else mattered except for curiosity value. Increasingly then, as the century approached an end, few men braved the consequences of surveying a region without also intensively examining its natural history. Few chorographical works produced after 1656 were able to achieve the standards set by their predecessors because many were written in the traditional chorographic mode established before Baconism or Dugdale came on the scene. Such chorography, in many senses, now became a distant cousin to the new regional natural historical works, even though the latter were indebted to their beggared relation for having originally established a tradition of regional study in Britain.

Just as the natural histories retained some of the ingredients inherent in the earlier chorographies, so, too, did the latter manifest certain qualities that one usually associates with works of natural history. Carew and other chorographers, for example, took notice of the flora and fauna, plotted geological strata, and so on. The difference lies in part in the degree of attention dispensed to natural phenomena as opposed to that given over to civil history and institutions. Natural histories always emphasized the former while chorographies usually stressed the latter – although in at least one chorography, Carew's *Survey of Cornwall*, both were roughly mixed.[16] Civil history and institutions, generally, played little part in regional study in the second half of the seventeenth century, although they are contained to a limited degree in some works, especially at the time when the writing of regional natural history was still in its infancy.

Today the term 'natural history' is often used as a synonym for 'botany' or for 'biology.' For the natural historian of the second half of the seventeenth century, however, the geology of the substrata, the productions of the soil, and the zoology of the locality were all proper objects of inquiry, as were all related natural phenomena. Observation and description of nature were therefore extended to limits beyond those of the chorographer, whose main interest lay in the topography, history, or antiquities of a region, or a combination of these. When a chorographer noticed the natural phenomena, his observations were of secondary importance to his main focus and were intended to complement the picture rather than form its centre.

The ancients were the first to record observations of the earth's natural constitution. Aristotle (384–22 BC) is often regarded as the father of natural history. He devoted, for example, a portion of his treatise on 'meteorics' to a discussion of earthquakes, and also speculated on matters such as volcanic activity, river systems, and the weather. The Greeks were also the first to grasp the true nature of fossils. Xenophanes (sixth century BC), it seems, observed impressions of small fish and marine shells; Pythagoras and Xanthus, meanwhile, are 'reported to have accepted the occurrence of these shells as an indication that the mountains were at one time under the sea.'[17] Early investigations such as these were well known among many men engaged in regional studies.[18] But as the seventeenth century progressed, the ideas of Francis Bacon began to take hold, and so the attitude of men towards the ancients was changing. Even Bacon refused to accept their dicta as authoritative and recognized the value of their findings only after these had been tested; ancient scientists, he held, should be regarded as 'counsels,' not 'dictators.'[19] But the veneration of the ancients died a slow death, and this adherence to antiquity still had some effect upon many of the more enlightened minds right up until the end of the century – upon minds of the calibre of John Ray or Robert Hooke, scientists who still supported their discussions of natural history with quotations from Pliny, Strabo, Plato, Aristotle, and others. Thus, one also finds that many of the lesser figures did not totally disregard the literary or classical sources either.

Natural history, as the term implies, dealt with the 'deeds and works' of nature, and for Bacon it was very broad, covering all the 'Phenomena of the Universe.'[20] Bacon, following to some degree Pliny, divides the subject matter of natural history into three states or conditions of nature: 'nature in course ... [that] is history of Creatures'; 'nature erring or varying ... [that] is history of Marvels'; and 'nature altered or wrought ... [that] is history of Art.'[21] 'Nature in course' is nature 'free,' following her ordinary course of development. 'Nature erring or varying' finds her driven out of her usual course 'by the perverseness, insolence, and frowardness of matter, and violence of impediments; as in the case of monsters [that is, wonders and freaks].'[22] 'Nature altered or wrought' is nature 'in constraint, moulded, and made as it were new by art and the hand of man; as in things artificial.'[23] In a similar fashion, then, nature may fall into the categories of a history of generations, of pretergenerations, and of arts – the last Bacon called 'Mechanical and Experimental History.'

Bacon believed that while the history of nature in course is extant, the history of nature erring and nature wrought 'are so weakly and unprofitably handled that they may be set down as deficient.'[24] Regional natural historians could not help, then, but be influenced by this call to explore 'these works of nature which have a digression and deflexion from the ordinary course of generations, productions, and motions,' especially if they are 'singularities of place and region.' Bacon certainly found there to be enough literature extant on marvels and the fabulous, but it was generally uncritical, unmethodical, and without 'due rejection and as it were public proscription of fables and popular errors.'[25] What some people today fail to realize is that such accounts of irregularities of nature were often incorporated into natural historical studies for this reason alone, that is, in response to Bacon's advice. The history of pretergenerations was vital for providing 'the most clear and open passage [from the wonders of nature] to the wonders of art.'[26] Of equal weight as far as the natural historians were concerned was Bacon's appeal to not neglect the examination of

> superstitious narratives of sorceries, witchcrafts, charms, dreams, divinations, and the like, where there is an assurrance and clear evidence of the fact ... For it is not yet known in what cases; and how far, effects attributed to superstition participate of natural causes, and therefore howsoever the use and practice of such arts is to be condemned, yet from the speculation and consideration of them (if they be diligently unravelled) a useful light may be gained, not only for the true judgment of the offences of persons charged with such practices, but likewise for the further disclosing of the secrets of nature.[27]

Narrations touching miracles of religion, not being true or natural, were considered by Bacon 'impertinent for the story of nature'; and generally this bit of advice was followed. This was not always the case, however, with his recommendation that 'those narrations which are tinctured with superstition be sorted by themselves, and not mingled with those which are purely and sincerely natural.' It was not always easy to distinguish between the two. Whatever the situation, the pursuit of 'nature erring' would enable man 'to lead and drive her afterwards to the same place again.'[28]

Bacon valued most of all the history of 'nature wrought or mechanical' in the various arts. He felt it necessary, however, to defend it, since 'it is esteemed a kind of dishonour upon learning for learned men to descend to inquiry or meditation upon matters mechanical.'[29] Including a history of the arts as a species of natural history is justifiable because

> the artificial does not differ from the natural in form or essence, but only in the efficient; in that man has no power over nature except that of motion; he can put natural bodies together, and he can separate them; and therefore that wherever the case admits of the uniting or disuniting of natural bodies, by joining (as they say) actives with passives, man can do everything; where the case does not admit this, he can do nothing. Nor matters it, provided things are put in the way to produce an effect, whether it be done by human means or otherwise ... Still therefore it is nature which governs everything.[30]

Furthermore, 'history mechanical' is 'the most radical and fundamental towards natural philosophy,' a natural philosophy 'as shall not vanish in the fumes of subtle or sublime speculations, but such as shall be operative to relieve the inconveniences of man's estate.' It will be of immediate benefit 'by connecting and transferring the observations of one art to the use of others, and thereby discovering new commodities,' but it will also 'give a more true and real illumination concerning the investigation of causes of things and axioms of arts.' Nature thus reveals herself best when subjected to the 'trials and vexations of art,' just as Proteus's many shapes are not revealed until he is 'straitened and held fast.'[31] The history of 'nature wrought or mechanical,' according to Bacon, may already be found in 'some collections made of agriculture and likewise of many manual arts,' but normally 'with a neglect and rejection of experiments familiar and vulgar.'[32] Bacon also castigated the scholastics, blaming their failure primarily on their neglect of nature and the observations of experience. Such Baconian ideas about the history of arts were especially to affect the reformers of the 1640s and 1650s, Samuel Hartlib's design for the history of trades being one prominent example.

A history of the arts, according to Bacon, was vital towards disclosing nature's component parts, whose external appearances or shapes often hide their true colours. The arts that have the greatest effect in this regard – that is, those that are most capable of displaying, altering, or preparing natural materials – include chemistry, agriculture, dyeing, and the manufacture of paper, enamel, glass, gunpowder, and so on. Those less 'useful' include weaving, carpentry, the making of instruments, and architecture. Bacon warned of the danger of overlooking experiments that do not have have an immediate effect on the arts, that is, those that at first glance or through habit may seem to be merely incidental.[33] This warning, however, had the added effect of occasionally directing some of his followers to the pursuit of experimentation that we would consider nonsensical or trivial, and even to the indiscriminate collection of data.

The second division of natural history, according to its use and end, was into narrative history or inductive history. Narrative history is extant, but is flawed by its reliance on 'fables, antiquities, quotations, idle controversies, philology and ornaments.'[34] Not surprisingly, Bacon's preference is for inductive history, which he finds wanting. Histories of pretergenerations and arts are lacking, and as for history of generations, 'only one out of five parts is sufficiently handled,' namely a history of the 'Exquisite Collections of Matter' – or 'Species.' And even then, writers have shown interest only in superfluous matters such as figures of plants and animals rather than enriching their work with 'sound and careful observations, which should ever be annexed to natural history.'[35] Bacon lists the other four parts that deserve further attention:

First, a history of the Celestial Bodies, exhibiting the actual phenomena simply and apart from theories. Second, a history of Meteors (including comets), and what they call the Regions of the Air ... Third, a history of the Earth and Sea (considered as integral parts of the universe), mountains, rivers, tides, sands, woods, islands, and the shapes of continents as they lie; in all these, inquiring and observing rather the laws of nature than cosmography. Fourth, a history of the Common Masses of Matter ... for I find there are no accounts of fire, air, earth, and water, with their natures, motions, operations, and impressions, such as to form a just body of history.[36]

In the *Parasceve ad historiam naturalem et experimentalem* (*Preparative toward a Natural and Experimental History*) Bacon established rules for the collection of natural historical material while emphasizing the co-operative nature of this type of venture. The first involved the elimination of antiquities, citations of authors, and scholastic-like rhetoric; the second, the omission of superfluity in the description of species; the third, the evidence of superstitious narratives and ceremonial magic in natural historical accounts.[37] In compiling a natural history one should also beware of disregarding familiar things previously thought to be superfluous or things that might seem base, filthy, trifling, or subtle – for any or all of these might prove of worth. Of prime importance is the need for accurate measurement, since 'Operation, not speculation, is the final end in view.'[38] Natural historians were also advised to closely consider three types of statements: those that are true, doubtful, or false. The first should present no difficulty and be framed simply; the second should carry a qualification, such as 'it is reported,' etc; while the third should be expressly rejected as untrue, the hope being that such fables will be weeded out of contemporary thinking. Anderson lists the five minor precepts that remain: first, questions as to fact, not causes, should be asked in order to provoke further inquiry; second, new experiments should be recorded as to the procedure so that the worth of the information may be determined by the reader, who might be encouraged to discover possibly a better method; third, statements of observation and experiment 'are to be written in truth with religious care,' without doubt, since 'This record is the book of God's works and – so far as there may be an analogy between the majesty of divine things and the humbleness of earthly things – is a kind of second Scripture'; fourth, occasional observations by the way may be interspersed, as well as general observations and comments on 'things that are not' – 'a star is never oblong or triangular'; and, finally, the opinions of the several philosophic sects should be passed over, whenever possible.[39]

Bacon then presents a list of subject titles of particular histories that are to be examined and written out, towards the goal of compiling a natural history of the 'Phenomena of the Universe.' There are 130 in all, some of which relate directly to the study of nature, some to the study of man, and others to the study of the nature-man relationship or to the mechanical arts. There was, of course, considerable overlap, since 'many of the experiments must come under more titles than one,' and Bacon advocated that there be no artificial division among any of the branches of knowledge themselves. Anderson reprints this 'Catalogue of Particular Histories by Titles' in whole. The following are some representative titles:

History of the Heavenly Bodies; or Astronomical History
History of Comets
History of Lightnings, Thunderbolts, Thunders and Curoscations
History of Rainbows
History of Hail, Snow, Frost, Hoar-Frost, Fog, Dew, and the like
History of ... Earthquakes ... Islands springing up de novo ...
Natural History of Geography; of Mountains, Valleys, Woods, Plains, Sands, Marshes, Lakes, Rivers, Torrents, Springs ... omitting Nations, Provinces, Cities, and such like Civil matters

Histories of the Greater Masses Follow Next
History of Flame and Ignited Things
History of Air, in Substance, not in Configuration

124 'Speculum Britanniae'

Histories of Species Follow Next
Histories of perfect Metals, of Gold, of Silver; and of the Veins, Marcasites of the Same ...
History of Fossils ...
History of Stones ...
History of the Magnet
History of Miscellaneous Bodies, which are neither wholly Fossil nor Vegetable; as Salts, Amber, Ambergris, etc
History of Plants, Trees, Shrubs, Herbs; and of their parts, Roots, Stalks, Wood, Leaves, Flowers, Fruit, Seeds, Guns, etc.
History of Birds ...
History of Quadrupeds ...
History of Serpents, Worms, Flies, and other Insects ...

Histories of Man Follow Next
History of the Figure and External Limbs of Man, his Stature, Frame, Countenance, and Features; and of the variety of these according to Races and Climates ...
Anatomical History ...
History of Excrements ...
History of Faculties; Attraction, Digestion, Retention ...
History of Sleep and Dreams
Medical History of Diseases, and their Symptoms and Signs
History of Cookery, and of the subservient arts, as of the Butcher, Poulterer, etc.
History of Honey
History of Sweetmeats and Confections
History of Ironworking
History of Pottery
History of Architecture generally
History of Printing, of Books, Writing, Sealing ...
History of Farming, Pasturage, Culture of Forests
History of the business of War, and of subservient arts ...
History of Navigation, and of the subservient crafts and arts
History of Athletics ...
History of Horsemanship[40]

Bacon's particular histories, on which he worked after 1620, had the dual aim of eliminating 'traditional theories and doctrines "by means of certain instances"' (this was not achieved), and 'to arrange the instances in given fields so as to form a "systematic catalogue" serving as a basis for the new philosophy.'[41] He came to formulate a theory 'on the function of "particular topics" as "places or directions of invention and inquiry in every particular knowledge."'[42] Bacon described the foci of investigation or 'topics' within the particular histories, and included both intellectual queries and suggested experiments: 'Since history and experiments very often fail us, especially those Experiments of Light and Crucial Instances by which the Understanding may determine on the true causes of things, I give injunctions touching new experiments contrived, as far as can be at present foreseen, to meet the special object of inquiry. And such Injunctions form a kind of Designed History.'[43] Bacon was well aware that a carefully selective approach to inquiry was required, that a random gathering of facts would prove of little use. As Purver points out, 'Bacon knew that the faculty of recognition or diagnosis was an essential part of inductive research,' and 'he saw this faculty, not just as an ingredient in the making of particular discoveries, but as a prime factor in the creation of Sciences.'[44]

Unfortunately, as Bacon realized all along, the task of producing histories in this manner was enormous. And as Rossi points out, in his last years Bacon, seeing the need for the accumulation of large amounts of information, was almost driven into an unsystematic collection of natural data in order to provide material for the particular histories. In essence, his history 'became more and more "literary" as he was reduced to making uncritical and indiscriminate use of an increasing number of traditional sources.'[45] In this dilemma he was not alone; more than a few future inquirers into natural history faced the same problem, to one degree or another. Bacon managed, however, to produce several histories, including 'The History of Winds'; 'The History of Life and Death,' centred on the degeneration of the human body with old age; 'The History of Dense and Rare or the Contraction and Expansion of Matter in Space'; and 'The History of Celestial Bodies,' although not all of these were published in his lifetime.

It is hoped that the above has made clear both the similarities and the differences between civil and natural history. In the seventeenth century few men, in any case, were thought capable of achieving the ultimate in the civil history sphere, namely the production of a perfect history – one covering 'Times,' 'Lives,' or 'Deeds'. Even Bacon's *Henry VII* fell somewhat short of the ideal, especially because of its uncritical use of sources and its overall derivative flavour. To be an antiquary, therefore, implied that one was engaged in a less than perfect activity. Bacon's pronouncements on the subject only tended to reinforce this line of thinking, so that by the second half of the century Thomas Shadwell, in his critique of the virtuoso, used the antiquary as a prime example. In Bacon's words, an antiquarian production involved 'work laborious indeed, but agreeable to men, and joined with a kind of reverence; and well worthy to supersede the fabulous accounts of the origins of nations, and to be substituted for fictions of that kind; entitled however to the less authority, because in things which few people concern themselves about, the few have it their own way.'[46] Nevertheless, as Shapiro makes clear in her work, 'It was antiquarian scholarship that produced most of the important historical achievement of the seventeenth century, particularly after the decline of perfect history in the 1620s.'[47] Among the reasons for this is the antiquary's reliance on 'facts' and 'things,' 'which was in keeping with the growing demand for accurate evidence and convincing proof. Antiquarian research tended to limit itself to topics where visible remains and documentary evidence were available.'[48] Usually conjecture, for the sole sake of philosophizing or embellishing the narrative, was outside the orbit of the antiquary, who preferred to limit himself to collecting and compiling, letting the facts speak for themselves. Of course, this was not always the case – for example, in instances dealing with the Brute legend or with family histories, which often required embossing or where extant documentary evidence was lacking or non-existent. The chorographer, then, as antiquary, was perhaps less prone to speculation not supported by evidence, be it literary or archaeological, than someone who considered himself a true historian. It should be noted here that for most chorographers archaeological evidence – had they used the term – was usually restricted to the combing of monumenta, and the like, for information derived from the copying of inscriptions, and with few exceptions was not 'archaeological' in the modern sense of the word.[49] Shapiro states, however, that seventeenth-century 'historians' were uncertain as to whether political history could be combined with detailed documentary research, at least in a readable form. One implication is that they therefore tended to avoid the latter, although after Dugdale this was probably less the case. One may also argue that some of the concerns of the antiquary would be considered 'historical' in the modern sense of the term in reference to the subject matter at hand – for example, Anglo-Saxon and other political institutions, feudalism – even if contemporaries labelled any work without a sustained narrative structure or central role of individual as 'antiquarian.'[50] If this is the case, then the voluminous amount of research and writings – on such subjects, and in this manner – produced by 'antiquaries' should be considered as having contributed much to our knowledge of history itself and not just to antiquarianism *per se*. The picture becomes even more confusing when we look at someone like John Aubrey; should Aubrey be considered primarily a historian, an antiquary, or a scientist? Perhaps calling him a 'virtuoso' is the safest approach. Although Shapiro says that many antiquaries, especially those involved in chorography, were connected to the scientific movement, perhaps it is closer to the truth to state that those individuals strictly interested in traditional forms of chorography went their own way, while the scientific investigation of countries and regions was undertaken by the men who were more closely associated with the Royal Society and/or Baconism in general. Still, it is true that

> it was the more painstaking antiquarian scholarship, with its closer links to natural history, the scientific movement, and philological specialization, which, more often than 'history,' produced a sensitivity to the pastness of the past, to problems of development, and to the uniqueness of various institutions and periods. 'Perfect history,' whose links with Renaissance historical thought are obvious, may in the end have contributed more to general social science than to history itself. For it was practitioners of perfect history who were particularly interested in causal explanation and political generalization.[51]

If science gained something from antiquarianism, then history also partook of empirical science, repudiating some aspects of the humanist rhetorical tradition, though in the eighteenth century 'the pendulum shifted back somewhat to a more literary conception of history.'[52] Shapiro has admirably described the ties between humanism – in particular, for our purposes, history – and science. These include, in both fields, the limiting of investigations to probable truths of physical observation and antiquarian analysis, leaving others to concern themselves about future use of their efforts. In addition, a good deal of collecting, collating, ordering, and so on, without serious analysis took place, and there occurred a movement away from premature generalization and theory. Men tended to shy away from distant goals, such as perfect history and natural philosophy, respectively, in order to focus on verifiable facts of matter. Firsthand observation was considered most reliable for historians (and chorographers), and 'truth' and impartiality were a concern to both historians and scientists.[53] There were other instances of commonality: historical and scientific theory emphasized probable rather than certain knowledge, and, as we shall see, membership in the two communities overlapped.

We have seen how some of the chorographers relied on collective fact gathering; the regional writers of the second half of the seventeenth century, such as those who participated in the 'Britannia' project, directed by Edmund Gibson, perfected this system. Since Bacon did not regard the two major varieties of history as mutually exclusive or totally distinct, natural historians who followed him, regardless of what they publicly may have professed, did not feel obligated to avoid completely discussion of matters that normally fell into the category of civil history. Bacon did separate man and nature, but not in terms of a dichotomy between history and nature; the historical and the human are not placed in opposition to the natural and the non-human. Many natural historians believed that natural history for Bacon was primarily restricted to the recording of matters of fact. But – and it is not inconceivable that here he may have had chorography in mind – these facts had to be sorted out of the literary quagmire in which bits of information on natural history were to be found; so that it could 'shrink into a small compass.'[54]

The study of natural history, of course, was not restricted to the regional writers; other virtuosi and scientists applied Bacon's dicta to their own particular researches. Moreover, Bacon's appeal for reliance on experiment, measurement, induction, documentation, and impartiality encouraged all scholars to develop methods free from their own prejudices, and to evaluate their data accurately and dispassionately. Bacon emphasized that the inductive and the quantitative method was critically important because, as he stated, the mind distorted the impressions of the senses, failing to distinguish between external evidence and its own subjective impulses by mixing 'its own nature with the nature of things.'[55] A natural history may have a twofold use, therefore; it may serve either as a narration of particular occurrences in nature of interest for its own sake, or as 'the stuff and material of a solid and lawful Induction.' Most members of the Royal Society accepted Bacon's belief that the natural world must be studied because discoveries bear visible fruit. Before Bacon, as John Aubrey put it, 'Things were not then studied. My Lord Bacon first led that dance.'[56] The Royal Society, therefore, did not inaugurate the study of 'things' to the exclusion of 'persons and actions,' but by emphasizing and intensifying it, established an intellectual climate that made such study respectable and that enabled it to flourish.

The increase in interest in natural phenomena and in natural history had more than a little to do with the continued influence of the Paracelsians – followers of Paracelsus (1493–1541) – dominated by the iatrochemists or chemical physicians.[57] This group criticized the Aristotelianism of the universities, even though many of the concepts they subscribed to were in fact Aristotelian in nature, and the traditional methods in medicine, which reached back to Galen. Their own philosophy was imbued with ideas taken from the Christian neo-Platonic and hermetic texts and called for a study of natural phenomena through study of scripture and the 'second book,' nature itself. The latter would centre on a method employing observation and experimentation. At the basis of all this was a concept of a chemical universe, in which the earth itself constituted a chemical laboratory – a concept that explained the origin of volcanoes, hot water springs, mountain streams, and the growth of metals.[58] New chemical cures were being sought for diseases; the application of chemistry to agriculture by Paracelsians resulted in correct theories about the benefits of manuring, which they postulated offered essential soluble salts to the soil; and questions about geological process were being debated.[59] But by the same token, 'iatrochemistry was born of superstition, for Paracelsus's view of nature was deeply imbued with magic even though he gave it an empirical dress.'[60] Thus, though what is termed rationalism in science had made considerable headway in the seventeenth century, traces of the magical element were retained, so that, for example, 'Respectable naturalists continued to credit the spontaneous generation of frogs and insects into the second half of the century'; many such examples are to be found in the works of the regional writers.[61]

In the macrocosm-microcosm world view of the Paracelsians, everything in the universe was in harmony, connected in one way or another with everything else, and correspondences existed between the celestial and the sublunary worlds to the extent that man (microcosm) was a reflection of the earth (geocosm), and in fact of the entire universe (macrocosm) in which he existed, embodying all of its constituent parts. A study of geocosm and macrocosm, that is, of the book of nature, could thus lead to a better understanding of man, and in general allow man natural-magical control over nature – hence the notion of the 'magus,' an element in the thought of neo-Platonists, most notably Marsilio Ficino and Pico della Mirandola. Medicinal cures were sought through the study of flora and fauna, which were impressed by 'signatures' that make them identifiable and that were usually connected with celestial objects in some way. In like manner, a study of astronomy could lead to an understanding of the relationship among parts of the celestial world, while a study of cosmography could do the same when it came to the relationship among earthly herbs and minerals; 'such a knowledge would equip him [the physician] to understand the structure and function of corresponding parts of the human body and the occurrence of diseases within it.'[62] Bodily functions were thought of in terms of chemical reactions, hence chemical – including alchemical – mineral remedies were eagerly sought for by the Paracelsians; they often came up against the Galenists, who were entrenched in the medical schools well into the seventeenth century. For Paracelsus, then, the visible world was a veneer for an invisible one that was injected by an active, to some extent magical, power. This neo-Platonic thinking 'justified the rejection of the element-theories of the ancients and dissolved the notion of matter into that of patterns of spiritual powers.'[63]

Connected with all this was the 'Corpus Hermeticum,' although Paracelsus took ancient hermeticism one step further by making the social-Christian ideal of the alleviation of man's lot an end goal of the knowledge of nature; this in itself goes a long way to explaining the interest of many sixteenth- and seventeenth-century medical practitioners in natural historical study. As Rattansi states, such ideas, being congenial to millennium utopianism, meant that Comenius and others of like mind based their calls for reform 'on the conviction that the millennium was at hand, and would be marked by the recovery of the knowledge of creatures that Adam had possessed in his innocence, and of the Adamic language which had given him power over all things.'[64] Sixteenth-century figures mistakenly believed the hermetic writings – Gnostic second- and third-century AD texts – to be Egyptian in origin, ascribed to Hermes Trismegistus, and pre-dating or contemporaneous with Moses. Hermes and the other ancient Egyptian priests or barbs were credited with being sorcerers of a sort, and it was thought that their knowledge of magic ultimately had its effect on Plato and the Platonists. The anti-Aristotelian tone of the neo-Platonists helped lead to anti-scholasticism, which further paved the way for the 'revolution' in science; as evidenced in the sixteenth and early- to mid-seventeenth centuries, before the mechanical philosophy took hold – if, in fact, it ever did completely. Mechanics, in fact, was viewed by certain Renaissance figures – John Dee, for example – as a branch of 'mathematical magic': in this sense hermeticism is seen as a direct stimulus for the growth of the mathematical mechanical sciences, and at least one observer regards Dee, through his emphasis on mathematics, as advancing the cause of science much more than did Bacon, who deprecated it.[65]

If the pseudo-sciences, or pseudo-philosophies, collectively helped give rise to what we call 'modern' science, they in turn succumbed to modern science. The process was gradual, so that even after the founding of the Royal Society the residual effects of what some contemporaries termed the 'occult' are still discernible in the thought and work of some British figures who often are regarded as true scientists:

The emphasis on empiricism and plainness could promote naivety, leading away from the intellectual exploration that is the true course of science; conversely, recognition that natural phenomena are more complex than scholastic philosophy allowed could end by making the world an unfathomable mystery. Science had lost much of its established intellectual discipline in the sixteenth century, and only acquired a new one in the second half of the seventeenth. Crudely considered, the experimental method tended towards indiscipline with its suggestion: anything is possible, therefore let it be tried.[66]

In the early days of the Royal Society, at least, it was difficult even for some sober-minded men to avoid being pulled in five different directions, so that, indeed, 'The route to complete rationalism in science was hard to follow.'[67] On the one hand there is Glanvill condemning hermetic alchemy and Sprat attacking astrology; on the other there is Ashmole becoming a disciple of the alchemist William Backhouse, and Sir Robert Moray a follower of the Jesuit hermeticist, Athanasius Kircher – who may also have held some sway over other members of the Society including Hooke, Oldenburg, and Aubrey.[68] But even a figure like Glanvill believed in spirits and in an *anima mundi*.

In England, references to Paracelsus can be traced to William Turner's account of the mineral waters at Bath (1557), and then to the writings of the physicians and surgeons. However, by the early years of the seventeenth century, the deeper occult elements of Paracelsian thought were of little concern to most Englishmen.[69] After 1640 the English Paracelsians were especially influenced by the work of J.B. van Helmont (1579–1644), whose works helped revive an interest in Paracelsus, and contributed to the attack on Galenism, which included calls for injection of chemistry into medicine. It was the identification of hermeticism with the more extreme sectarians, the failure of the chemical physicians to halt the plague of 1665, along with the generally more scientific climate of the day that began the discrediting of extreme Paracelsian thought among the English.[70] Baconian science, though it was in part an outgrowth of pseudo-science, was winning the day by the late 1640s because it seemed compatible with a defence against hermetic sectaries. The mechanical also appealed to many because it prevented a return to Aristotelian doctrine, with its inherent danger to Christian belief. The decline of the pseudo-sciences can also be attributed to Boyle's *Sceptical Chymist* (1661).[71] It should be noted, however, that even Paracelsus did not believe in transmutation, usually regarded as at the core of alchemy, which 'to him was the search for the active essential in natural substances which would serve in medicine more effectively than the unrefined Galenic soups.'[72]

How does Francis Bacon fit into this scenario? In some ways he presents us with a paradox; one may regard him as emerging out of the hermetic tradition, while at the same time attempting to dissociate himself from its extreme forms.[73] Like the magus, he conceived of science as power and of man as having the capacity to exercise dominion over nature. Unlike the magus, he possessed a modern outlook, 'if the Adamic mysticism behind the Great Instauration is not emphasized.'[74] Bacon, of course, advocated a reliance on experience and experimentation, though he detached it from its magical connection, but at the same time he rejected the mixing of the secular and religious spheres, eschewing the Paracelsians for pretending 'to find the truth of all natural philosophy in the Scriptures' and in the human body's correspondences with the universe.[75] Rattansi put Bacon's position in the proper perspective:

Despite Bacon's zeal to distinguish his own work sharply from the spirit and methods of the Hermetics, many of their ideas were reflected in his own writings. He ridiculed the alchemists for deriving chemical records from pagan myth; but his own works were pervaded by a belief in a pristine knowledge which he associated, above all, with Solomon, and he attempted to unveil natural secrets conveyed by the ancients in the guise of myth and parable. In the natural histories which were to serve as the foundation of his physics, Bacon was forced to rely on authors like Pliny, Cardano, Paracelsus, and Porta. To waste labour on the first principles of nature, whether elements or atoms, 'can do but little for the welfare of mankind,' and Bacon, therefore, had to resort a great deal to the 'spirits' which alchemists, natural magicians, and iatrochemists had discussed as the carriers of celestial powers. Finally, his *New Atlantis* (1627) bore great similarities to the Hermetic-Utopian societies depicted in such works as J.V. Andreae's *Reipublicae Christianapolitanae descriptio* (1619) and Campanella's *De Civita Solis* (1623).[76]

Hartlib could believe that the secrets of a metal-transformation tincture could be unlocked, while the death of Thomas Vaughan, a believer in the true philosopher's stone, was attributed by some to the accidental inhalation of the fumes of mercury.[77] Beale's interest extended to Pico, Bruno, and the Rosicrucian Robert Fludd; like Bacon he 'deplored the worst effects of astrology as a vulgar art, but felt nevertheless that it was "a serious affair," one worthy of reform.'[78] One has only to read Hunter's interpretation to see the great extent to which John Aubrey believed humans 'are governed by the planets,' although at the same time he realized that supernatural phenomena were especially susceptible to charlatanry.[79] Robert Plot showed an interest in wizards and witches, an 'elixir of life,' and in alchemy in general. His manuscript papers contain transcriptions from alchemists like van Helmont, Basil Valentine, and others, and in a 'Petition to the King' he sought royal patronage for a proposal to establish a college to promote hermetic studies.[80] The interest in hermeticism and in 'vulgar Baconism' extended to Isaac Newton – which might lead one to speculate that the interests of the 'great' scientist, in some ways at least, corresponded closely with those of men usually associated with 'lesser' intellectual traditions. In fact, one may argue that Newton's justification for the mathematical study of forces 'reached back [through the Cambridge Platonists] to the Florentine Platonist vision of a continually active and creative God whose love for His creation was imperfectly mirrored in human love as it was, at a lower level, in the cosmic sympathy, immanent in a *spiritus*, which bound the universe together.'[81] Perhaps the pseudo-sciences had come full circle.

Bacon adopted and adapted the hermetic doctrine of the transmutation of metals – achieved by purifying a metal, that is, by introducing into a base metal the 'spirit' of another. He believed generally in the possibility of transmutation from one substance to another, though not through some mystical process but by uncovering through analysis the irreducible qualities of a substance. For Bacon magic was 'the science which applies the knowledge of hidden forms to the production of wonderful operations; and by uniting (as they say) actives with passives displays the wonderful works of nature.'[82] Yet as Rossi points out, Bacon tried to separate himself from the alchemical and scientific traditions of the Renaissance by establishing the model of the mechanical arts, which included a collaborative research system.[83] In this set-up learning is not restricted to the illuminati, and the new science is to serve not the needs of the individual but of all mankind. If magic is to be retained at all, according to Bacon, it is basically for its value as a model, one that disclosed how nature can be dominated and improved.

Ironically, as Yates states, this very attitude on Bacon's part may have contained undercurrents of cosmic mysticism, in that his conception of the regenerated Mosaic Adam, in close communion 'with nature after the Great Instauration of the Sciences, seems to bring us back into an atmosphere which is after all not so different from that in which the magus lived and moved and had his being.'[84] Is it any wonder that some of the virtuosi followers of Bacon retained elements of this atmosphere? For us there is certainly the danger of reading too much into the pseudo-scientific tradition(s) of Bacon and the intellectuals of the seventeenth century. But if one interpretation is to be accepted, part of the difficulty is that we are dealing with the muddied intellectual waters of two phases of the Scientific Revolution: 'the first phase consisting of an animistic universe operated by magic, the second phase of a mathematical universe operated by mechanics. An enquiry into both phases and their interactions, may be a more fruitful line of approach to the problems raised by the science of to-day than the line which concentrates solely on the seventeenth-century triumph.'[85]

There certainly was enough interaction between the two phases, to the extent that Webster in a recent work defuses the idea of a major gulf between neo-Platonism and early-modern science.[86] Paracelsus, for example, is taken as stretching 'across a broad spectrum lying between traditional magic and experimental science' and, if anything, more akin to the latter than the former; and Newton has his own niche at the end of this Paracelsian road, in which we find 'the harmonization of knowledge derived from the Bible, Ancient Tradition and Nature.'[87] Patrick Curry states that with the arrival of the new science, astrology – and one might add the other strands of pseudo-scientific thought – 'declined drastically in prestige but it lived on to assume new meanings for different people in the changing world.'[88] The fact is that the regional natural historians, like the members of the Royal Society and other virtuosi, tended to draw their intellectual strength from several sources, not all of which can be classified as 'modern.' This does not necessarily counter the argument posed in this study, that it was Baconism that generally seemed to be the most influential; and the confines or constructs of Baconism itself, as we have seen, to a degree incorporated what may be called differing or opposing world-views.

Francis Bacon's scholarly reputation and his contribution to the realm of human knowledge have commanded the attention of keen minds to this day. For the most part, it seems one either staunchly admires the Lord Verulam, seeing him as a man of wide learning and a regenerator of science, in England at least, or one holds him in apparent contempt, viewing him as a 'scientific pretender,' a detractor of Copernicanism, uncritical, or at least lacking in the cohesion of thought required of one who formulates the new learning. There are elements of truth in all these opinions, if one examines them closely enough. Over the past 150 years especially, his contribution to modern science has been disputed by many. Only relatively recently have his supporters started making themselves heard once again. But the fact of importance to this study is his impact on the 'scientists' or 'virtuosi' of the second half of the seventeenth century, when, in England at least, he was often quoted and admired. Whether or not his loose 'system' of experimental science was defective did not seem to matter to many, who instead tended to focus on the valuable aspects of this system, or at least on those aspects relevant to them. Into this group fall many of the members of the Royal Society in its early years, as well as many regional natural historians who often had connections with the society. The 'Baconism' of these virtuosi was often acquired indirectly, for example, through diluting agencies such as the utilitarians of the Interregnum, or through individual interpretations of just what Bacon said. The picture is further complicated by the fact that few of the regional natural historians bothered to theorize about, or at least openly communicate, their speculations about the details of Bacon's 'philosophy' – that is, his views on science and nature. But it is also abundantly clear that underlying the work of most of these men is an adherence to a method that they usually perceived as Baconian, one that is implicit throughout their own work. This is not to say that there were no other influences; Cartesianism and hermeticism – the latter perhaps transmitted *through* Bacon – for example, affected their views to some degree. The empirical method, as they understood it, was at the basis of their work. Invariably, most credited Bacon with being its founder, even if today it is clear that even Bacon relied on the work of others, and that some credit for the role of empiricism should be given also to 'non-Baconians' such as Descartes.[89]

CHAPTER SEVEN

Metamorphosis

But by far the greatest obstacle to the progress of science and to the undertaking of new tasks and provinces therein, is found in this – that men despair and think things impossible. For wise and serious men are wont in these matters to be altogether distrustful; considering with themselves the obscurity of nature, the shortness of life, the deceitfulness of the senses, the weakness of the judgement, the difficulty of experiment and the like.

Francis Bacon *Novum Organum*

NO ACCOUNT of regional study in the mid-seventeenth century can afford to overlook the work of the circle gathered around Samuel Hartlib, from 1628 on an enthusiastic supporter and often originator of schemes for educational, social, and scientific reform. Indirectly, at least, Hartlib's outlook on such reform influenced the work of many of the regional writers, not the least of whom was John Aubrey, who corresponded directly with Hartlib. Although apparently none of these reform schemes were realized, 'they accelerated the foundation of the Royal Society by conditioning men's thoughts to the ready acceptance of such a project.'[1] In this respect, Dugdale aside, the Interregnum was not as barren in its contribution to regional study as might appear at first glance.

Until Henry Oldenburg took over by 1660, Hartlib served as the chief English correspondent of Continental natural philosophers, and even in England itself 'before 1660 virtually every new recruit to Baconian science made a point of soliciting Hartlib's assistance.'[2] In his ideas and his work he combined social idealism or utopianism with utilitarianism. In England, of course, a variety of the former may be traced to Sir Thomas More, especially the ideal that knowledge of nature would alleviate the lot of humanity on this planet. More's Utopians regarded 'the knowledge of yt [physical science] amonge the goodlieste, and mooste profytable partes of Philosophie. For Whyles they by the helpe of thys Philosophie searche owte the secrete mysteryes of nature, they thynke that they not onlye receaue thereby wondefull greate pleasur, but also obteyn great thankes and fauour of the auctore and maker thereof.'³ Sir Francis Bacon wrote of another Utopia – Bensalem – in his *New Atlantis*, emphasizing how science has as its end the 'enlarging of the Bounds of Human Empire ... the effecting of all things possible,' and he followed a similar line in the *Novum Organum* as well as in some of his other works.⁴ Hartlib was a great exponent of Baconian thought, arguably the chief purveyor of it to the Royal Society. Similarly, by the 1620s, 1630s, and into the Interregnum, he became an exponent of proposals to establish ideal societies or communities, variously called 'Antilia' or later 'Macaria,' schemes that never did come to fruition. Although in one sense utopian, such schemes were also seen by Hartlib as practical propositions that might be adapted in one form or another to the English scene. Here is the Baconian influence in its utilitarian aspect. Jones, in *Ancients and Moderns*, has attempted to demonstrate that the Royal Society was largely concerned with the improvement of agriculture, commerce, industry, and other practical schemes. Although, more recently, the members of the Royal Society have been discussed in the light of their other interests, it seems not unfounded to attribute at least part of the 'practical' element in the work of the Royal Society to the Baconism of Hartlib's circle and its influence: 'the actual nature of much of the work of the Royal Society makes it plausible to suggest that the Baconianism which did inform the activities of many of the early members had been filtered to them largely through the preoccupations of the group of educational, social, and intellectual reformers which had gathered round Samuel Hartlib in the 1640s and 1650s.'⁵ Hoppen goes on to state that partly 'through the distorting filter of Hartlib and his circle' the Royal Society came to regard highly the Baconian emphasis on 'the benefit of mankind, the stockpiling of information, reticence in the erection of systems, and the superficial lack of discrimination evidenced by the *Sylva Sylvarum*.'⁶

Webster's *The Great Instauration* focuses on Hartlib as the ante of a group of reforming 'Puritans,' at the same time drawing notice to Bacon's emphasis on providential and millenial themes.⁷ Webster thus links the development of Interregnum science with a social-religio-intellectual movement.⁸ The intent here is not to argue for the Puritan-science or even religion-science connection.⁹ Nor is it necessary here to identify the Hartlib circle as encompassing or even influencing all of the leading scientists of the 1640s and 1650s, for this clearly was not the case. But it is essential to point out that in one sense the interests of Hartlib and his associates and correspondents reflect the utilitarian concerns of gentlemen throughout England, some of whom, at least, were introduced to such schemes by their reading of Hartlib's publications on husbandry, John Beale's *Herefordshire Orchards, A Pattern for All England* (1657), Gabriel Plattes's *A Description of the Famous Kingdome of Macaria* (1641), and other such works associated with Hartlib in one way or another. During the English civil war the movement of armies hindered antiquarian pursuits in many ways. However, it also had the effect of promoting knowledge of regional differences, and encouraging a lot of practical efforts to improve the land. In the long run the diversion of gentlemen's energies into practical enterprises urged them in a more scientific direction, which coloured regional study in the second half of the seventeenth century. The gentry were running their estates and exchanging practical ideas about farming with one another; at the same time some of them were investigating their country's antiquities, especially local ones. In other words, the gentry had strong interests in practical agriculture and scholarly writing, which gradually, in the course of the senventeenth century, trained them in a fresh approach that seemed much more scientific.

It is important to note that the writings associated with the Hartlib circle did not constitute in themselves an agricultural revolution, and that the innovations in agriculture and husbandry in particular, though promoted by the Hartlib circle, built upon edifices already present in these fields: 'Since they were climbing on a bandwagon that was already rolling and, in a time of increasing flexibility in the ownership and management of land could appeal to the profit instincts of owners and farmers, it is perhaps unsurprising that their writings seem to have had some effect.'[10] But this should not diminish the propagandist efforts of the 'Puritan' reformers of the Interregnum to promote useful knowledge. Joan Thirsk points out that innovative agricultural experiments 'lay with single individuals, who were goaded by economic and/or intellectual interest, and had the determination and persistence, zest, energy, and money to experiment.'[11] Thirsk traces agricultural improvements back to 1500 at least, and notes that all sixteenth-century English authors of agricultural textbooks came from the gentry, who as a class wished to see their estates efficiently run. John Norden is noticed as preserving and raising the standards of such writing, exhibiting – as noted earlier in the present study – 'a keen, observant eye,' and a 'practical cast of mind.'[12] By the 1650s Hartlib was acting as the principle go-between for English agricultural improvers, and acted as disseminator of information by publishing the works of Richard Weston, Cressy Dymock, John Beale, and others. A full take-off in agriculture was deterred by a number of factors – for example, the fact that many men were content merely to eke out an existence,

Metamorphosis 139

resistance by unwilling labourers, the high financial risks of innovation, etc.[13] But a determined drive in this direction was spearheaded by 'the bookish circle of improving farmers' of the Interregnum, informally and loosely organized, led by Hartlib, and often building upon – when not actually influencing – the effort of Continental intellectuals and upon those of 'smallholders, gardeners, and cottage labourers.'[14]

The interests of the scientifically oriented men of the Interregnum, therefore, tended to steer many of them away from the 'unproductive' amassing of genealogical facts and barren observations of topography, to 'practical' agricultural, botanical, and natural historical subjects that could prove to be more relevant to their own means of livelihood, and perhaps, collectively, to the greater well-being of the nation and even humanity as a whole. Thus, the example was set for Restoration science in general, and for future regional writers in particular. In this regard, at least, the influence of the Interregnum figures is carried well into the period after 1660. But it would be incorrect to attribute this influence as springing solely from the puritan ethos that emerged in the Interregnum; the early members of the Royal Society, including those involved in regional study, display a wide heterogeneity of intellectual traditions, some of which had pre-1640 origins, and which were developed independently of the influence of Puritanism.[15]

Aside from Hartlib and Gerard Boate, the Interregnum figure who was perhaps the most influential on future regional study was John Beale, the 'Cidar Man.' As Webster states, 'his "plain and unpolished account of our Agriculture in Herefordshire" might be regarded as an application to an English region of the methods pioneered by Boate in *Irelands Naturall History*.'[16] Like Hartlib and Boate, one of Beale's main concerns was the application of science to the benefit of economic and even social planning, and he believed, therefore, that through agriculture the practical value of Baconian science could be illustrated. His *Herefordshire Orchards* was intended to promote the 'publick welfare in the peace and prosperity of this Nation.' His focus in this work, which was also an economic regional survey, was a bit narrower than that of Boate, not simply in the area it covered but also topically. Beale's main focus was on arboriculture. Fruit trees and the uses of fruit and cider manufacture were covered in a full manner, but other commodities and natural resources were given scant, if any, treatment. This is not to disparage the nature of his work, which as Webster points out exhibits a movement 'towards a scientific explanation and experimental elaboration of his observations,' a feature common in future regional study.[17]

Beale first communicated with Hartlib in 1656, following the publication of Hartlib's *Legacy of Husbandry*. Some future editions of this work incorporated

Arnold Boate's 'Interrogatorie, Relating more particularly to the Husbandry, and Natural History of Ireland,' and this – along with Beale's own long-standing practical experience in the field and interest in Bacon – perhaps served as the immediate inspiration for Beale's interest in writing about the agricultural techniques of Herefordshire.[18] Hartlib put Beale in touch with other like-minded men, including Ralph Austen, a skilled gardener and author of a *Treatise of Fruit-Trees* (1653). This was typical of Hartlib's role – to solicit letters and information from an informal circle of friends and acquaintances outlining their ideas or projects, often editing them for publication.[19]

On the surface, then, the apparently random collection of information relating to natural phenomena was part of the intellectual baggage associated with the Baconian aim of gathering facts. Unfortunately, outside of the Hartlib circle and perhaps among individuals associated with science at Gresham, Oxford, and Cambridge, such 'Baconian' attitudes were slow to gain acceptance, and hence we find most regional study during the Interregnum still being written in its chorographical, or quasi-chorographical, form. Beale's case is especially interesting. After Hartlib put him in correspondence with Henry Oldenburg and John Evelyn, whose interests coincided with those of Beale, Beale became a member of the Royal Society in 1663 and thus established a primary link in the scientific chain leading from Hartlib to the society.

It would appear, therefore, that the Baconian interest of the Royal Society was partly, if not mainly, due to the interests and activities of the Hartlib circle in the immediate past, even if, as Webster has argued, the reforming, public-spirited aims of the Puritan natural philosophers generally came to an end with the Restoration.[20] (Although Webster acknowledges some lasting links, these are seen to be cut off from any radical reform of church, state, or society; the tendency of the majority was to shy away from the democratization of society and politics.) At the same time, one should not overlook the fact that many members of the Royal Society, including those interested in regional study, became identified with various branches of experimental or natural science, 'in which were many of those problems of importance from which "the integrated worldview of the puritan reformers deflected attention away."'[21]

Aside from acting as an intermediary, Hartlib promoted a more concrete scheme for a state-backed 'Office of Address' to put into effect the Baconian program, acting as sort of a clearing-house for scientific information and as a labour exchange.[22] Virtuosi including William Petty, Robert Boyle, and John Evelyn were drawn to Hartlib, and, in fact, these three men took on the compilation of histories of trade of the type advocated by Bacon, while another

Metamorphosis 141

of Hartlib's friends, Theodore Haak, was instrumental in establishing the London group of scientists – some of whom went on to help form the Royal Society – whose prime interest was in natural philosophy. This group was intact from 1645 to 1648. Hartlib was also familiar with the later activities of the 'Invisible College' centred on Robert Boyle and pledged to social action, which probably included the Boates.[23] He was also involved with the Oxford club of scientists, another prime source of early Royal Society members.[24]

A major reason for Hartlib and Dury's call for a general education that would include the lower classes was their desire to see a growth in the contribution of artisans in particular to the compilation of Baconian natural histories, and to advance technology and the economy.[25] For them as for many Puritan reformers, the emphasis on human toil as a way to a saintly life also gave added esteem to the manual arts. The resultant downgrading of 'unproductive' scholasticism, with its emphasis on ancient literary, Ptolemaic tradition, tended by itself to depreciate the value of desultory chorography in the minds of some potential regional writers. Only the millennial expectation of a dominion over nature – implying a knowlege of nature – could lead to spiritual salvation. Also, because Bacon referred to the technical manual arts as a model for the reform of natural philosophy, Hartlib's proposed Office of Address was geared toward placing a history of trades and technology in the forefront of his social program, along with a utilitarian and experimental attitude.[26] As for natural history, Hartlib's conception of its role was highly influential during the Interregnum, but as a basis for a scientific approach to social and economic planning its influence tended to decline as the seventeenth century progressed, while scientific investigation for its own sake tended to become slightly more important.[27] The feature of Hartlib's interpretation that remained constant and highly influential was his emphasis on classification and on organizing natural histories on a regional basis. In this, of course, Bacon indirectly played a role, but it is not stretching the point too much to turn to the earlier chorographies – as geographical frameworks, at least – as exemplars. The major fruits of Hartlib's calls for regional histories, which would facilitate the compiling of the histories of trades, are seen in the efforts of Beale, the Boates, and William Petty. As far as regional natural history is concerned, what followed after the Restoration, although directed to practical aims and the accumulation of scientific knowledge, often did not achieve the same high standards.

To return briefly to a theme mentioned earlier in this chapter, Joan Thirsk has recently discussed the seventeenth century in terms of its being a 'golden age for English books of husbandry.'[28] In this, foreign literature on the subject played an important role, helping to persuade a select group of English land-owners, scholars, or parsons in particular to extend their 'interest beyond bookish reading to practical experiments on farming and its improvement.'²⁹ It has already been noted that Hartlib and many of his correspondents fall into this potpourri of domestic and foreign influences, while, once again, the local gentry became involved in techniques of husbandry. Aside from foreign writings, by the time of the Interregnum native Britons – through 'the wisdom of their practical experience and their reading of others' – turned to producing works on agriculture: 'Humanism and the ideals of the conmonweal blended in support of a political objective: writers believed that their fellow men would get a better living from the land if new ways of farming and new crops from abroad were better publicized.'³⁰ All this, of course, should still be seen against the backdrop of the greater, if not always enunciated, goal of promoting one's country in the eyes of the rest of the world. Multiple occupations were also becoming the norm, a portent of the coming of the virtuosi, who firmly entrenched themselves in the annals of Restoration intellectual life. The commonality of interests between Interregnum and Restoration science is evident, to cite one example, in the interest of Walter Blith, William Dugdale, and John Aubrey in schemes for draining fen lands.

The agricultural writers would not, however, have been nearly as influential if it had not been for favourable economic and political conditions, combined with current intellectual interests and individual zeal to encourage improvements.³¹ The introduction of many new crops and developments in new farming systems made much headway by the Interregnum, and also came into play here: 'The right agricultural environment had to be found for these crops, their labour requirements had to accord with the social resources of the neighbourhood, and the financial rewards had to be sufficient to make the crop more profitable than other crops competing for the same land.'³² Hartlib and others added to the impetus through their contacts with government officials, and by their promotion of enlightened socio-economic policies. The latter, especially, would have the added effect of alleviating the condition of the poor – for example, through employment connected with the growing of labour-intensive crops – in addition to enriching the landlords. The English civil war itself had a positive effect in that it 'placed the need for agricultural improvement higher on the agenda than ever before, for it inflicted great financial losses on land-owners of both parties, and those who had earlier shown an interest in land improvement naturally turned back to these possibilities in the quieter days of the 1650s.'³³ After the Restoration, by which time 'virtually all the agricultural improvements of this [the seventeenth] century were known in general terms,' falling rents coupled with falling prices, periods of general depression, and an 'altered climate of social intercourse' worked to inhibit further rapid agricultural change; 'Gentlemanly interest in improved agriculture was [hence] channelled formally after 1660 through the Royal Society,' which publicized many of their ideas in the *Philosophical Transactions* of the society.[34] But in the preceding period the tone for what was to follow was set to a large degree by agricultural writers and others, including a Parliament that encouraged self-sufficiency in agricultural production – the need to overcome dislocation of trade during periods of conflict with Spain and the Dutch being one reason for this – and including Hartlib and his circle. This is true not only in the realm of agriculture *per se*, but also in the field of regional study, where the 'practical' was combined with the scientific, and sometimes the not-so-scientific, to overhaul the manner in which men studied their localities.

The well-known figures William Petty and Robert Boyle should not be overlooked. Petty was a stand-out in the history of trades and in Irish land adventurism of the 1650s; according to Aubrey he 'was beloved by all the ingeniose scholars,' and apparently Thomas Hobbes in particular 'loved his company.'[35] Hartlib recommended Petty to Robert Boyle in November 1647 as 'a most rare and exact anatomist,' and Petty and Boyle soon were conducting experiments together at an anatomical laboratory in Dublin.[36] Later, once Boyle had secured his Irish estates following the Cromwellian settlement in 1652, Hartlib sought information from him on topics relating to husbandry, and solicited his assistance in providing information towards a second edition of Boate's natural history of Ireland. Boyle's friends in Ireland were also to be recruited for this effort, which involved the use of 'interrogatories' – a series of queries arranged alphabetically under headings relating to the natural history of Ireland, which were contained in editions of Hartlib's *Legacy*.[37] Although the plan did not bear the sort of fruit hoped for, the important point here 'is that the metamorphosis that agriculture was undergoing at that time was largely due to gentlemen like Boyle, who farmed their land and were sufficiently educated to appreciate the potential of the new methods.'[38]

In the Baconian spirit, Boyle – at one time a patient of Gerard Boate – 'determined to let experimentally confirmed facts alone dictate his conception of nature,' although he placed more emphasis on the use of reason than did Bacon, realizing that the collection of data had to be directed towards a goal, with reason being the guide.[39] Boyle was a connecting link between his friends in the Invisible College and Hartlib's circle of more loosely organized reformers, of whom Petty was one. He was an enthusiastic supporter of various proposals for the production of histories of trade by Petty and others, and John Evelyn even dedicated to him his *Sculptura; or, The History and Art of Chalcography and Engraving in Copper* (1662). As early as 1647 Hartlib had sent Boyle a design for a history of trades authored by Petty, which Boyle no doubt communicated at least verbally to members of the Invisible College and to other virtuosi.[40] And as late as 1671 he wrote an apology for such a history, pointing out its practical applications and the indebtedness of scientific virtuosi to craftsmen and tradesmen; this he did at a time when the mechanical arts were becoming secondary to the study of natural philosophy *per se* within the orbit of the Royal Society's activities – despite a brief period of revival of interest in the mid-1660s.[41]

Although in the late 1640s and 1650s Baconians such as Boyle put more stock in the history of trades as a utilitarian study than in the study of plants, animals, or the natural products of the earth, others did the opposite, while still others of whom Petty is a good example, seemed to combine the two spheres. A study of the history of trades seemed better suited to the confines of the 'laboratories' or shops of the practitioners of the mechanical arts, or to 'Solomon's House' – that is, a scientific college – as Bacon hinted. Such a study, of course, was part and parcel of Bacon's wider idea of a new natural history, but in a more restrictive sense, relating as it did more precisely to man's activities alone, going beyond 'nature in course.'[42] Some regional natural historians of the post-Restoration period tried to emulate Petty's example as well as Bacon's dictum in their attempts to incorporate discussion of the mechanical arts or the history of trade in their studies. But their works fell somewhat short in this regard.

It would have been difficult, in any case, for them to attain the achievements of Petty, who was, according to Aubrey, a 'Person of so great worth and learning and haz such a prodigious working witt that he is both fitt for and an Honour to the highest Preferment.'[43] In the context of the history of trades, Petty's *The Advice of W.P. to Mr. S. Hartlib for the Advancement of Some Particular Parts of Learning* (1648), along with ideas on educational reform, is a guiding light listing the 'profits and commodities' accrued from this type of study.

In 1652 Petty arrived in Ireland as chief medical officer of the English army stationed there, although he soon gave up his medical practice to pursue scholarly and business pursuits. In December 1654, Petty was given responsibility under Surveyor-General Benjamin Worsley for undertaking the Down Survey, a topographical survey of Ireland that entailed the production of general maps of that land.[44] This was a major component of a series of at least nineteen different surveys conducted between 1653 and 1660. The task itself was the basis of the redistribution of lands forfeited by Irish 'rebels' to English adventurers during the Cromwellian resettlement;

here again the advancement of learning was based upon utilitarian concerns, coupled with political expediency. As Webster states, the surveyors – 'being advocates of Baconian natural history, histories of trade and political anatomy' – were therefore able 'to transcend the immediate practical purposes of the survey, to prepare the ground for an elaborate exercise in economic geography.'[45]

Through his use of a large force of soldiers – many of whom had a trade – in surveying the land, Petty was able to demonstrate the worth of Bacon's collaborative approach to science, utilizing humble men who, with the exception of a limited group of overseeing surveyors and other technical professionals, were not well educated, let alone experts in any specialized field of science.[46] Through such techniques a vast amount of detailed economic and geographical information was collected and quantified. Future regional natural historians easily identified with the overall concept, the exception being their general disdain for – or inability to undertake – the quantification stage of it. With notable exceptions, such as Edward Lhuyd, 'collecting' seemed more important to them than 'evaluation.'

By 1659 Petty was back in London, where as a member of the Royal Society he expanded his intellectual pursuits into the realm of political arithmetic – upon which his fame mainly rests – naval engineering, and so on. His lands and interests in Ireland, however, continually commanded his attention, and from 1676 to 1680 he resided in Dublin. Petty became the first president of the Dublin Philosophical Society (1684) and watched over its progress until his final departure from Ireland in June 1685. In his 'Advertisements,' which set out the society's objectives, the members were directed to 'chiefly apply themselves to the makeing of experiments' – these being preferred to 'Discourses, Letters and Books' – and were admonished to 'not contemn and neglect comon, triviall, and cheap experiments and observations.'[47] This attitude, however, did not deter Petty from occasionally procuring chorographical accounts of Ireland in manuscript, to serve as sources for his folio atlas of that land. But this fact does not detract from Petty's contribution to regional study. Unfortunately, the virtuosi who came after, who are for the most part associated with the Royal and other scientific societies 'were incapable of commanding the disciplined labour needed for the collection of substantial amounts of well-organised quantitative data or detailed cartographical records. After 1660 the successful natural histories were basically the work of individual enthusiasts such as Robert Plot. Although Baconian in general spirit, these works tended to reflect the idiosyncratic interests of their authors rather than pursuing the systematic programme adopted by the compilers of *Irelands Naturall History*.'[48]

CHAPTER EIGHT

'To Serve the Commonwealth of Learning'

> ... to make faithful Records, of all the Works of Nature, or Art, which can come within their reach: that so the present Age, and posterity, may be able to put a mark on the Errors, which have strengthned by long prescription: to restore the Truths, that have lain neglected: to push on those, which are already known, to more various uses: and to make the way more passable, to what remains unreveal'd ... And to accomplish this, they have endeavor'd, to separate the knowledge of Nature, from the colours of Rhetorick, the devices of Fancy, or the delightful Deceit of Fables.
> Thomas Sprat *The History of the Royal Society* (1702 ed)[1]

THE ROYAL SOCIETY was founded in 1660 and granted a royal charter in 1662. This is not the place to retell its entire story, for, as in the case of Bacon, this has been done admirably by several historians. Its origins are traced to a series of regular gatherings of scientifically oriented men in London and Oxford, going as far back as 1645, if not earlier.[2] But, as Hoppen states: 'We are still far from achieving a satisfactory understanding of the complicated web of traditions, sources, and intellectual systems that provided both an inspirational dynamic for the work of natural philosophers such as those in the Royal Society and patterns of expression through which their preoccupations could be articulated.'[3] In the mishmash of often conflicting intellectual, religious, social, and philosophical currents within the membership of the society, it is difficult to locate a single, unifying concept of the nature of the scientific enterprise underlying its early aims.

Baconism or elements of it are often seen as the inspirational heart of its activities, with its stress upon a natural philosophy based on natural history and an empirical research program and its general quest for the advancement of learning in order to reap the practical fruits thereof. Recent historians have emphasized the distilling effect that the Hartlib circle had on Baconian science as perceived by the early members of the Royal Society. Many of the latter did, in fact, see Bacon as 'the simple utilitarian, the apostle of mere fact-gathering, hag-ridden by fears of hypotheses.'[4] Some regarded this image as a negative one, while others saw it as the model to follow. Whatever brand of Baconism existed, it was mixed in the society with mystic, chemical, neo-Aristotelian, Cartesian, and other views, or with individual idiosyncratic varieties or combinations of these. Not only the lesser figures, such as our regional natural historians, but also the society figures usually designated as 'great,' such as Newton or Boyle, were influenced to one degree or another by some of these 'unscientific' as well as by 'scientific' pursuits; sometimes two seemingly diametrically opposed traditions are visible in the thought of a single individual.[5] There are many instances of members who displayed no affinity for any particular intellectual current, but were accepted into the society on the basis of their general interest in natural philosophy, their reputation simply as 'collectors,' their personal or social connections with current members, and so on. Within this milieu one may, like Hoppen, regard the Royal Society as a deeply serious institution: 'If at times many of its activities may seem to have involved no more than a random gathering of "curious" materials, this is a profoundly misleading impression. But the pursuit of the serious tasks which the society set itself does not imply rigid intellectual exclusiveness, and it is consequently unhelpful to discuss the concerns of the early Royal Society in simplistic terms of black and white, of "ancient" and "modern," of "new science" and "old learning."'[6]

The Royal Society certainly had its failings, being somewhat of a gentlemen's club, too much London oriented, amateurish, and so on.[7] But all this does not negate the fact that it was a vibrant, if not the chief, centre of scientific investigation, at least in its early years. Furthermore, as Hunter says, 'the Society met a real need in organising science of the day – in gathering information and stimulating, coordinating and publicising the research of scientists who, despite the Society, continued to work on their own.'[8] Henry Oldenburg, the society's secretary (1662–77) was especially instrumental in guiding the provincial virtuosi in the compilation of natural histories, as well as acting as a sort of clearing-house for other types of scientific information.[9]

It is evident that there was indeed a substantial overlap between 'real' and 'amateur' science in the second half of the seventeenth century, and this overlap is also reflected within the membership of the Royal Society itself. It is also the case in the still smaller group comprising the regional natural historians; for every Edward Lhuyd we find ten or more 'lesser' minds. As Hunter points out, difficulty lies in first establishing the definition of 'serious scientist.' And the tendency to emphasize the speculative or mathematically inclined minds of the period, in historiography, has at the same time played down the 'levelling effect of Baconism,' whereby 'The demands of induction gave minor observers a role they had been denied when deduction a priori dominated intellectual life, as under the old scholastic regime from which the new scientists were so eager to extricate themselves, and it conferred significance on lesser thinkers whom one might otherwise dismiss.'[10] The far-flung provincial types acted as sources of 'invaluable observations for London scientists to consider and use as a basis for synthesis and theory,' and as popularizers of the new philosophy.[11]

For Hunter, there is a connection between the scientific reformers of the 1640s and the scientists of the Royal Society; both groups followed Bacon's belief that not so much speculators as practical men of experience would lead the true natural philosophy. For example, Oldenburg and Sprat were in part, at least, heirs to Hartlib's merging of science and technology; one might add to this list Aubrey, Plot, Boate, and other regional natural historians, even if their observing and data collection outran attempts at actual experimentation or involvement in the mechanical arts. But Hunter warns of 'reading the appeal to utility too narrowly and assuming that it referred exclusively to practical, everyday needs.'[12] Perhaps more regional writers than is generally realized, like Boyle, tended to put at least as much emphasis or more on knowledge of the natural world for its own sake and for religious enlightenment.[13] The intellectuals, it seems, were often frustrated as well in their attempts to master various crafts, techniques, or inventions. Sometimes, even when they were able to accomplish this, there existed the problem of reaching the illiterate masses.

Hunter sees less connection between science and religion and between science and political ideology than some other historians of science do, although he sees commonplace attitudes loosely defined as 'Puritanical' and, later, certain latitudinarian ideas as conducive to the development of scientific thought. Several interests and schemes that originated in the Interregnum retained an interest – if often only in modified form – such as plans for educational reform, greater enlightened central government planning, and efficiency in the economic and social sectors. As with Hartlib, science and technology, of course, were to be applied in the pursuit of these goals, though with less of a 'left wing' bent than Hartlib had in mind. But lack of adequate income, opposition by sectional interests, erratic government patronage, political distrust of the zeal for reform, and the wars of the 1690s prevented the Royal Society, as well as other like-minded individuals, from successfully implementing many of these schemes. Still, although collectively scientists may have conformed politically to the straight and narrow, pragmatically and prudently supporting the government of the day, it is difficult to accept too rosy a picture of their wholesale withdrawal from political affairs. One only has to look as far as the large number of writings produced by scientists/virtuosi – even if one sometimes has to read between the lines – involving the ideological applications of the new science, or to consider the repressive measures threatened or taken by the authorities of the Restoration – as in the case of Oldenburg's religious orthodoxy – to see that the situation needs to be assessed more carefully, or on an individual by individual basis.

By the late seventeenth century, many scientists and intellectuals adopted natural philosophy as one way to approach religion; the former would soon become 'virtually self-justifying – a psychologically satisfying substitute for the former pursuit of religion, theology and personal salvation.'[14] The mix of religion and natural philosophy was nothing new, stretching back at least to Vesalius, if not Paracelsus and earlier. In England, Bacon's defence of the study of nature on pious as well as on practical grounds tended to shield such study from interference by theologians, and contributed to the division between science and religion – though at the same time proclaiming their common end.[15] Bacon encouraged the study of the 'second book,' that is, natural science, as a means to uncovering the greatness of God.[16] But this in itself tended to shift intellectual interest away from theological disputes, including those generated by the radical reformers of the 1640s and 1650s. Discussion of the Almighty was now increasingly found explicitly or implicitly within the less volatile context of discussion of His works of nature, generated by non-theologians. After 1660 this was also keeping with the reaction against the excesses of the English civil war when many – like Sprat – sought a 'Rational Age' (although the excessive Anglican High Church policy made such a search more difficult than it could have been). Here one may in fact detect the latitudinarian spirit of the early founders of the Royal Society.[17]

A large group of the natural philosophers, while separating the study of God and the study of nature, did not regard the Bible and science as contradictory. A sort of 'priesthood of nature' developed, although this is more evident in the work of men like Boyle and Browne than in that of some of the regional natural historians, the latter being usually more purely utilitarian or Baconian in outlook.[18] Boyle also went further than most other natural philosophers in insisting on the affinity between mechanistic (that is, scientific) and teleological explorations of a structure or process, and in advocating the importance of functional hypotheses to research into the mechanical operations of structures.[19]

There was added incentive to believe in an active God, namely, in order to avoid an accusation of atheism, which ranked with popery and sectarianism as an apparent threat to the fabric of English society. Bacon himself was attacked by some for his rejection of final causes. This further stimulated the interest of the virtuosi in the gathering of data with the aim of illustrating the existence, as well as the greatness, of God. The 'scientist as priest' idea according to Hunter, spread 'not least in the provinces, where virtuosi espoused "the great and worthy designe of an Universall History of Nature to be borne out of her owne bowells by faithfull experiments for the confusion of growing Atheisme, and Philosophy falsely soe called."'[20]

While Hunter has examined Restoration science within its social context, other historians of science have argued for its ties, or for its lack of them, with Renaissance hermeticism and occultism, or the 'supernatural.' Keith Hutchison, for example, has discussed some versions of the mechanical philosophy in terms of their role in protecting 'a radically supernaturalistic ontology against the naturalism of the Renaissance. The Mechanists' conception of matter as totally barren was used to offer a guarantee that supernatural activity was ever-present in the universe.'[21] The comparison is made to Boyle's belief that some supernatural force, at least, was required to set matter into motion. Even Luther and Calvin are brought into the picture as exponents of radical supernaturalism. I shall return to this topic later in this chapter.

The men of the Restoration and, later, the regional writers began increasingly to include antiquities alongside natural history in their studies; thus they made it clear that these two disciplines 'were regarded as coming more within the purview of the natural scientists than of the historians.'[22] This meant that antiquities were coming under the wing of empirically oriented Baconians, who preferred field investigation to literary authority. If it is true that the scientific revolution in England centred on forty or fifty men, many of whom were academic professionals like Hooke or Newton, it is equally true that the sum total of the contributions of non-academics or 'dabblers of science' is also important.[23] These virtuosi, such as Aubrey and a great many of the regional writers, helped to popularize science, especially in the early decades of the Royal Society, and 'to these men in whom a love of natural science sometimes mingled with connoisseurship in art, a taste for history, or a desire to explore remote parts of the globe, must be ascribed much of the breadth of the scientific movement in the second half of the seventeenth century.'[24]

It seems that recent historiography has tended to emphasize the contributions of the virtuosi. However, the question raised by Houghton remains with us today: to what extent were the virtuosi 'Baconian' in their approach to investigation of the natural world? Houghton argued that many, in fact, were quite 'un-Baconian,' citing the example of Evelyn: 'Nowhere ... does he show the slightest concern with what to Bacon was the main raison d'être of the study of nature or mechanical art – the discovery of law.'[25] Evelyn was, in other words, interested in rarities solely as an admirer of the wondrous, not in order to explain them. Through his call for a history of marvels to be included in the study of natural phenomena, Bacon is held partly responsible for an attitude like Evelyn's, even though Bacon did warn against such admiration as an end in itself. Again, 'It is clear that in the tradition of natural magic Bacon was attracted to the 'natural,' the virtuosi to the "magic."'[26] All in all, one may agree with Houghton that among the motives spurring the virtuosi was the ancient one of romantic sentiment for communing with the heroes of the past, as well as a desire for reputation – motives that impelled men in antiquarian as well as in scientific directions.[27] But as far as the regional natural historians are concerned, there is clear evidence that some of them, at least, went beyond Evelyn in their Baconism, endeavouring, like the Royal Society, to 'know,' rather than merely to 'admire.'[28] In this they were more like Wilkins, 'lying between the real natural philosophers, men like Boyle and Hooke, Ray and Newton, and the mob of gentlemen who played with science, as they wrote, with ease.'[29] For they regarded natural historical investigation as more than a recreation or a game of magic, rather as a serious, utilitarian – that is, Baconian – pursuit. For this reason, the virtuosi as a group were recognized, if in a somewhat back-handed manner, in comedies of the time as neither pseudo-scientists, charlatans, nor fools, and even Thomas Shadwell's famous comic parody, *The Virtuoso* (1676), drew not upon observations of 'astrologers, alchemists, or witchmongers,' but upon 'facts gleaned from observations of the new philosophers [virtuosi].'[30]

The early Royal Society was not alone in treating, or claiming to treat, Bacon as the paragon of scientific method; 'for over two centuries Englishmen complacently regarded Francis Bacon ... as the 'Legislator of Science.''[31] His role in illustrating the importance of 'new' scientific methods and his attack on the ancients even found some admirers abroad, Pierre Gassendi being but one notable example.[32]

Empiricism, that is, scientific method based upon experimentation, experience, and observation, played a vital role in the program of the Royal Society. This is often reflected in the words of society members. Oldenburg wrote that it was not the concern of the society 'to have any knowledge of scholastic or theological matters,' its 'sole business' being the cultivation of 'knowledge of nature and useful arts by means of observation and experiment.' Boyle advocated experimentation, stating that members 'were not come so far as to frame Systems'; and Hooke openly opposed the society's meddling in 'any hypothesis, system, or doctrines of principles of naturall philosophy ... nor [should the society] dogmatically define, nor fix axioms of scientificall things.'[33] Some virtuosi, including some involved in regional study, in fact adopted a kind of naïve empiricism that restricted them solely to experimentation, observation, and the collection of data, but this approach often did contribute useful information of a scientific nature.[34] Yet as M.B. Hall points out, 'However much the Society as a body might hesitate to favour hypothesis, its aim was to establish something more than a mere collection of random experiment.'[35] This led members of the society, including some of the most openly vocal opponents of hypothesis, into this realm. The society was not averse to hypotheses '"grounded in fact," and subject to confirmation by (empirical) facts,' and individual 'Fellows might hold what views they liked, and did so – Cartesian, Baconian, Epicurean, mystic, Aristotelian, medical [and] chemical.' While as a body the society desired to avoid a priori systems, 'those based upon empirically untestable tenets or principles,' some of its members, for example, Hooke, did stray into the sphere of a priori hypotheses.[36] For others, like Newton, however, 'experiments offered proof, and once they had proved an hypothesis, it has not to be set aside in favour of another just for the intellectual exercise.'[37]

Within the membership of the society, one finds conflicting views of the role of experimentation as opposed to rationalization – though the two are not necessarily mutually exclusive – and as Hall points out, empiricism itself 'wears many faces.' Bacon felt that multiplicity of experimentation could lead to conclusion, and the Galilean ideal of the closely controlled experiment, adopted by some society members, could also lead to the same result. For Bacon experiments are to be applied for the purpose of discovering the natural cause of some effect, not simply to produce a particular effect. For many of the regional natural historians, 'vulgar Baconism' was the rule; 'theory was to be eschewed, but utility was not.'[38] This is also not to say that none was involved in the emerging rationalist and mechanical traditions in natural philosophy of the later seventeenth century, as was Hooke, but for many their interest in this direction was limited. In the society there were 'Those preoccupied with the compilation of natural histories [who] were prepared to enlist anyone willing to contribute however marginally to this endeavour,' and it is clear that they had some success in this regard.[39]

At the same time, in the work of, say, Lhuyd, Plot, Sibbald, and others, we find evidence of a striving to attain something more than mere 'vulgar

'To Serve the Commonwealth of Learning' 153

Baconism.' Even if many regional writers may be classified as such, the point is that the 'greater' scientific minds of the time did not remain totally uninfluenced by their correspondence and activities; nor were they viewed as having tarnished Bacon's grand design. But some of the regional natural historians exhibit a growing tendency – though restricted in scope, and sometimes unwittingly – of becoming probabilistic in their approach, swaying from a strict Baconian quest for certainty, even when they still engaged in an essentially Baconian research program.[40] These regional natural historians display an affinity with Shapiro's 'family' of 'moral certainty, probability, and hypothesis,' even where the mere accumulation and examination of data superseded an interest in actual experimentation and hypothesizing.[41]

All these elements coalesce, of course, in Plot's chemical experimentation or Lhuyd's theorizing on the origin of fossils, but not to the same degree as in the works of Boyle or Hooke. The Baconians made attempts at scientific classification and nomenclature and contributed substantially to these. They were often 'followers' of Lister, Ray, Willis, and others. Their prime aim was one of collecting and examining data in order to provide the natural historical basis for Bacon's pyramid of knowledge. It is in this light that one should refer to Shapiro's statement that within the society as a whole:

Baconian optimism ... faded, and with it the confidence that data collection and analysis would yield general axioms or the forms of nature. Metaphysical principles were rarely if ever discussed in the context of empirical science. Although the building metaphor, with its 'strong foundation' and 'firm base,' was endlessly repeated, no one seemed willing or able to explain how substructure and structure were related, or the process by which natural history became natural philosophy. The Royal Society, to a very considerable extent, quietly abandoned Bacon's philosophy of science while continuing to pursue Baconian projects.[42]

It almost seems that the material collected and even the material classified by the regional natural historians was too enormous in terms of number, breadth, and variety, for the intellect to comprehend, and derive certain axioms from it. The most one could hope for, more likely, was conjecture and informed opinion, as Shapiro indicates. Yet for a few the more data, or evidence, they accumulated the greater the hope they held of attaining knowledge, or at least of arriving at a higher level of probability, if not mathematical certainty. Again, since most of the regional natural historians did not engage in detailed philosophizing about the degrees of certainty that were attainable, it is difficult to determine where each individual stood on this matter, and even more difficult to ascribe to them as a group a drift away from the Baconian position.[43]

Available evidence indicates that like other scientifically oriented men, the members of this somewhat arbitrary grouping at best took the stance that one could ascertain that matters of fact were 'highly probable' or even 'morally certain.'[44] This could lead to Baconian axioms, but many of our practitioners never reached this latter stage; and if they ever did they abandoned the 'philosophical goal of comprehending the forms of nature for the more limited knowledge of appearance and phenomena,' while 'borrowing from fields such as history, law, and religion, which were long accustomed to dealing with matters of fact.' Since 'the status of facts might change as more and better information was provided,' the regional writers could anticipate and participate in a 'more accurate collection of natural data.'[45] Whatever the case, most of them did not express openly a concern about not being able to attain certainty. Yet they did participate in the general drift from the rhetorical to a more scientific view of probability that is also perceived in other fact-related fields such as civil history and law. They placed importance of evidence based on experimentation and observation above that based on 'authority,' be it ancient or otherwise, and usually above hypothesizing, which possessed a lower degree of probability. Their emphasis was on the empirical approach. Data could and did lead sometimes to hypotheses in the course of their investigations of the natural world. There was little effort made to go beyond this in order to acquire the kind of knowledge of causes advocated by Bacon. For many the task at hand consisted solely of gathering more data, before even a cursory attempt at hypothesizing could be made. For those who did venture that far, the nature of their hypotheses was closer to that of Boyle than that of Spinoza or Leibniz; that is, they believed hypotheses must be demonstrated by empirical evidence, rather than logically and 'rigorously' from a priori, certain principles.[46] On the whole, the regional natural historians felt that hypothesizing was best left to their successors as well as to others. And even by the end of the century, they tended to disregard any possible strict application of Newton's demonstrative, mathematical system to their own work; this was best left to those more involved in mathematics *per se* or in the mechanical or corpuscular studies.

So there are two sides to the story. On the one hand, Baconism is evident in the work of the regional natural historians; it is seen in their ongoing interest in its utilitarian aspects, in their observation and data collecting, and to a lesser degree in their interest in experimentation. On the other hand, there is some drift away from the strict Hartlibean-like interest in the history of trades, exhibited during the Interregnum, a recognition of the unlikelihood of their arriving at certitude, at least in their lifetime, and in some quarters a greater – though still rare – tendency to hypothesize.[47] The difficulty lies in the gist of Shapiro's comment that 'the tension between Baconian and non-Baconian tendencies was not always resolved or even explicit.'[48] All this leads one to assume further the heterogeneous nature of the early society, and that in many instances a notion of exactly what natural philosophy was differed from one individual to the next. While the early society was a major proponent of Baconism, recent studies have pointed out the value to experimental philosophy of Harvey's, Gilbert's, and Continental traditions. From this perspective Baconism is often seen as more important for its popularizing and psychological functions than for its relevance to the scientific problems posed by the evolving mechanical philosophy. The worth of the Royal Society itself is seen primarily in its capacity as an organizer of science, especially through the efforts of men such as Oldenburg. There is much truth to all this, although at least some of the 'vulgar' Baconians would have given Bacon much more credit for establishing a practical, workable method of science.

The Royal Society, therefore, although officially adopting an essentially Baconian research program, encompassed virtuosi not all of whom were Baconian purists. The scene is further complicated by the fact that the Royal Society certainly does not represent the be-all and end-all of the scientific revolution as it manifested itself in Britain, or even England, and its effect, or lack of effect, on those workers outside its immediate domain requires closer study; some of the 'provincial' naturalists covered in this study fall into this category.[49]

It may be argued that in its early years the Royal Society paid little attention to mechanics.[50] Before Newton, mechanics – attributable to the influence of Galileo – and metaphysics of science – stemming mainly from Gassendi and Descartes – were 'as yet imperfectly assimilated.'[51] Nevertheless, more and more attention was being given to the 'mathematico-mechanical' natural philosophy, which was associated with the experimental method, and even the universities were not immune to the new ideas. Cartesianism was stronger than Baconism at Cambridge, at least, even if the traditional Aristotelian premisses predominated learning as a whole there.[52] It is noteworthy, for our purpose, that the medical faculty in particular was busy promoting the systematic study of natural history and breaking away from the authority of Hippocrates and Galen.[53] Natural history and natural philosophy, therefore, were being promoted outside the orbit of London, in areas where Baconism was perhaps not as directly influential in their development. There we find the imposing figures of John Ray and Wallis becoming imbued with a thirst for natural philosophy, stimulated through medical inquiry. Equally important are the lesser lights, little-known medical practitioners who upon graduation dispersed to all parts of England, not only to practise medicine but also in many instances to make incursions into the study of the natural history of their new locales. This process was well under way by 1658, when Hartlib commented to Boyle that at Cambridge 'We have divers fellows of Colleges who have made excellent progress in anatomy ... They have also much travelled in botanics, and have got together many hundreds of plants in several gardens here.'[54] Natural philosophy, then, was being touted in several quarters of Cambridge, although there, thanks to Henry More and his circle, it was equated more with Cartesianism than Baconism, at least after the Restoration.

To digress for a moment, the Royal College of Physicians also had a role to play. Until 1662 it was the only learned society in England outside the universities with a royal charter. Physicians, 'the only students of any natural science who had any organization of their own,' provided the Royal Society with the illustrious Goddard, Petty, and others who were members of the College.[55] It was not long before the new society eclipsed the old in terms of dedication to scientific research. Perhaps this was the case even in terms of the attention devoted to advancements in the medical sciences, and it was the junior society that cultivated connections with medical men overseas and sought information about exotic remedies.[56]

At Cambridge, Henry Power provides an interesting example. Sir Thomas Browne allegedly wrote *Religio Medici* for his young friend Power, and encouraged him to seek the knowledge one needed to become a physician not simply by reading books, but through observation and experience.[57] Soon Power was 'herbalizing,' in his own words 'going out three or four miles once a weeke have brought home two or three hundred Hearbs.'[58] Another Cambridge medical graduate relevant to our study is William Stukeley, a fellow of both the Royal Society and the College of Physicians, whose interests ranged from natural history to antiquarianism; he was one of the revivers of the Society of Antiquaries. A.H.T. Robb-Smith says that at Cambridge 'the Neoplatonism of Henry More and the Cambridge enthusiasm for Descartes was an intellectual philosophy with a cool restraining influence to it; they sought to interpret rather than to study nature. Very different was the insurgent inquisitiveness of the Oxford experimental school, with, as Boyle put it, "an unsatisfied appetite of knowledge."'[59]

The Cambridge Platonists adapted the fundamental mechanical formulation of Descartes, partly as a bulwark against the Aristotelian-inspired spread of 'atheism' and against the sectaries, the latter basing heresy on 'personal inspiration.' But it is generally thought that the Cambridge group, which passed on some important ideas to Newton, differed from Descartes in that its members believed in the continual active intervention of God in the operations of nature and the universe, whereas Descartes relegated this sort of active divine intervention solely to the beginning of time, or Creation. The Cambridge men thus set the foundation of the 'plastic virtue' or 'plastic spirit' principle, wherein 'plastic virtue' constituted the active, subordinate agent of God. It injected an 'incorporeal' element into an otherwise spiritually dead or inert universe of matter, in itself a restatement of the neo-Platonic world soul, though free of magical or hermetic overtones.[60] Again we return to the idea that at least a few mechanical philosophers did in fact adopt a view of matter that was designed 'to protect a radically supernaturalistic ontology against the naturalism of the Renaissance,' in contrast to the Aristotelian conception that the universe operated without any supernatural activity at all.[61] Boyle, for example, believed that God has continually sustained the motion of matter ever since the Creation, and actively directs every event in the material universe.[62] Fundamental to this argument of a supernatural element in the mechanical philosophy is Descartes's own attitude that since matter lacked inherent activity, God 'had to create the whole universe afresh every instant.'[63] A Cambridge Platonist such as Ralph Cudworth, too, is seen as introducing the notion of the 'plastic nature,' because he could not quite accept that God was so completely involved with the details of the universe, especially with regard to its defects. The 'plastic nature' thus acts as 'a naturalistic intermediary between God and mechanism, functioning to *reduce* the need for the supernatural permeation of the universe which he saw the mechanical philosophy in its pure form as demanding.'[64]

For many regional natural historians such considerations lay outside their realm of intellectual activity, where 'collection, description and classification' were still the dominant themes.[65] There were those, of course, like Robert Plot or Edward Lhuyd, who speculated, for example, about the nature of 'plastic virtue' and of the mechanical philosophy itself, and some followed in the tradition of Boyle, adopting this from Bacon and that from the mechanical philosophy.[66] In general, there were few intelligent men, even 'Baconians' through and through, who did not find their curiosity aroused by the diverse scientific theories, or philosophies, of their day, including that of Descartes. But as G.A.J. Rogers has pointed out: '... it is all too easy to assume that because somebody agreed with Descartes's conclusions about, say, the corpuscular philosophy, that they therefore also agreed with Descartes's scientific methodology.'[67] For some, the Cartesian model itself offered hope of attaining absolute certainty with regard to knowledge of the physical universe. Though Descartes was more of an empiricist than some critics believe, as well as a rationalist, for many English virtuosi he was too much of a theoretician, whose philosophy of science held little relevance to actual experience, experimentation, and investigation, and for whom the collection of natural historical information was essentially not of value. In most natural history of the time, even the 'great systematizing and collecting naturalists ... were only incidentally men of ideas; for the most part they were content merely to know,' 'for the grey, sober truth is that the only practicable first approach to the study of living things was the Baconian one of collection and comparison. Living processes are too varied in their manifestations and too complex in their chemical character to yield to a priori reasoning.'[68]

Like the short-lived Elizabethan College of Antiquaries, the Royal Society stressed loyalty to the search for truth and viewed itself as a body of researchers, not of authorities. In addition, both societies found it necessary to fend off charges that their respective efforts might somehow damage the universities. Perhaps the major difference between the two societies, as far as regional study is concerned, is best seen in the papers dealing with antiquities that were presented to the Royal Society. As Joan Evans succinctly states, these papers 'are far closer to modern field-work than are the essays of the Elizabethan College of Antiquaries [Society of Antiquaries], for they are concerned with objects, not documents, and attempt their interpretation in the modern spirit.'[69] And, as Stuart Piggott points out, the empirical approach that one associates essentially with the Royal Society 'made possible the liberation of archaeological studies from a dependence on literary sources, so many of dubious validity, and from any entanglement in the quest for antecedents or the hunt for respectable ancestors.'[70]

The new ideas did not take hold overnight, and the Royal Society should certainly not be taken simply as the cynosure of science. There is an interval of almost half a century between the period when Bacon published most of his works and the date Plot issued his *Oxfordshire* (1677), the book that was most influential in firmly establishing local natural history. But still, the first regional natural history published in English, Gerard Boate's *Irelands Naturall History*, was written in 1645 and published in 1652, long before Plot undertook his study. It would therefore at first glance seem rather surprising that it took so long for the type to become popular. For almost a decade before the founding of the Royal Society, a relatively recent model was there for all to see. The truth of the matter is that in many circles there were few firm signs that the Baconian philosophy was taking hold. In William Dugdale's *Warwickshire* (1656), the unimportance of archaeological investigation is clearly seen, as is the disregard for natural history. Dugdale's close friend Sir Thomas Browne, author of the famous *Hydriotaphia, or Urne-Burial* (1658), and the equally famous *Religio Medici* (1642), has often been portrayed by connoisseurs of the scientific revolution as a major link between Bacon and the Royal Society or even as the first of the moderns.[71] Browne went further than many of his contemporaries in rebelling against the relative unreliability of many past writers upon natural history, especially in the first book of his *Pseudodoxia Epidemica* (commonly known as *Vulgar Errors*), first published in 1646.

Evans draws attention to the fact that the engravings that illustrate the *Hydriotaphia* are the first to be published of Anglo-Saxon pots, and that his treatise, *Sepulchrall Urnes lately found in Norfolk*, constitutes 'the first English excavation report.'[72] However, on the contrary, this particular work exhibits little of the interest of the Royal Society in artefacts in their own right. In some of his works, Browne's sources were primarily literary. Thus, for the most part, as an antiquary he fits into the climate in which there was still little attempt made to apply the systematic method of inquiry to the material remains of the past. In his reliance on literary, even on occasion classical authority, Browne did not go far beyond the work of his contemporaries. And yet it cannot be denied that he had a stake in establishing the stage for the antiquarian investigations of the Royal Society. In one short treatise written in response to a query by Dugdale, for example, Browne discusses the origin of the 'artificial' burrows found throughout diverse parts of the country, suggesting that there should be an 'ocular exploration' of one of these, that is, it should be opened up and any remains found there examined and even dated by the presence of 'distinguishing substances.'[73] So, although Browne may not have been involved in the actual field-work himself, it has to be recognized that he was one of the first men to formulate the idea of doing it.[74] Although this idea may have been based on general Baconian principles, he deserves credit for setting out the first concrete proposal for an archaeological application of these to a specific situation. The fact that Browne's treatise is wholly free from philosophical speculation and literary parallels suggests in itself the direction in which it was pointing – towards the modern tradition of archaeological investigation later propounded by the Royal Society. Therefore, it seems that the varying nature of Browne's studies may indeed place him in the category of a transitional figure, if not a modern. It is the use of reason and experience that primarily associates Browne with the moderns, even if his use of authority sometimes tends to place him with the ancients.

This brief notice of Browne's work brings us back to an earlier theme. The connection between medical study and natural history went back, of course, to Galen and the ancients. In Restoration England, we find cross currents of medical thought, everything from Paracelsianism and Galenism to the more modern or 'scientific' influences, such as those exacted by the examples of Vesalius and Harvey. Towards the end of the century 'British medicine differed markedly from what it had been at the beginning,' with the Royal Society leading the rout of ancient medical theory that was associated with Galen.[75] Medical practitioners – even Galenists – had a head start on the other natural historians in that their work always had centred on empirical activity – for example, observation based on anatomical dissection – as well as on reason – reflecting on information gained empirically.[76] Unfortunately for the Galenist, 'as empirical observations multiplied and vast areas of new perceptual experience cried out for explanation, the Galenic ideas were no longer adequately adaptive,' and 'the data of chemistry and physics could be explained better by revising the notions of elements and the status of qualities.'[77] The mid-century reformers called for reform in medicine, advocating a 'chemical' understanding of – significantly – both medicine and nature, and for the replacement of what they viewed as a haphazard Galenic method, which had done little to advance the study of herbs.[78]

But before 1640, as Webster states, the medical faculties at the English universities were conservative in their approach to medical education, with emphasis on Galen and Hippocrates and almost none on dissection.[79] An interest in natural philosophy had already sprung up at the universities, especially at Oxford, where the earl of Danby had endowed a botanical garden. But scholasticism still prevailed in the curriculum, and after the 1635 plague, religious and eventually civil strife proved a handicap to further development.[80] With the return to a bit more normalcy after the English civil war a new spirit of inquiry is evident in the medical faculties, even if Galenic and Hippocratic studies still predominated. At Oxford, for example, Petty and others gave greater attention to anatomy and physiology; looking to lay enthusiasts such as Boyle or the local royalist physician, Thomas Willis, the students generated a momentum; the collection of the botanical garden there was expanded, and medical disputations assailed tradition and reflected a growing interest in iatrochemistry and the ideas of van Helmont and Harvey.[81] At Cambridge the religious sanction given to natural history by the Cambridge Platonists drew medical students and naturalists together, John Ray being a prominent example of a Platonist scholar convinced that 'the study of nature is essentially a religious duty,' since natural history was the best method of exhibiting God's works.[82] Natural theology also permeated the work of Henry Power and Sir Thomas Browne.

Bacon's urging of the compilation of medical data also had its effect. In *De Augmentis Scientiarum* he claimed that through using the methods of the new experimental science new drugs could be developed and diet as well as husbandry improved, thus leading to longer and healthier life. In the hermetic tradition he emphasized mineral and metallic cures.[83] This in itself contributed to a growing interest in geology, mineralogy, and associated types of natural historical investigation, even in men whose main interest otherwise resided in medicine. Knowledge of the healing virtues of plants and herbs was already common throughout the rural communities of Britain, particularly among those who had no recourse to professional medical help and therefore had to provide their own care, or defer to the only educated figure in the community, the clergyman. The other educated person likely to be a purveyor of medical information was the gentleman, and knowledge of plants and herbs, it has been suggested – and seems logical to conclude – was essential to the education of the gentry.[84] John Aubrey was especially curious to uncover bits of medical folklore. Apparently, the interest in practical botany did not extend go the College of Physicians, which took little trouble in encouraging the systematic translation of medical works into the vernacular; it seems that the apothecaries took a greater interest in matters related to botany.[85] But the physicians, generally, helped to direct regional study away from chorography and toward natural philosophy. This brings to mind William Harvey, and dynamism as contrasted with staticism.

Harvey's reputation as a socially prominent court physician, a benefactor of the College, and a respected anatomist 'legitimated, in a direct and concrete way, the role of physician as scientist.'[86] Robert G. Frank, Jr, has brought attention to the '1645 Group,' one of the progenitors of the Royal Society; of the nine members named by John Wallis, five were physicians and four knew Harvey intimately.[87] One of these, Christopher Merrett, is cited for his attack on 'bookish' or literary methods of education in medicine, and for including in his natural history references to plants growing in the Oxford region.[88]

The Commonwealth period saw the College of Physicians become the focus of the physician-scientists. If its members' interest did not lie in botany, it certainly can be seen in zoology. Besides indulging in attempts to confirm Harvey's theory of the circulation of the blood, almost to a man they were 'intimately acquainted with,' or ready to 'atomize,' any 'Fish, Bird, or Insect ...' and also knew the medicinal virtues of most European and American drugs, and understood minerals.[89] These interests must be viewed against the backdrop of Bacon's 'Solomon's House,' or ideal of cooperative investigation.

The scene was not much different at Oxford, beginning in the 1650s at least, except that there virtuosi played a greater role in scholarly activities. William Petty, a friend of Hartlib, is presented by Frank as a shining example of the experimental physician-scientist, and there is also Boyle, schooled in anatomy. By the time the Royal Society was formed in 1660, 'the new intellectual traditions of English medicine had [thus] been firmly established.'[90] A good segment of its members were physicians, many of whom almost eulogized Harvey in paving the way for their own activities, even if he did not exactly plan a revolution in research.[91] But also:

the rise of a new role for English physicians, that of investigator or virtuoso, was a complex process in which the intellectual majesty of scientific ideas was reinforced by personal ties. New physiological concepts and experimental methods were learned, not primarily through direct communication with a dominant scientific accomplishment (a paradigm if you will), but by having such innovations in ideas reinforced by the social prestige of institutions and the approval of role models.

English physicians found ... that investigation, once begun, whets an appetite for knowledge beyond the strictly medical sciences.[92]

Without the input of the medical men natural history, including the non-biological sciences, would not have advanced to the degree it did in the 'scientific revolution.' Indeed, science in general would have suffered. As Hall has noticed, 'certainly the largest single class in the scientific movement was that of the medical men, preponderantly physicians, in all countries.'[93] Out of their ranks came many of the regional natural historians, men like Gerard Boate and Robert Sibbald, who epitomize the interest in wide-ranging scientific investigation, and yet who through their work helped to 'compartmentalize' knowledge, that is, contributed to the foundation of disciplines such as geology, palaeontology, and archaeology.

Of course, for individual reasons, not every physician engaged in regional study adopted the methods propounded by the Royal Society. If we look to Robert Thoroton, for example, author of the chorographic study of Nottinghamshire, these reasons might include first, the great influence his chorographer friend, William Dugdale, had on him, persuading him to put into systematic form the information Thoroton had already compiled for his own personal pleasure, and second, his inability, because of the time demands of his practice, to go out and study the natural history of the county. As has been pointed out also in the introduction to the 1972 reprint of Thoroton's book, his ability to investigate the archaeology of the county was severely restricted by 'the enormous task of elucidating the manorial history of every

'To Serve the Commonwealth of Learning' 163

village,' which 'made him abandon attempts at more varied and [even] topographical description.'[94]

Although John Aubrey was not a physician, his polymathic interests likewise exhibit the inconsistency found in the work of the Royal Society during its early stages, which was characterized by a mix of pre-scientific as well as modern investigation. The first volumes of the society's *Philosophical Transactions*, for example, contain many contributions that today would be dismissed as pre-scientific nonsense, hermetic magic, and so on. For instance, one finds a consideration of the alleged healing power of a stone taken from a snake's head in Java.[95] Harvey and the other members of the society, including its first president, Henry Oldenburg, shared a belief in astrology.[96] Robert Plot, of course, was much involved in alchemical speculation while Robert Boyle searched for experimental proof of transmutation and witchcraft.[97] Even John Ray, who rejected the 'Doctrine of Signatures' – which held that some intrinsic property of a plant suggested to man its use as a specific cure – still was able to write that 'one Observation I shall add relating to the Virtues of Plants, in which I think there is something of Truth, that is, that there are, by the wise Disposition of Providence, such *Species* of Plants produc'd in every Country as are most proper and convenient for the Meat and Medicine of the Men and Animals that are bred and inhabit there.'[98]

But as time went by, and as Thomas Sprat and others assailed the occult, a new tone became apparent in the *Philosophical Transactions*. The society began to commit itself seriously to a mixture of natural and civil studies, with increasing emphasis on what today we describe conventionally as natural history. Martin Lister, who exerted a considerable interest upon the regional writers, contributed articles on topics such as plant and animal life that he had examined.[99] Between 1663 and 1703 several papers on pre-Roman Britain and on other archaeological topics were included in the *Philosophical Transactions*. So, 'natural history' came in time to be one of the premier concerns of the society. What we know today as 'archaeology' was of interest to the members of the society – in fact, to a considerable degree – but they generally referred to it as 'antiquities,' a label more general than 'archaeology' in that it did not always, especially at first, include matters such as serious excavation.

Of equal importance to the regional writers were the proposals and the questionnaires sent out by members of the Royal Society for the systematic study of natural phenomena. A 'Georgical Committee' consisting of thirty-two members was appointed by the society in March 1664, and in connection with the design of the society to collect 'Histories of Nature and Arts,' it proposed the drawing up of 'Heads of Enquiries' in the form of a questionnaire to be sent to 'experienced Husbandmen in all the Shires and Counties of England, Scotland and Ireland.'[100] This was done in the earnest desire that the recipients would '... from their owne and their knowing friends observation and experience, give as full and as punctuall answers thereunto, as they could; that thereby it might be knowne, what is knowne and done already, both to enrich every place with the aides, that are found in any place, and withall to consider, what further improvements may be made in all the practise of Husbandry.'[101] It was also decided to collect 'out of Mr. Hartlib's *Legacy*' certain inquiries as should occur to Dr. Croone. Also, Sir William Petty in Ireland and 'Mr. Awbrey' [Aubrey] in Wiltshire, and others, were to be responsible for obtaining reports on Husbandry-practise' and the like, throughout Britain. The 'Heads of Enquiries' were duly printed in *Philosophical Transactions* 1 and consisted of questions on the types of soil found in various parts of the country, on how these might be improved, on the methods of agricultural production, and so on. These first inquiries generally met with a low level of response. But this attempt to compile descriptions of agricultural practices from all parts of Britain 'is in itself,' according to Reginald Lennard, 'a significant fact both in agricultural history and in the history of English science,' because it showed 'how keenly interested the scientific researchers of those days were in matters of practical utility.'[102]

Soon afterwards Boyle advocated a similar plan, one specifically related to the interests of the regional writers.[103] Boyle's 'Heads' established the nature of the required research, for example:

1. To the First sort of Particulars, belong the Longtitude and Latitude of the Place ... and consequently the length of the longest and shortest days and nights, the Climate, parallels, etc ...

2. About the Air may be observ'd, its Temperature, as to the first four Qualities (commonly so call'd) and the Measures of them: its Weight, Clearness, Refractive power: ... What duration the several kinds of Weather usually have: What Meteors it is most or least wont to breed ...[104]

Boyle also stressed the importance of investigating 'the Earth itself' for example, its topography, and the specific nature of the soil, and also of noting the appearance and important characteristics of the inhabitants:

4. [In addition to] Productions of the Earth, there must be a careful account given of the Inhabitants ... settled there: And in particular, their Stature, Shape, Colour, Features, Strength, Agility, Beauty (or the wont of it) Complexions, Hair, Dyet, Inclinations, and Customs that seem not due to Education. As to their Women

(besides the other things) may be observed their Fruitfulness or Barrenness: their hard or easy Labour, etc. And both in Women and Men must be taken notice of what diseases they are subject to, and in these whether there be any symptom, or any other Circumstance, that is unusual and remarkable.[105]

The compilation of material on the 'External Productions of the Earth' (grasses, grains, timber, animals) was advocated: and 'Subterraneal observations,' that is, those regarding types of minerals and quarry stones, also found a place in Boyle's 'Heads.'

The proposals of the 'Georgical Committee' and those of Boyle seemed to unleash a flood of similar ones. Thus we find John Hoskyns advocating 'a physicall survey, e.g., what soyle, what temper, how early things ripen, how healthy the inhabitants and ... what fruits the land is most given to,' etc.[106] Later John Woodward formulated *Brief Instructions for Making Observations in All Parts of the World* (1696). John Aubrey was especially receptive towards the idea of a geological survey of Britain, one that was also suggested by Lister in a paper read to the Royal Society, which contained a proposal for a new soil map.[107] Aubrey 'oftentime wished for a mappe of England coloured according to the colours of the earth: with markes of the fossils and minerals.'[108] Another one of his projects included aiding John Ogilby in drawing up a set of printed queries.[109]

The interest of Englishmen in natural history soon extended to all parts of the earth. The New World received its share of attention. Plot's *Oxfordshire* served as a model for John Banister's unfinished 'Natural History of Virginia'; Richard Blome added a description of laws and customs to *A Description of the Island of Jamaica*, which also contained extensive natural history; Thomas Glover's account of Virginia primarily concentrated on the natural phenomena; and the list goes on and on.[110] The casual traveller was also drafted into the service of the Royal Society. R.W. Frantz has shown admirably how the society impressed upon the traveller the inestimable value of accurate observation and furnished him with a set of directions to guide him towards this goal.[111] There was an added incentive; frequently a traveller obtained permission to print his findings in the Society's *Philosophical Transactions*, alongside those of the famous scholars of the day. Such scientifically based findings received considerable critical praise.

And so scientific activity was stimulated on all fronts.[112] No longer would it be considered sufficient for any serious scholar merely to perambulate the region that engaged his attention, gathering information to a considerable degree on the basis of visual observation and local hearsay alone. No longer would it be deemed a worthwhile task for him to design a study around the genealogies or pedigrees of this or that family. No longer, also, would the eclectic selection of information from literary sources alone be considered appropriate: investigation of the natural phenomena of the local environment would also be required of him. The desire to write natural histories, then, was part of the Baconian aim of the scientists of the age. It 'received a new impetus in the late seventeenth century, gaining a topographical application that allowed it to benefit from the enthusiasm for antiquarian topography of the previous century,' by which time Kendrick's 'fogs' were rapidly dissipating.[113]

During the period between the publication of Gerard Boate's *Irelands Naturall History* in 1652 and Robert Plot's *Oxfordshire* in 1677, regional study was still in a state of flux. In the minds of men natural history and to a lesser degree chorography seemed to compete for position as the most useful method of describing a region. These two opposing – if related – forces are perhaps most vividly exhibited in the work of two men, Joshua Childrey and John Aubrey. These two figures are rated among the regional writers most closely associated with the activities of the Royal Society in its formative years. It is not entirely surprising, therefore, that they displayed a keen interest in natural history, to the point where it largely superseded chorography in their work. It is their rather haphazard though ingenious work that serves as a major bridge between the chorographers and the practitioners of natural history.

Joshua Childrey's *Britannia Baconica: or, the Natural Rarities of England, Scotland and Wales* excluded pedigrees because they had been 'copiously handled' previously by Lambarde, Camden, Dugdale, and Philipot, among others.[114] This book, as Hunter so aptly puts it, 'illustrates well how pure description could be seen by the ardent Baconian as the only proper concern of the scientist.'[115] Childrey (1623–70) was educated at Rochester Grammar School, and entered Magdalen College, Oxford, in 1640, taking his degree of BA in 1646. Until the Restoration he spent his time running a school at Faversham in Kent. From 1660 on he made his living as a man of the cloth, successively holding the posts of chaplain, archdeacon, and rector in towns throughout England. Sometime during this period he came into contact with men of science, as attested by his contribution to the Royal Society's *Philosophical Transactions* and by his letters to Henry Oldenburg, which have been recently printed.[116]

Hunter regards *Britannia Baconica* as 'a rather more superficial earlier attempt of the same kind as Plot's,' because it mixes literary sources and empirical observation to greater degree than do Plot's two main regional works.[117] Despite his stating that the work of earlier chorographers holds little significance for his own study, Childrey seeks out information in their works and utilizes it in his own. For example, he quotes Camden's account of certain spring waters near Flamborough Head, and cites the work of Giraldus in the section on Cheshire.[118] On another occasion he enlarges upon Camden's postulation that at one time England was physically joined to the Continent.[119] Also, Childrey justifies his description of the caves in Wiltshire on the grounds that he finds no mention of them by any of the antiquaries.[120] But he goes even further than this; he believes the collection of materials is the main function of natural philosophy, deploring the fact 'that men first fancy Opinions and Axiomes to themselves, and then by the help and art of Distinguishing, wrest and fit particular Instances and Observations to them.'[121]

Childrey gave Bacon full credit for establishing the study of natural philosophy, and based the title of his own book on the hope that it would serve as 'part of several Histories' in Bacon's Catalogue at the end of *Novum Organum*. In the preface Childrey tells us that he intended his work to benefit the 'Vulgar,' to instruct them not to scoff at the gentry; and that they not 'mis-believe or condemn for untruths all that seems strange, and above their wit to give a reason for.' The gentry could also be beneficiaries, by seeing that 'England is not void of those things which they admire abroad in their travels.' That the public –and most virtuosi, for that matter – were not yet fully aware of the value of the study of natural philosophy is reflected in his confession '... that such kind of writing is a little too bold yet, before the Histories of Art and Nature are compleatly done.'[122] One has to remember that *Britannia Baconica* came out about the same time that the Royal Society was founded, and therefore was not as influential at first as it later became.

The 'rarities' were described county by county in Childrey's book. We find here chorographic-like accounts of the physical qualities of the population of Devon and Cornwall, who were 'active in wrestling, and such boisterous exercises'; an exact description of the medicinal wells at Tunbridge, and of a small underground rivulet of Medway, in Kent.[123] He also reports on phantasmal events, such as the horrible and fearful groaning heard many times by fishermen off a shore in Yorkshire when the sea was calm.[124] This last example is indicative of the continuing interest in any phenomenon that was apparently inexplicable. Childrey speculated about the influence of the planets on the weather, and at one point even suggested to Oldenburg that the society investigate a report that it rained blood near Apsam in Devonshire.[125] But his interest in such matters – unlike that of the majority of the chorographers – was tempered by religious scruples.[126]

Experiment and inductive methods played some role in Childrey's work.

On several occasions he forwarded to the Royal Society material of the type found in *Britannia Baconica* for which he was highly praised by Oldenburg.[127] In this fashion the society received detailed information from Childrey on experiments such as one whereby quicksilver, if properly 'fixed,' would take off the impression of any seal in wax:

1. Salis Albi lib. unam. Floris aeris finely pulverized uncias sex. Argenti vivj uncias tres. First take a new Frying pan, and fill it full of clean water. First hold it over the fire till it be hot. Then adde the salt, and afterwards the Verdegrease, and let them boile, till it' come to a purple colour. Then adde ye Quicksilver, and let it boile. Scum off the froth, and at last poure off ye water, and put the compounded quicksilver into a stone Morter and with a pestle agitate it, till it become very cleare, and well purged. Wash it with the liquor still pouring on more, till it be perfectly cleansed. Then forme the masse into the fashion of seales. And if at anytime the masse grows too hard, it is but adding a little more quicksilver, and it will be fit for operation. This was given me by a very ingenious Gentleman (now dead) in manuscript.[128]

Oldenburg, for his part, enlisted Childrey in the society's project of collecting information on various matters concerning husbandry through the use of its agricultural enquiries.[129]

As Childrey tells us himself, he 'first fell in love with the L. Bacons Philosophy in ye yeare 1646, and tried severall Experiments (though such as I now reckon not to be of any moment) in 1647. 1648. 1649. and 50.'[130] He culled the material for his *Britannia Baconica* from such experiments, and from several of his notebooks, which contained observations on the physical environment. These he intended to bequeath to the society.[131] In *Britannia Baconica* we find the localities where snake-stones, star-stones, cockles, periwinkles, oysters, scallops, and mussels are preserved in solid stone. Thus, today Childrey is best known as the earliest British writer to refer frequently to fossils.[132] His prime importance to this study is based on the fact that Robert Plot later to a large extent patterned his studies on Childrey's *Britannia Baconica*. (The major link between Childrey and Plot – namely John Aubrey – is the topic of chapter 9.)

Childrey's admonition for others to follow Bacon's example is worth quoting, because it succinctly states the views of like-minded men of the time. Childrey's desire is thus

to serve the Commonweath of Learning, which much wants such [natural] Histories as this to be written, and laid as a sure Foundation, whereon to build those Axiomes that make us true Schollars, and knowing men in Philosophy. I have as nearly (as I could) followed the Precepts of my Master, the Lord Bacon, and (by way of acknowledgement, from whom I received my first light into this way) have given my Book the Title of *Britannia Baconica*; and the rather, because it will serve for a part of several Histories in his Lordships Catalogue, at the end of his *Novum Organon*. I have not at all meddled with matter of Antiquity, Pedigrees, or the like, those being copiously handled by several of our Countrymen already; as the learned Cambden in his *Britannia*, Mr. Dugdale in his Description of *Warwickshire*, Mr. King in his *Vale Royal*, Mr. Lambert in his *Perambulation*, Mr. Philpot in his *Villare Cantianum*, and others. Only I ventured at the description of the Caves in Wiltshire, because I find it mentioned by none of our Antiquaries. I have here and there attempted to give the Causes of the Rarities I relate, having the example of my Lord B. for my authority ... And though I cannot but confess, that such kind of writing is a little too bold yet, before the Histories of Art and Nature are compleatly done; yet possibly I may in some, hit upon the true Reason by chance; and unless men were more forward (then I see they are yet) in collecting such Histories, these kinds of confidences must be dispensed with. Indeed, had those men that have spent so much time & pains in writing voluminous Comments on Arist.[otle] but labored as diligently in writing Coments upon Nature, & (with that self denial and indifferency, which becoms ingenuity in the dark) in trying to render a reason of such and such odd appearances in things, though some of them had been but false Positions; doubtless the Philosophical part of Learning would have been at a much better pass, and Inquisition a great deal more happy and thriving then it is at this day. The pest of Learning is, that men first fancy Opinions and Axiomes to themselves, and then by the help and art of Distinguishing, wrest and fit particular Instances and Observations to them.[133]

Despite the fact that Childrey had the right idea, his work left much to be desired when it came to natural historical description. For example, his account of 'Radnorshire' did not transcend that of the chorographers. It consisted solely of the following: 'This Shire hath sharp and cold air, because of the snow lying long unmelted under the shady hills, and hanging Rocks, whereof there are many.'[134] In mixing 'superstition with a critical and scientific spirit, as R.F. Jones stated, Childrey 'presents an excellent example of the struggle between the budding scientific spirit of the age and the inherited chaos of superstition and fear'; hence, 'nature is still mysterious, but there is a disposition on his part to assign all phenomena to natural causes. His work furnishes a clear picture of the impact of Bacon's philosophy on a representative mind of the age.'[135] What this meant, however, was that natural history still had a distance to go before Bacon's plan could be fulfilled.

CHAPTER NINE

'This Secret Call'

> Here indeed was the birth of serious field studies. Lhwyd was studying antiquities for themselves and not merely as one element in the cultural landscape; and so was John Aubrey, justly the most famous of the seventeenth-century antiquaries ... Aubrey, and perhaps Lhwyd as well, blazed a trail which leads from Leland and Camden to Stukeley, Colt Hoare, Cunnington, and to the great field archaeologists of the last fifty years, such as William Freeman, Crawford and Fox.
>
> Glyn Daniel *A Hundred and Fifty Years of Archaeology*

AS WE HAVE SEEN, the seventeenth century in Britain witnessed a substantial increase in the production of county studies by country gentlemen and scholars alike. Joshua Childrey was but one of many virtuosi who, by the 1650s and 1660s, began to give more thought to the inclusion of natural phenomena in their work on counties or regions. Bacon, Hartlib, and like-minded men encouraged thought and activity in this direction. One of the earliest English writers – if not the earliest – to exhibit this change in interest, stimulated largely by the influence of Baconian science, was John Aubrey (1626–97).

Although Aubrey is famous as a literary figure of the seventeenth century, and as somewhat of a mystic, less had been said of his overall contribution to regional study. Beginning in the 1650s, Aubrey collected into notebooks, often haphazardly, his observations on antiquities and natural history. Yet as a scientist, he apparently left much to be desired even by the standards of the day. Having read one of Aubrey's manuscripts, the naturalist John Ray wrote him: 'I think ... that you are a little too inclinable to credit strange relations,' and White Kennett said that Aubrey was known as 'the Corruption Carrier to the Royal Society.'[1] He has also been called a backward-looking dilettante.[2] Aubrey's two-sidedness is evident if we consider that he was able to compile a work largely devoted to archaeology, the 'Monumenta Britannica,' which was thoroughly in tune with the modern scientific antiquarianism, while his *Miscellanies*, a work on folklore, ghost stories, magic, astrology, and so on, demonstrates his credulity and involvement with hermetic lore.

The debate among historians and scientists over whether Aubrey should be counted with the ancients or the moderns has not yet been settled – although in the context of current historiography this is not a pressing issue.[3] Aubrey was Baconian in his scientific writings, mainly due to his desire to uncover practical knowledge of the type that would allow man to establish some control over his life. He believed that the temperament and character of the people were affected by the air they breathed, the type of soil on which they lived, and their diet. In his 'Naturall Historie of Wiltshire,' for example, he observed that regional differences of pronunciation 'must proceed from the earth, or aire or both'; or that 'At Huntley in Gloucestershire, the nature of the people breakes with the soile.'[4] This type of environmental determinism was mixed with astrology as a basis for many of his explanations. The desire for practical knowledge, according to Hunter, also led Aubrey to write his *Brief Lives*, his most famous study: 'One of his aims in collecting the biographical data that he used in *Brief Lives* was to collate human life with its astrological circumstances.'[5] Hunter believes that it was a 'streak of practicality' that led Aubrey to value the magical element in the *Miscellanies* as well as the natural phenomena in his 'Naturall History of Wiltshire.'[6] Hunter's view, overall, is that Aubrey was a serious intellectual, actively and enthusiastically involved with the scientific movement of his time.

The appreciation of Aubrey's genius is a recent development, however; the typical eighteenth-century view that he was credulous is summarized in a piece in *The Gentleman's Magazine* that referred to Aubrey as a 'gossiping anecdote-monger,' one who makes it difficult to 'discriminate the gold from the dross, the truth from the lies,' and who 'transmitted a variety of tittle-tattle.'[7] Despite Ray's mild rebuke, many of Aubrey's contemporaries held him in high esteem. His membership in the Royal Society was public recognition of excellence, and his friendships with the leading figures of his day were a further honour. His contemporary reputation as an important investigator is reflected in the fact that Charles II commissioned him to survey Stonehenge and Avebury; like most of the people who knew Aubrey personally, the king tended to overlook Aubrey's quirks.

The story of Aubrey's life is well known. He was son of a gentleman estate owner, Richard Aubrey, at Easton Pierse in Wiltshire, where 'the Indigenae, or Aborigines, speake drawling; they are phlegmatique, skins pale and livid, slow and dull, heavy of spirit ... they are generally more apt to be fanatiques: their persons are generally plump and feggy.'[8] Aubrey at first received a private education under Robert Latimer, Thomas Hobbes's preceptor, then proceeded to Oxford, entering Trinity College in May 1642. Smallpox and civil war interrupted his stay there, so that he left Oxford in 1643, but by the spring of 1646 (after another brief stint at Oxford), he read law briefly at the Middle Temple, although he was never called to the bar. During this period he had read Browne's *Religio Medici*, which had just been published, and contributed a plate of Oseney Abbey to Dugdale's *Monasticon*. In 1652 Aubrey's father died, leaving him with an estate that was to involve him in several lawsuits. These and some matrimonial predicaments eventually ruined Aubrey financially and affected him emotionally, but he found some consolation in directing his energies to the pursuit of antiquarian knowledge.

That Aubrey might not have been as backward looking as some historians suppose may be deduced from the following comment. Until about 1649, when the Royal Society had its embryonic beginnings at Wadham College, Oxford:

... 'twas held a strange presumption for a man to attempt an innovation in learning; and not to be good manners to be more knowing than his neighbours and forefathers. Even to attempt an improvement in husbandry, though it succeeded with profit, was look't upon with an ill eie ... 'Twas held a sinne to make a scrutinie into the waies of nature; whereas Solomon saieth, *Tradidit mundum disputationibus hominum*: and it is certainly a profound part of religion to glorify God in his workes.

In those times to have had an inventive and enquiring witt was accounted resverie (affectation), which censure the famous Dr. William Harvey could not escape for his admirable discovery of the circulation of the blood. He told me himself that upon his publishing that booke he fell in his practice extremely.[9]

Aubrey frequently refers to the earlier, pre–Royal Society period of science as the 'darke time.' He was elected to the society in its infancy; before that he belonged to the club of 'Commonwealth Men' at Oxford. Yet compared to the high standards of investigation exemplified in the Royal Society after its foundation, the scientific concerns and projects espoused at Commonwealth Oxford stand up rather well.[10] It was at this time that Aubrey was first exposed to the new methods of empirical investigation, and to the ideas concerning the practical applications of new discoveries and invetions, such as William Petty's sowing and harrowing engine, Christopher Wren's balance and machine for double writing and so on.[11] It was also during this period that Aubrey met Samuel Hartlib, who impressed Aubrey with his practical scientific programs.[12] The Oxford group encouraged Aubrey to attempt a minor Baconian effort of his own, and so in 1656 he began compiling material for a natural history of Wiltshire. He apparently needed little prompting, for he enjoyed undertaking the type of work such a compilation required. But he preferred going out into the countryside, talking with local inhabitants, observing historical trends in architectural design, and collecting and evaluating artefacts, over what he considered the more arduous conventional type of historical-genealogic research found in most contemporary regional studies. To be sure, he realized the value of archival research and utilized documentary material in many of his works. However, he found 'this searching after Antiquities is a wearisome Taske,' and stated 'of all studies I take the least delight in this.'[13]

Aubrey added a new dimension to John Leland's original goal of surveying the entire country, as revealed in Aubrey's dictum that: 'Their is no Nation abounds with ... [more] varieties of Soiles, Plants, and Mineralls, than ours: and therefore it very well deserves to be surveyed.'[14] However, one of the first regional studies that he undertook, the 'Essay towards the Description of the North Division of Wiltshire,' was directly modelled on Dugdale's *Warwickshire*; it was based on the same kind of heraldic and genealogic information, derived from similar documentary and monumental sources. Thus it largely resembled the earlier chorographies. Dugdale and Aubrey were close friends, and they exchanged various pieces of antiquarian information over the years. But the immediate impetus for the compilation of the 'North Division' came from a meeting of gentlemen at Devizes in 1659, where 'it was wish't by some, that this County (wherein are many observable Antiquities) were surveyed in Imitation of Mr. Dugdales Illustration of Warwickshire.' As such a survey was too large a task for one man, 'Mr. Yorke ... advised, to have the Laboure divided. I would undertake the North.'[15] To be truthful, the manuscript of the 'North Division' makes no pretension to chorography. It consists, in reality, of Aubrey's notebooks, where the arrangement is irregular. Still, it does hold information of considerable chorographic value. Aubrey began his notebooks in 1659, and lingered over them most of his life, although he thought they were close to completion about 1671. His hope was that they would one day fall into the hands of some antiquary who would properly assemble them.

Aubrey's preface to the 'North Division' contains an apologia for the reconstruction of the past from fragmentary relics, in which his disdain for Puritan zealots is not hidden:

In former daies the Churches and great houses hereabout did so abound with monuments and things remarqueable that it would have deterred an Antiquarie from undertaking it. But as Pythagoras did guesse at the vastnesse of Hercules' stature by the length of his foote, so among these Ruines are Remaynes enough left for a man to give a guesse what noble buildings, etc. were made by the Piety, Charity, and Magnanimity of our Forefathers ... so here, the eie and mind is no less effected with these stately ruines than they would have been when standing and entire. They breed in generous mindes a kind of pittie; and set the thoughts a-worke to make out their magnificence as they were when in perfection. These Remaynes are 'tanquam tabulata nuafragii' (like fragments of a Shipwreck) that after the Revolution of so many yeares and governments have escaped the teeth of Time and [which is more dangerous] the hands of mistaken zeale. So that the retrieving of these forgotten things from oblivion in some sort resembles the Art of a Conjuror who makes those walke and appeare that have layen in their graves many hundreds of yeares: and represents as it were to the eie, the places, customs and Fashions, that were of old Time. It is said of Antiquaries, they wipe off the mouldinesse they digge, and remove the rubbish.[16]

The preface also contains a summary history of England, which is one of Aubrey's finest pieces of sustained writing. He begins: 'Let us imagine then what kind of countrie this was in the time of the Ancient Britons,' and goes on to describe the ancient inhabitants as 'almost as savage as the Beasts whose skins were their only rayment.'[17] Relying on Camden, Aubrey tells us how the Romans were able to subdue and civilize them, then the narrative follows the successive waves of invaders until the story is brought up to Aubrey's time.

Aubrey's strong interest in both architectural history and religion is brought to light in this opening section, where the two subjects are interwoven. Some of his information was obtained from Dugdale, who held similar interests, but in the main Aubrey based his remarks on his own personal observations, for example: 'When I came to Oxford Crucifixes were common in ye glasse in the Studies window: and in the Chamber windowes were canonized Saints ... But after 1647 they were all broken: Down went Dagon. Now no religion to be found.'[18] Aubrey's main interest was in coats of arms, window paintings, and inscriptions, many of which he quoted verbatim in his study. Although this interest was common to other chorographers, in Aubrey it seems to have been connected with what could almost be termed a type of religious mysticism.

In the heart of every antiquary beats a nostalgia for *temps perdus*; from his eye courses a furtive tear as he stands amid bare, ruined choirs in an age when the world is turned upside down, religion with it. This was truer of Aubrey than of most; in his inordinate interest in the architecture of churches and other religious houses and in their artefacts, he found it almost impossible to

'This Secret Call' 175

discuss anything without some allusion to the religious element. An exponent of a sort of natural theology, Aubrey sought to glorify God in his works. In an abbreviated description of Haselbury Quarry, he invokes 'the old men's story' that it was upon the direction of St Adelme that the quarry was miraculously discovered.[19] The countryside through which he journeyed in search of his material delighted him. Of Down Ampney House he informs us: 'This is a very noble seate ... situated with great convenience for pleasure and profits: By the house runnes a fine brook, which waters these gallant meadowes on the west side, where depasture a great number of cattle: – thirty milk-mayds singing.'[20] Neither does the 'North Division' lack its share of anecdotes. Riding through Garsdon he recalled the tale of a footman to Henry VIII: Henry, tumbling from his horse, 'fell with his head into mudde, with which, being fatt and heavie, he had been suffocated to death, had he not been timely relieved by his footman Mody.'[21] Mody was later granted Garsdon Manor for his assistance. Another tale recounts how Aubrey could not read certain inscriptions on high windows, 'for want of a short telescope,' so that the tuckling of the charges on the shields does not always agree with his written description of them. This is yet another example of why Aubrey found research so physically difficult.

Aubrey's personal manner adds vitality to his work, but also makes it difficult for the reader to approach it in a serious manner. He praised much that had been lost in the past, but at the same time admitted that everything was not rosy in the days of old. In discussing another favourite topic, the social climate of earlier times, he describes the English court as unpolished and unmannered; that of James I, for example, was so uncivil to women that the queen herself could not pass by the king's apartment without receiving some affront. Learning from the time of Erasmus until about 1650 is called 'downright pedantry'; the gentry and citizens were generally unschooled, and 'the conversation and habits of those times were as starcht as their bands and square beards: and gravity was then taken for wisdom.'[22] Aubrey's lively personal style extends to his other sources. The anecdotes quoted on the authority of 'My Grandfather Lyte,' or 'the parish clerk's wife,' often give local flavour.

Aubrey's 'Perambulation of Surrey' was written mainly in 1673 and resembles the 'North Division' in two respects.[23] First, the work consists of notes of a perambulation of the county and was never intended to serve as a formal topographical history. Second, it contains very little natural history, concentrating on genealogy and heraldry, even though Aubrey, writing in 1691, claims to have mixed antiquities with natural history in this work.[24] There is a third similarity, this one not in the nature of the work itself, but in the fact that once again Aubrey began it in co-operation with others but was later left to his own devices. 'Surrey' was part of a project by John Ogilby, the king's cosmographer and geographer, which aimed at the description of 'Britannia,' and which was to be part of Ogilby's larger project, 'A Description of the Whole World, viz. Africa, America, Asia and Europe.'[25] Having fallen on lean times, Aubrey was eager to take up his appointment as Ogilby's deputy for Surrey, after he was recommended to Ogilby by Christopher Wren. As in the case of his predecessor, John Leland, Aubrey thus was ensured the assistance of all the justices of the peace, mayors, vicars, and other responsible civic officials in his search for antiquities.

Aubrey began his perambulation with zest in early July 1673, 'hoping that the delicate aire and diversion of Surrey will ease my ... spirit.' But by October his role was ended by Ogilby, who, learning that Robert Plot at Oxford was formulating a scheme for a series of regional studies along rather different lines, discontinued his own work in this direction. Ogilby now preferred to use 'what scraps he can get out of bookes or by heere say.'[26] Aubrey naturally was disappointed, and he could not conceal his frustration. Referring to Ogilby, he said: 'God deliver me from such men.'[27] After so much work, Aubrey experienced several misadventures including the theft of his horse at Esher, and the loss of his money; hence he was reluctant to lay aside his notes on 'Surrey.' He continued to add material to the 1673 base for many years afterwards.

Aubrey begins 'Surrey' with a study of the Thames River. Searching into etymology he quotes Edward Lhuyd on the origin of the name 'Thames,' from one of Lhuyd's letters to him.[28] From there on, coats of arms and transcriptions of memorial tablets dominate the text. Included in the study were the 'Queries In Order to the Description of Britannia,' which he had earlier considered with Ogilby and others. The first of these pertained to the type of material normally found in chorographic literature, for example, cities, their antiquity, government, fairs, and so on. But some also had featured the kind of information found in natural histories, which Aubrey himself generally disregarded in 'Surrey.'[29]

Despite his disappointment over the actions of Ogilby, Aubrey was nevertheless pleased to hear of Plot's intended 'Natural History of England'; 'I am right glad to heare of Dr. Plot's designe, it agrees so much with my Humour. I can much assist him in it: having a good Penus naturales [store of natural history] of my owne Collecting: which I dayly augment.'[30] From this time on, Aubrey took an interest in Plot's work, and the two men aided each other in their respective efforts, to their mutual benefit. Aubrey tells how in 1675 he 'came contracted and acquainted with Dr. Rob. Plott, who ... had

'This Secret Call' 177

then ['upon the Loome'] his *Naturall History of Oxfordshire*: wch, I seeing he did performe so excellently well.'³¹ Aubrey went on to describe how he had sent Plot his own papers, including 'Surrey,' and offered Plot assistance in undertaking a natural history of Wiltshire. But Plot had already been invited into Staffordshire, 'to illustrate that Countrie.' He accepted the invitation, finishing his work in December 1684, when Aubrey 'importuned him again to undertake this County [Wiltshire]; but he replied, He was so taken up ... of the Musaeum Ashmoleanum.'³² Plot by that time was determined to meddle no more in county natural histories, with the exception of his native Kent; he encouraged Aubrey to finish and publish his collections on Wiltshire lest they perish or be attributed at some future date to someone else.

The result of Plot's encouragement was the compilation, most of it in 1685, of Aubrey's notes into the 'Naturall Historie of Wiltshire,' his only regional study that incorporated a large amount of information on natural phenomena; he claimed it to be the first essay of its kind in the nation.³³ As he did to his other regional studies, he appears to have added material to this production for several decades; he began work on it in 1656, and in 1691 submitted two draft volumes (Aubrey MSS 1 and 2) to the Ashmolean. (A separate draft volume was submitted to the library of the Royal Society.) According to Aubrey it was a 'secret Call' that had originally drawn him to pursue this kind of study:

I was from my Childhood affected with the view of things rare; which is the beginning of Philosophy: and though I have not had leisure to make any considerable proficiency in it; yet I was carried on with a strong Impulse to undertake this Taske. I know not why; unles for my own particular private pleasure.: Credit there was none: for it getts the contempt ... of a mans Neighbours. But I could not be quiet till I had obey'd this secret Call: Mr Camden, Dr. Plott and Mr Wood confess the like same.³⁴

In 'Wiltshire' Aubrey describes the country under numerous headings: 'Air,' 'Springs Medicinal,' 'Rivers,' 'Soils,' 'Minerals and Fossils,' 'Stones,' 'Formed Stones,' 'Plants,' 'Beasts,' 'Fishes,' 'Birds,' 'Reptiles and Insects,' 'Diseases and Cures,' 'Gardens,' and so on. This study then, represented a novel approach in the field of regional study, contrasting with the earlier chorographies in the greater detail with which the natural phenomena of regions were described and arranged. Many of the topics found in 'Wiltshire' were quite similar to the concerns of the Royal Society.³⁵ Nevertheless, it is clear that Aubrey as county historian cannot be referred to as a 'natural historian' in the strictest sense of the term. He is now known more for his archaeological investigations than for his natural historical studies.

Aubrey had great respect not only for Dugdale but also for Carew, Owen, and other early giants of chorography.[36] For example, in order to discover the proportion of the downs of Wiltshire to the vales he divided Speed's map of the county, 'with a paire of cizars, according to the respective Hundreds of Downes and Vale, and I weight them in a curious ballance of a Goldsmith.'[37] He was also deferential to the 'curious enquiry' of John Stow. But although he admired Camden also, he was misled once because he 'could not in modesty, but jurare in verba tanti Viri,' 'trust in the words of such a man.'[38] Among the other antiquarian works that Aubrey utilized, Weever's *Ancient Funerall Monuments* holds a prominent place.

Aubrey named the preface of 'Wiltshire' the 'Chorographia super-et subterranea naturalis' ('natural chorography above and below ground'), which consisted of an account 'of what parts of England I have seen, as to the soiles.'[39] In this context Aubrey used the term 'chorography' solely to denote the new type of soil or 'land' studies now being carried out by many members of the Royal Society, while dispensing with the usage of the term that had been accepted until then. The 'Chorographia' was first sent to Martin Lister in 1684 in connection with the paper read by Lister at the Royal Society that advocated a geological survey of Britain. Because Lister was perhaps the first Englishman to envisage such a scheme, his specific proposal for the project of a geological map is worth quoting in whole:

We shall then be better able to judge of the make of the Earth, and of many Phænomena belonging thereto, when we have well and duely examined it, as far as human art can possibly reach, beginning from the outside downwards. As for the more inward and Central parts thereof, I think we shall never be able to confute Gilbert's opinion thereof, who will, not without Reason, have it altogether Iron. And for this purpose it were advisable, that a Soil or Mineral Map, as I may call it, were devised. The same Map of England may, for want of a better, at present serve the Turn. It might be distinguisht into Countries, with the River and some of the noted Towns put in. The Soil might either be coloured, by variety of Lines, or Etchings; but the great care must be, very exactly to note upon the Map, where such and such Soiles are bounded. As for example in Yorkshire (1.) The Woolds, Chaulk, Flint, and Pyrites, etc. (2.) Black moore; Moores, Sandstone, etc. (3.) Holderness; Boggy, Turf, Clay, Sand, etc. (4.). Western Mountains; Moores, Sand-stone, Coal, Iron-stone, Lead Ore, Sand, Clay, etc. Nottinghamshire; mostly Gravel Pebble, Clay, Sand-stone, Hall-playster, or Gypsum, etc. Now if it were noted, how far these extended, and the limits of each Soil appeared upon a Map, something more might be comprehended from the whole, and from every part, then I can possibly foresee, which would make such a labour very well worth the pains. For I am of the opinion, such upper Soiles, if natural, infallibly produce such under Minerals, and for the most part in such order. But I leave this to the industry of future times.⁴⁰

Lister was interested in geology for its own sake, but his proposal contained little if anything relating to the process of stratification or to the order of superposition of the soils. But among Aubrey's 'many ... clever remarks on Geology, [made] long before the principles of that Science were systematically laid down' were ones that indicate that he had a considerable knowledge of the major trends in English stratigraphy, as well as of the soil patterns.⁴¹ For example, like George Owen half a century earlier, Aubrey noted that 'east from Bridgeport in Dorsetshire to Dover in Kent, runnes a veine of Chalke: and the like south and north from Merton in South Wilts., near to Calne in North Wilts'; or, 'The Isle of Wight is Chalke: the Needles at the west end are pricked Rocks of Chalke.'⁴² Aubrey's geological observations were those of a practical field researcher; he described the structure of a deposit of 'thunder-stones' (pyrite), noting that its distance from any navigable river made it commercially unviable to exploit.⁴³ He personally has tested a mineral vein thought to hold iron pyrite, only to discover that it contained a deposit of marcasite.⁴⁴

The new scientific antiquarianism that is primarily exhibited in some of Aubrey's other works does, however, have its place in 'Wiltshire,' even if the terminology is not exactly modern, and especially in the 'Chorographia' section. Aubrey displayed an exceptional interest in 'petrified shells' or fossils: 'As you ride from Cricklad to Highworth, Wiltsh., you find frequently roundish stones ... which (I thinke) they call braine stones, for on the outside they resemble the ventricles of the braine; they are petrified sea mushrooms.'⁴⁵ In Aubrey's day the term *fossil* had not yet assumed its modern meaning, that is, those remains – whether an impression, trace, or actual fabric – of animal and plant life from a previous age that are found embedded in the earth's crust. In the seventeenth century fossils were still denoted as 'all bodies whatever that are dug out of the Earth.'⁴⁶ The sole distinction made was between those bodies native to the earth and those that were 'adventitious' – also referred to as 'foreign' or 'extraneous' fossils – ones that were exuviae or remnants of animals.

Much of what Aubrey had to say about the organic adventitious fossils is contained under the chapter title 'Formed Stones' – which, incidentally, indicates the broad usage of the term *fossil* as compared to its modern, narrow meaning.⁴⁷ Yet he was also keenly interested in anything else dug out of the earth, including the 'Mineralls' (metals). He described these in detail in another chapter, giving their locations and at times referring the reader to other sources of information.[48] Two other fascinating and informative chapters, one on soil types and one on stones, comprise the remainder of the geological portions of 'Wiltshire.' It was here that Aubrey referred to Bacon's experimenting with different types of soils to produce several sorts of plants.[49]

This reference brings us to yet another aspect of Aubrey's investigations – his use of the geological outline as a foundation for natural history; that is, his attempt to come to practical conclusions through correlations between geology and vegetation. He had hoped that certain trees could be identified as the products of distinguishable 'marks of minerals,' and similarly believed that a great variety of 'earths' would result in many sorts of plants; that red earth bears good barley; that 'at Bradfield and Dracot Cerne is such vitriolate earth ... [that it] makes the land so soure, it beares sowre and austere plants.'[50] In a less scientific manner, however, he extended this type of analysis to his general hypothesis that 'according to the severall sorts of earth in England (and so all the world over) the Indigenae are respectfully witty or dull, good or bad.' As we have seen, Aubrey applied this sort of reasoning to the particular context of his fellow Wiltshiremen.[51]

The first part of 'Wiltshire' also included 'An Hypothesis of the Terraqueous Globe. A Digression,' a review of the current palaeontological theories of the origins of the earth. Thomas Burnet's *Telluris theoria Sacra*, presenting the most famous of these, strove to combine old and new science in a sort of justification of God's ways with respect to the Creation and the workings of the universe. According to Burnet, in the Creation God surrounded the earth with a cover of water with oil floating on top, and superimposed on this a globe of dust-filled air. The dust descended into the soil creating a smooth crust, which became a global paradise.[52] Of course, all this was done in accordance with natural providence, and correlated with the biblical phenomenon of the Noachian flood, then commonly known as the 'Great Deluge.'

Aubrey found much to criticize in Burnet's eschatology, arguing that even though the occurrence of 'petrified fishes' shells gives clear evidence that the earth had been totally covered with water, the present surface of the earth is the result also of a series of earthquakes and not the flood alone. 'As the world was torne by earthquakes,' he observed, 'as also the vaulture by time foundered and fell in, so the water subsided and the dry land appeared.' He continued: 'Then, why might not that change alter the center of gravity of the earth? Before this the pole of the ecliptique perhaps was the pole of the world.'[53] As Britton has noted, Aubrey was not inclined to read the sacred writings too literally on this subject, preferring to avoid conflict with official church doctrine on the earth's creation, as revealed by his referral of the reader to Père Richard Symond's then-radical belief that the scriptures might in some places be erroneous when it came to philosophy, even though the doctrine of the church by itself was correct.[54] Hunter believes that Aubrey's position on the matter was unusually radical even among the most advanced thinkers; Aubrey held that the world was older than was commonly supposed in his time, something that he thought could be proved by the study of stratigraphy, while most of his contemporaries unflinchingly accepted the authority of biblical chronology on this topic.[55]

The remainder of the first part of 'Wiltshire' contains material closely related to the kind found in the earlier chorographic literature, especially in the descriptions of rivers and flora and fauna. But even here Aubrey manages to endorse eagerly schemes that were part and parcel of the intellectual baggage shouldered by the fellows of the Royal Society. He was particularly fond of the idea that if the Thames and the North Avon rivers 'were maried by a canal between them, then might goods be brought from London to Bristow ... which would be an extraordinary convenience both for safety and to avoid overturning.'[56]

Of less scientific value are Aubrey's views on some other matters. In the chapter 'Of Men and Women' he displays that typical Restoration fascination with freaks and the improbable, in his recitation of occurrences in Wiltshire of remarkable longevity, monstrous births, etc: 'Mr. William Gauntlett, of Netherhampton, born at Amesbury, told me that since his remembrance there were digged up in the churchyard at Amesbury, which is very spacious, a great number of huge bones, exceeding, as he says, the size of those of our dayes. At Highworth, at the signe of the Bull, at one Hartwells, I have been credibly enformed is to be seen a scull of a vast bignesse, *scilicet*, half as big again as an ordinary one. From Mr. Rich. Brown, Rector of Somerford Magna.'[57] Then there is his preoccupation with astrology and with the supernatural. Aubrey included various observations on occult phenomena in his writings on natural history. Magic, witchcraft, apparitions, and 'accidents' all had their place alongside the accounts of natural phenomena, particularly in the chapters on 'Air,' 'Accidents,' and 'Fatalities of Families and Places.' Sometimes Aubrey provided a rational explanation for a phenomenon generally considered to fall within the realm of the supernatural, but this was not usually the case. It is possible that he attempted to understand the world of the supernatural for the same reason that he looked for answers in the natural world – he wished to manipulate it for utilitarian purposes. However, the fact that Aubrey eventually removed much of the occult material originally included in 'Wiltshire' and put it into his *Miscellanies* might indicate that he was more critical and that he utilized solid criteria of selectivity to a considerable degree.

The second part of 'Wiltshire' contains chapters on such miscellaneous topics as 'worthies' of the county – taken largely out of Fuller's *Worthies of England* (1662) – 'Learned Men who received Pensions from the Earles of Pembroke,' architecture, gardens, agriculture and the commodities of the county, fairs, and markets, 'hawks and hawking,' and so on.[58] These also contain practical information and advice, such as 'Some Excerpta, out of John Norden's Dialogues. Though they are not of Wiltshire, they will doe no hurt here: and if my countrey-men know it not, I wish they might learn.'[59] Aubrey had inherited also a strong interest in mapping and surveying, though not at a professional level, and quite often he used mathematical instruments in his field-work. On other occasions, however, the descriptive elements in his work – also based on an eagerness to record anything that might prove to be useful knowledge someday – was strictly the result of unaided visual observation, as when in 1660 he came upon Wilton House: 'There was then remaining on the south side some of the walles of the great gate; and on the north side there was some remaines of a bottome of a tower; but the incrustation of freestone was almost all gone; a fellow was then picking at that little that was left. 'Tis like enough by this time they have digged all away.'[60]

Like most of Aubrey's works, 'Wiltshire' did not reach the printer during his lifetime. In a letter to Wood he stated that he had completed the final chapter of this study on 21 April 1686.[61] In 1691 John Ray came into possession of the manuscript, and read it 'with great pleasure and satisfaction ... the book cannot but take with all sorts of Readers.'[62] He added his own notes to it and shortly thereafter returned it to Aubrey in the hope that the latter would speed it to the press. Although Aubrey had already formulated some vague plan along these lines, which perhaps involved Edward Lhuyd, it came to nothing. He shared the predicament of many of his fellow antiquaries – the inability to secure the kind of effective patronage that would allow him to realize his publishing plans. Aubrey was honoured by the Royal Society, which had a transcript made of 'Wiltshire,' but on the whole his study could not and did not attract a large following.

Some of the material found in Aubrey's regional studies is also interspersed in his 'Monumenta Britannica, or A Miscellanie of British Antiquities,' although for the most part this work evolved separately. Written mainly between 1665 and 1693, it contains considerably more examples from the rapidly developing field of scientific antiquarianism than does 'Wiltshire', and constitutes the first work entirely devoted to systematically recording material artefacts and other such finds. It is primarily responsible for Aubrey's widespread reputation today as the first investigator to use modern archaeological methods: 'Historians, chroniclers, and topographers there had

'This Secret Call' 183

been before his time; but he was the first who devoted his studies and his abilities to archaeology, in its various ramifications of architecture, genealogy, palaeography, numismatics, heraldry, etc.'[63] The appearance of Inigo Jones's *The Most Notable Antiquity of Great Britain Vulgarly Called Stone-Heng ... Restored* (1655), and several other books on Stonehenge in the 1650s, encouraged Aubrey to investigate this monument for himself. Except for Aubrey's 'Monumenta Britannica' these works display few signs of modern archaeology. Jones and the writers stimulated to respond to his book argued the origins of Stonehenge primarily from a historical perspective, utilizing literary sources such as Speed, Verstegan, Stow, Raleigh, and Leland as their authorities.[64]

The 'Monumenta Britannica' opens with a section entitled 'Templa Druidum,' which contains most of the archaeology and which Dugdale urged Aubrey to publish.[65] It also holds the first account of the megalithic remains of Avebury, and the first detailed study of Stonehenge. Aubrey's 'discovery' of Avebury would on its own merit have given him a spot in the annals of British archaeology. It was in 1648 during a hunting excursion on the 'Grey-Weather' Downs by Marlborough that he first encountered Avebury:

... the chase led us (at length) through the village of Aubury, into the closes there: where I was wonderfully surprized at the sight of those vast stones: of wch I had never heard before: as also at the mighty Bank and graffe [ditch] about it: I observed in the Inclosure some segments of rude circles, made with these stones, whence I concluded, they had been in the old time complete. I left my Company a while, entertaining myselfe with a more delightfull indaegation: and then (steered by the cry of the Hounds) overtooke the company, and went with them to Kynnet, where was a good Hunting dinner provided.[66]

The direct prompting for Aubrey's study came from the king in 1663. Charles had learned of Aubrey's antiquarian efforts from the president of the Royal Society, Lord Brouncker. Having 'admired [observed] that none of our Chorographers had taken notice of it [Stonehenge],' he commanded that Aubrey be presented to him, and proceeded to instruct his honoured subject to survey and to describe in writing both Avebury and Stonehenge.[67]

In his survey of Stonehenge, Aubrey was so thorough that in his plans of it he even showed depressions in the ground that subsequently disappeared from view because of their relative shallowness. (These were relocated in 1921 and are now called the 'Aubrey Holes.') Generally, his drawings and notes of the site are extremely accurate and indicate the actual positions of individual stones and earthworks. Furthermore, he was probably the first to label a barrow – one of those features common to Wiltshire and to other areas in Britain – as originating with the ancient Britons, even if he did not find much support for this identification.[68] Refuting earlier theories that attributed the construction of Stonehenge variously to the Romans, Danes, Phoenicians, or Saxons, Aubrey assigned their origin to the Druids, a conclusion that proved erroneous. The fact that he postponed the publication of his findings until he had investigated all other monuments across the land that were comparable to Stonehenge and Avebury demonstrates, according to Glyn Daniel, that Aubrey was following a procedure in conformity with that of the modern field archaeologist.[69]

As Hunter has argued, beginning with Aubrey the collection and interpretation of field antiquities was given a new direction through the study of natural history. Observations of the natural environment in other works often aided Aubrey in evaluating artefacts from the past. A study of geology, for example, led him to note that the pebbles used on Roman roads were not usually local.[70] In this respect Aubrey pointed the way for regional studies to follow: by linking the two fields.

His piecemeal collecting was not comparable to the more systematic classification of the rigorous scientists like Ray, Lister, Lhuyd, or even Plot. Nevertheless, as Hunter makes clear, Aubrey's work, or methodology, 'represented an approach very different from that of the earlier county surveys, contrasting with them in the greater detail with which the natural and artificial phenomena of counties were described and arranged in chapters according to type, air, stones, beasts and so on.'[71] Aubrey, therefore, broke new ground. But, as mentioned previously, his haphazard collections and his disinclination to quantify or test most things means that the quality of his work fell short of the high standard set by men who later adopted a more precise systematic classification of natural phenomena.[72] This is not to take anything away from the pioneering nature of Aubrey's own investigations. Jon Bruce Kite, in his excellent work on Aubrey, states:

Wrong though he was in his conclusions about the Druids, Aubrey was utterly modern and right in his methods of surveying, in his objectivity and honesty as a field archaeologist, and in his desire to see all the data collected by others on comparable sites before he advanced his own erroneous conclusion. The humility and tentativeness with which he presented his Druid theory are human qualities which may not be either modern or old-fashioned, but which signal an open and enquiring mind ... [73]

Perhaps, then, if one values such distinctions, Aubrey may be seen as a 'transitional' figure in the annals of science and scholarship in general, with a foot in both camps, ancient and modern.

CHAPTER TEN

Remedying 'Chief Defects'

Naturall Philosophy, next to God's Word, is the most Sovereign Antidote to expell the poison of Superstition; and not only so, but also the most approved food to nourish Faith.

Joshua Childrey *Britannia Baconica*

THE FIRST PRODUCTION in the English language that can legitimately claim the title of regional natural history was, in fact, written by the Dutchman Gerard Boate (or Boet). Boate's seventeenth-century work entitled *Irelands Naturall History* is the earliest English-language specimen of a natural history that was not a translation from either classical or Continental literature. The author, educated at Dutch schools where the development of regional natural history preceded similar activity in Britain, was among the first to exclude from regional study accounts of local marvels, colourful digressions, supernatural occurrences, and citations from classical authors.

Boate (1604–50) is representative of one of the many ties between Continental and British scholarship. Like Thomas Browne and Robert Sibbald, Boate studied at the University of Leyden, obtaining an MD degree in 1628. No doubt he was profoundly influenced by the researches of the scientific community there. In medicine, Leyden had a high reputation, and such medical training imprinted itself on the study of botany as well. Men such as Otto Van Heurne, Paul Hermann, Albert Kyper, and others ensured the high quality of the scientific education at Leyden.[1] It is significant that this university, and the Dutch in general, had now begun to attempt the systematic natural history of their equatorial colonies. Field-work was carried out, notably in Brazil from 1637 to 1644, and the results were published. Such early research into natural phenomena had its effect on Boate, and his work in Ireland was of a similar type.[2]

In 1630 Boate left Holland and settled in London, where he found employment as royal physician. Boate the physician decided to become Boate the natural historian after he became a contributor to the fund, under the English Act of Parliament of 1642, that admitted the Dutch to subscribe finances for the reduction of the Irish. Since repayment was to take the form of a grant of forfeited Irish lands, he decided to undertake the compilation of a work that would post information on the rewards to be derived from Ireland's resources with the practical aim of augmenting the interest of 'adventurers' in obtaining land in that country.

Although Boate was appointed a doctor to the hospital at Dublin in 1649, he wrote *Irelands Naturall History* before ever setting foot in Ireland. Materials for his study were furnished primarily by his brother Arnold and by other planters familiar with that land. Thus Boate's book is based on the careful and exact observations of those who knew the area well.[3] At least one of his major informants, his brother Arnold, was also imbued with the emerging scientific attitude that placed so much importance on experiment and observation. According to his brother, Boate wrote *Naturall History* in 1645.[4] Between the writing of the manuscript and its publication, Gerard went to Ireland, where he died not long after his arrival.[5]

Judging from a statement by his brother to Samuel Hartlib Boate's manuscript seems to have been mislaid and regarded as lost during the intervening period.[6] As it turned out, Gerard Boate's papers came into the hands of Hartlib in London, who, with the assent and help of Arnold Boate, proceeded to publish the *Naturall History*. 'It is indeed revealing,' as Hunter states, that this work 'reflects Hartlib's practical preoccupation with husbandry and with mining iron and minerals, which he urged should be employed "For the Common Good of Ireland, and more especially, for the benefit of the Adventurers and Planters therein."'[7] As we have seen, Hartlib shared with the Boates the enthusiasm for the teaching of 'the true experimental natural philosophy; as also whatever is most needful and noble in the Mathematics, viz. Arithmetic, Geometry, Cosmography, Geography, Perspective and Architecture.'[8] He was an important contributor to scientific education and he advocated specialized agricultural instruction in *An Essay for Advancement of Husbandry-Learning* (1651) and in some of his other works.[9] Hartlib viewed Boate's *Naturall History* as a scientifically based study, useful for his own promotion of Ireland as an opportune location for agricultural experiment: 'I lookt also somewhat upon the hopefull appearance of replanting Ireland shortly, not only by the adventurers, but happily by the calling in of exiled Bohemians and other Protestants also, and happily by the invitation of some well affected out of the Low Countries, which to advance are thoughts suitable to your noble genius, and to further the settlement thereof, the Natural History of that countrie will not be unfit, but very subservient.'[10]

Boate arranged his book around a systematic plan, presenting Ireland under a series of headings covering such topics as climate, topography, minerals, waters, and so on. Genealogy and heraldry had virtually no place here. The *Naturall History* is divided into twenty-four chapters. The first chapter describes the situation and shape of Ireland and establishes the provinces, the counties of the English Pale, and the principal cities and towns. In this part of the book Boate's remarks are no more germane than anything found in the earlier chorographies; for example, the town of Galway is described simply as 'the head-citie of the Province of Connaught, to be reckoned, as well for bigness and faireness, as for riches; for the streets are wide, and handsomely ordered, the houses for the most part built of free stone; and the inhabitants much addicted to trafick, doe greatly trade into other countries, especially, into Spain, from whence they used to fetch great store of wines and other wares every year.'[11]

Nearly one-third of the book is devoted to a consideration of Ireland's coastline. Here Boate reveals himself to be a pioneer observer of the face of the earth, long before studies in earth-sculpture – or geomorphology, as the science is today known – became common in regional descripton. What is especially remarkable is the vivid and detailed manner in which these features are presented to the reader, as if the author himself had made notes of these in the field, and then incorporated them into his work. Boate paid attention to the kind of physical formations which either were given scant treatment by the chorographers, or escaped their notice altogether. Principal promontories, 'hilly sheares,' capes, sandbanks, and offshore rocks were all meticulously described, and the havens, 'for the most part so fair and large, that in this particular hardly any land in the whole world may be compared with this,' are afforded 'particular rehearsall.'[12] John Norden and other earlier seventeenth-century British chorographers had commented on the sea and its ravages, but not in as much detail. Norden, for example, merely noted that the Cornish rocks were weather-resistant enough to withstand shoreline erosion.[13] George Owen often referred to the sea off Pembroke as 'dealeinge so unkindely with this poore Countrey as that it doth not in anye where seeme to yeld to the lande in anye parte, but in everye corner thereof eateth upp parte of the mayne.'[14] John Speed took note of the Irish Sea, 'whose rage with such vehemency beateth against her bankes, that it is thought and said, some quantity of the Land hath been swallowed up by those Seas.'[15] The actual process involved in such erosion, however, awaited the serious investigation and speculation of the natural historians. Wave action was considered to be one culprit, and Joshua Childrey added another, tidal scour, when he wrote of the Cornish coast: 'The cause of the devouring of the Land by the sea, I conceive to be its being a Promontory lying open to the merciless stormes and weather, and withall, lying in a place where two currents meet and part; I mean the Tide as it comes in, and returnes out of the Sleeve, or narrow Seas, and the Irish Seas, and Seavern.'[16]

With Boate and Childrey leading the way other virtuosi soon directed their attention to the relationship between sea and shore, a concern not totally unexpected among a seafaring folk. John Wallis, for one, in upholding an earlier theory that claimed that islands were not part of the original Creation – for 'Almightie God the cause and conductor of Nature, in creating the world did leave no parte of his woork imperfect or broken' – investigated how a supposed one-time land bridge between England and France had been breached solely by tidal action.[17] In an elaborate argument Wallis went on to indicate how Romney Marsh, the Low Countries, and the plains surrounding the Thames estuary had all been the result of tidal scour, which was accompanied by erosive debris worn from the neighbouring isthmus. Such a process fitted nicely with the concept, then current, of a state of equilibrium among the natural processes. Whether or not that theory was true, the importance here is that such features of earth history were gaining the attention of more and more natural historians.

Although some of Boate's information on the coastline may have been derived from charts and Rutters, which in itself would have been an innovative step, most of it probably came from his friends and acquaintances among the sailors of the United Provinces, since in his day a considerable portion of the overseas trade of England was conducted in Dutch bottoms. On occasion Boate found it useful to search the literary sources for similar information, perhaps utilizing John Speed's maps for locating anchorages, or briefly quoting Speed, Giraldus, and Camden in his attempt to emphasize the tempestuous nature of the Irish Sea.[18] He was not as credulous as were many of the chorographers, and did not therefore take everything he found in the literary sources at their face value; he scoffs, for example, at the account of Giraldus Cambrensis of some strange fluctuations in the ebb and flow of tides.[19] Because Boate's description of the coastline takes up so much of the book, it is worth quoting a typical entry:

The next great Harbour upon this coast ... is that off Knocfergus, being a great wide Bay, the which in its mouth, betwixt the Southern and the Northern point, is no less than ten or twelve miles broad, growing narrower by degrees, the farther it goeth into the land, the which it doth for the space of fifteen miles, as far as to the Town of Belfast, where a little river called Lagon (not portable but of small boates) falleth into this Harbour. In this Bay is a reasonable good Road before the Town of Knockfergus (seated about nine miles within the land), where it is good anchoring in three fathoms, and three and a halfe. On the North Side of the Bay, somewhat neer the sea, under a Castle called Mouse-hill, is a sandbay, where it is good anchoring for all sorts of ships.[20]

Exact description of this kind was extended to the inland areas of the country. After discussing springs, fountains, rivers, lakes, and mountains, Boate displays an abiding interest in agriculture. The many aspects of agricultural life are fervently investigated, including the nature and fruitfulness of the soil, its suitability for tillage or pasture, the advantages of improving the land through the use of assorted types of manures (limes, dung, sea sand), the usage of marl, and so on. Generally speaking, Boate goes into much greater detail than the chorographers when it comes to agriculture. He explains the reasons why lands differ in richness: '... the best and richest soil, if but half a foot or a foot deep, and if lying upon a stiff clay or hard stone, is not so fertile, as a leaner soil of greater depth, and ... [lying] upon sand and gravel, through which the superfluous moisture may descend, and not standing still, as upon the clay and stone, make cold the roots of the grass, or corn, and so hurt the whole.'[21] Boate eventually arrives at the conclusion that, in general, Irish land should be better for pasture than tillage, because the country 'hath it a more natural aptness for grass.'

As for the heaths and bogs, Boate believes that their origin is recent, and that with proper drainage they could easily be made suitable for agriculture. According to Boate, the bogs could be classified systematically as watery, red or dry, grassy, hassocky, or wet, the latter known as 'Moones.' He contends that the English inhabitants are expert in drainage techniques, and have thus reaped great profits from such ventures, but he has only caustic remarks for the negligence of the Irish, who have been relatively inept when it comes to drainage.[22] In speculating about the origins of the bogs, Boate continues:

Very few of the wet-bogs in Ireland are such by any naturall property, or primitive constitution, but through the superfluous moysture that in length of time hath been gathered therein, whether it have its originall within the place it self, or be come thither from without. The first of these two ... taketh place in the most part of the Grassie-bogs, which ordinarily are occasioned by Springs; the which arising in great number out of some parcel of ground, and finding no issue, do ... soak through, and bring it to that rottenness and springiness, which nevertheless is not a little increased through the rain water comming to that of the Springs.[23]

Boate's fascination – almost obsession – with anything to do with water is clearly perceptible in his account of the inland drainage system. The lakes (loughs) are classified according to size, and the islands situated in them command considerable attention.[24] There is also an interesting example of the author's determination to discredit wild stories. He investigated the belief, long established throughout Christendom, that the 'Suburbs of Purgatorie' are located in a cave situated on an island in Lough Dug. He tells of the efforts of the earl of Cork and the 'Viscount of Elie,' who in the last two years of the reign of James I 'sent some persons of quality to the place, to inquire exactly the truth of the whole matter.' These investigators, 'descending down to the very Purgatorie and Hell,' found that this 'miraculous and fearfull cave' was 'but a little cell, digged or hewen out of the Rockie ground, without any windowes or holes.'[25] Boate was of the opinion that this cell, which the good earl and the viscount caused to be demolished, was once used for the devilish purposes of the friars, whose charges, after fasting, were likely to witness macabre apparitions in the depths of the cell.

Boate also debunks the alleged ability of Lough Neaugh to turn wood into stone. In this case, his brother had investigated the matter; he lived near the site but had never spoken to anyone who had witnessed such a transformation. Nevertheless, Arnold Boate affirmed 'that here and there upon the borders of that lough are found little stones ... [which] seem to be nothing else but wood, and by every one are taken for such, until one comes to touch and handle them.'[26]

In dealing with Ireland's forests and woodlands, Boate details their destruction, comparing the present scene with that described by Giraldus.[27] The contemporary scarcity of wood for fuel and of timber for construction is not surprising, he notes, when one considers that from Dublin to Tredagh (Drogheda), Dundalk, the Newry, and as far as Dromore, there are no woods worth speaking of. Only in the west and south are some large woods extant in the counties of Kerry and Tipperary, this despite the destruction wrought by the English. Boate ranges himself with those who argue that the obliteration of the woods was due more to economic factors than to warfare, the principal cause being the use of wood as a source of the charcoal used in iron smelting, an industry begun by the English who had settled in Ireland since the end of the Elizabethan wars. 'The trees and woods having been so much destroyed in Ireland, as heretofore we have shewed ... the inhabitants are necessitated to make use of other fuel.' Boate meticulously describes these other fuels, which included turf and sea coal, which was also imported from England and elsewhere. Because of the cheapness of imported coal, few attempts had been made to locate local coal deposits, although a deposit was accidentally found in the county of Carlow. It is apparent from Boate's analysis that there was an increasing reliance on imported coal, especially in the eastern counties, because of the limited amounts and high cost of turf and wood brought from the interior. In a similar fashion, Boate 'has an analysis of the factors governing the economic location of an iron industry which is quite in the tradition of modern geography.'[28]

Boate thought it highly unlikely that there were any gold mines in Ireland. He does provide a rare account of the occurrence of this metal there. A credible person informed him that a colleague stated 'that out of a certain rivulet in the county of Nether-Tyrone, called Miola ... he had gathered about one dram of pure gold.' Thus Boate concluded 'that in the aforesaid [Slew-galen] mountains rich gold mines do lye hidden.'[29] But Boate's main interest lay in the iron mines.[30] He identifies the various sorts of iron ore found in Ireland as bog, rock, pin, white, and shell ore, and in the tradition of George Owen, describes outcrops of rock that contain these ores. Silver and lead ores also have their places in the text, as do building and ornamental stones and the glass industry. Boate's account of the surface deposits is among the first detailing this branch of geology. He recognized a series of surface deposits that today are known to have been laid down by the last ice age.

The rest of the *Naturall History* deals with the same topics as the natural histories written later in the century: observations on temperature, quality of the air, snow, hail, hoarfrost, dew, thunder, lightning, winds, earthquakes, and other natural phenomena.

Boate's 'long distance' speculation and explanation regarding natural phenomena based on descriptions by others – such as his account of Irish bogs – were not inconsistent with the sentiment that scholars take advantage of the accumulated experience and knowledge of artisans, gardeners, husbandmen, travellers, and so on, to compile histories of trade and nature – a sentiment ultimately founded on the views of Bacon.[31] Boate was careful, therefore, to base his statements on observation and, at least in most cases, on verifiable fact. His informants were likewise firm believers in 'the true experimental natural philosophy,' anxious to be accurate.[32] In Boate's commentary on the location of gold in Ireland, it is significant that he emphasized the credibility of the source of his information, and, not having access to further details on this subject, was unwilling to speculate on its possible occurrence elsewhere in the country; Wicklow, which later became an important gold-mining area, was not even mentioned by Boate as a potential source of the mineral. This kind of reliance on verifiable fact and observation builds upon the methods of the chorographers, who relied on literary sources and monumental inscriptions for their information; it thus distinguishes the *Naturall History* as 'a

Regional Geography of quite exceptional merit,' one that also 'represented a major development in economic geography' in its utilitarian aim of encouraging a new Protestant plantation by accurately describing the natural wealth of Ireland.[33]

As noted earlier, the Boates were at the centre of a group of scientifically oriented Protestant settlers in Ireland who were more or less led by Hartlib in London. This group constituted an 'Invisible College,' dedicated to propagating experimental philosophy, especially as it applied to Ireland.[34] Hartlib supplied the inspiration and ideas, and was the driving force behind the attempt to complete the *Naturall History* after Boate died; Boate had planned three more parts, dealing with plants, other living creatures, and the 'old fashions, lawes and customes' of the natives of Ireland.[35] In this task Arnold Boate was succeeded by Robert Boyle and Robert Childe, both of whom were eminently suited to the task by their concern for husbandry and its improvement. Ultimately, however, their efforts reaped little success, at least as far as the completion of Boate's book is concerned. Little more was done toward the study of Ireland's natural history until in 1682 William Molyneux, an ardent Baconian, undertook to write the natural history of the country for Moses Pitt's great *English Atlas*.[36] Molyneux carefully sought out Irishmen versed in the scientific techniques of regional study, and from such inauspicious beginnings was formed in 1683 the Irish counterpart to the Royal Society, the Dublin Philosophical Society. The emphasis had shifted, meanwhile, to surveys, such as the celebrated Down Survey, headed by William Petty, of the mid-1650s. Such surveys, however, owed much to Boate, providing information of the type valued by him; the surveyors were advocates of Baconian natural history; Petty's aims were extraordinarily similar to those expressed in the *Naturall History*; and in general the Down Survey was constructed on Boate's 'tentative qualitative economic geography of Ireland.'[37] And so 'the Boates were important because they occupied a pivotal position, linking the earlier stirrings of interest in science in the 1630s with the later activities of the Interregnum.'[38] Overall, therefore, the *Naturall History* went far towards remedying many of the 'chief defects for which the Truths of Naturall Philosopie and the products thereof ... are so imperfectly known.'[39]

CHAPTER ELEVEN

'Learning So Much Neglected'

[The] genial father of County Natural Histories in Britain.
R.W.T. Gunther *Early Science in Oxford*[1]

LIKE HIS FRIEND John Aubrey, Robert Plot (1640–96) was interested in promoting useful knowledge, emphasizing how his own work would contribute 'to the great benefit of Trade, and advantage of the People.'[2] Also like the famous Aubrey, he was interested in the supernatural and included accounts of occult phenomena in his natural histories.[3] His *Natural History of Oxfordshire*, was published at the end of a lengthy period when natural history was still experiencing some difficulty in firmly superseding the chorographic element in the field of regional study, and was chiefly responsible for popularizing regional natural history. It was deliberately intended by its author to supplement the 'Civil and Geographicall Historys' that up to that time still exerted an influence on the field as a whole.[4] Plot, therefore, was one of the first regional writers to discard many of the methodologies and interests of the chorographers, preferring rather to investigate natural history scientifically.

Robert Plot descended from a 'genteel family at Borden near to Sittingbourne in Kent,' the only son of another Robert Plot.[5] Educated at the Free School at Wye, Kent, he matriculated at Magdalen Hall, Oxford, in July 1658, where he later became vice-principal and tutor. He graduated BA in 1661, MA in 1664, and BCL in 1671. About the year 1676 he left Magdalen and entered as a commoner at University College, where he resided until his marriage in 1690. A firm believer in Bacon's dictum that natural history 'is used either for the sake of the knowledge of the particular things which it contains, or as the primary material of philosophy and the stuff and subject-matter of true induction,' Plot intended to make a personal survey of the whole of England and Wales in order to compile their natural history.⁶ He recorded this intention in an interesting letter (c 1673) to Dr John Fell, dean of Christ Church. It is here that Plot proposed to follow the examples of William Camden and John Leland:

As often as I have reflected on the very great and no less commendable Service done to the Common-Wealth of Learning at home, and the Reputation of the Nation abroad, first by the indefatigable Travels of John Leland, and upon his Foundation a Superstructure added by William Camden Clarentieulx, and others; and that notwithstanding their great Industry not only considerable Additions might be made to whatever they have touch'd on, but a fair new Building erected (altogether as much to the Honour of the Nation) out of Materials they made little or no use of: so often have I thought with my selfe, provided I be judg'd a fit Person, the Design agreeable, and the Encouragement proportionable, that I might also in some measure deserve of my Country, if I would reassume their Labours, and once more take a journey at least through England and Wales, to make a strict search, and give a faithful Account to such as shall encourage me of all such Things (worthy notice) which they have wholly pass'd by, or but imperfectly mention'd.⁷

Besides building on the work of Leland and Camden, Plot mentions in his letter to Fell his intention of rectifying the defects he found in Sir Henry Spelman's 'An Interpretation of Villare Anglicum.'⁸ According to its preface, dated 31 October 1687, this gazetteer was 'made by the appointment of Sir Henry Spelman, out of Speed's Mappes.'⁹ He also hoped to add to Weever's *Ancient Funerall Monuments* information 'on all the other Dioceses in the same manner as he [Weever] has done the Dioceses of Canterbury, Rochester, London and Norwich.'¹⁰ Thus Plot indicated that he believed that the concerns of the virtuoso-scientists included the examination of inscriptions and similar sources. Conceiving his enterprise as a serious scientific project, however, Plot held it to be a history of 'Naturall Bodys, and manual Arts, found or practised within the Kingdom of England and Dominion of Wales.'¹¹ In one sense, it may be said that in his envisaging a series of county studies covering all the counties, he emulated Leland, Camden, and Norden, the major difference being that the content of Plot's volumes was natural history rather than chorography. Plot sought a royal commission to travel through all parts of the country, similar to the one held by Leland. He also armed himself with the following testimonial, signed by the principal dignitaries of Oxford:

These are to signifye to all whom it may concern that Robert Plott, Doctor of Laws, and now of Magdalen Hall in the University of Oxford, being studious to make search after the Rarities both of Nature and Arts afforded in the Kingdome for the Information of the Curious and in order to an Historical account of the same, by him promised hereof to be given, Wee whose Names are subscribed doe approve of that his ingenious undertaking and doe recommend him to the Courteous furtherance of such persons of whom he shall have occasion to make enquiry in the procedure of that Affair.[12]

In his work Plot relied on printed sheets of queries like those used a decade earlier by the Georgical Committee when it regarded agriculture in different regions; he supplemented them with his own. It is little known that Plot put his name to two sets of these, both of them more systematic than those drafted by Ogilby.[13]

Aubrey was one of the first to collaborate with Plot, sending him the results of all his years of collecting material on Surrey, Wiltshire, and several other shires.[14] Plot quoted 'Mr. Aubrey's notes' in his *Oxfordshire*, and Aubrey continued to provide additional information for years.[15] For example, in 1684 Aubrey transcribed and then forwarded to Plot the notes he had made on the flyleaves of his copy of *Oxfordshire*.[16] Unfortunately Plot seemed unwilling to acknowledge fully his debt to Aubrey. Despite his reference to 'Mr. Aubrey's notes,' he generally played down Aubrey's role, as is evident in one of his letters to Aubrey, where he mentions finding 'many things ... much to my purpose' in one of Aubrey's works, but adds that this did not apply to his study of Oxfordshire.[17] This may have been, however, a natural reaction of Plot to Aubrey's growing mistrust of the use of his (Aubrey's) materials by his colleagues. As Hunter indicates, Aubrey may also have been developing an inordinate pride about the value of his own work.[18] It is also possible that Plot merely did not desire to reveal to the whole world the extent of his debt to a man floating on the fringes of Oxford's intellectual community. Whatever was the case, the example of Aubrey's work did in fact guide Plot to a considerable degree. This is especially true of *Oxfordshire*, for this book, begun in 1674, was intended to demonstrate the methods Plot hoped to apply to the entire country.

From Plot's notebooks one can gain a clear picture of exactly how he went about his self-appointed task. He commenced his field-work in the parish of Cropredy and then, riding on horseback along the lanes or perambulating the fields on foot, he visited the northern parishes. Next, during the summer of 1674, he studied the countryside between the rivers Evenlode and Thames, completing the field-work the following summer when he toured the western sector of Oxfordshire beyond the Evenlode and then the eastern sector beyond the Thames. He used the rivers to divide the shire into five distinctive tracts. This method differed from that used by many of the chorographers in the past, who, as we have seen, arranged their narratives to follow the rivers and described the places in sequence along them. Plot was interested in the river system only in so far as it marked off the county into conveniently sized portions.[19]

In the address that prefaced *Oxfordshire*, Plot advised his readers that this work would aid in the 'advancement of a sort of Learning so much neglected in England [ie, Nature or Arts],' and in the promotion of trade.[20] Later in the book he claimed that in the account of the natural things of Oxfordshire he 'treated only of such as eminently ... were some way or other useful to Man.'[21] The opening address contains a commentary on the county map that Plot had researched and drawn up. Its accuracy, he maintained, 'far exceeds any we had before,' especially because 'it contains all the Mercat Towns, and many Parishes omitted by Saxton, Speed, etc,' and since 'it shews also the Villages, distinguished by a different mark and character, and the Houses of the Nobility and Gentry, and others ... and all these with their Bearings to one another, according to the Compass.'[22] However, the map was 'not so perfect' because Plot could not provide distances that were 'Mathematically exact.' Yet he was confident that all his placings were fairly accurate. The systems of house reference seemed to have considerable priority on the map, perhaps so as to compensate for the usual disdain of the natural historian for mere genealogy. 'This Map is so contrived,' he proudly pointed out, 'that a foreigner as well as English-man ... may with ease find out who are the Owners of most of them [ie, houses] ... And all this done by Figures ... placed in Order over the Arms in the limb of the Map.' He also saw the border of arms as not only useful as a reference to owners of houses depicted on the map, but also ornamental and acting as an 'Encouragement to the Gentry to keep their seats.'

It is certain from other features in the legend that Plot intended this volume as a forerunner to a county series; the symbols for ancient ways, fortifications, and sites of religious houses were therefore designed to apply to 'all following maps as well as this.' A village, for Plot, consisted of an assembly of more than ten dwellings, 'under which number I seldom think them worth notice.' Plot's rather aloof attitude is reflected in his plan for the incorporation of corrections. 'Gentry,' he serves notice, were expected to bring details of mistakes in the map directly 'to the Porter or one of the Keepers of the Bodleyan Library, who will be ready to receive them.'[23]

Throughout his book Plot concentrated his attention on natural features or practical problems, so there was no danger that he would become muddled in a matrix of genealogy and pedigrees. His method was to survey the county in each of its elements, thereby allowing the natural divisions to show up of their own accord. Or, as he put it:

I shall consider, first, Natural Things, such as ... Animals, Plants, and the universal furniture of the World. Secondly, her [the county's] extravagancies and defects, occasioned either by the exuberancy of matter, of obstinacy of impediments, as in Monsters. And then lastly, as she is restrained, forced, fashioned, or determined, by Artificial Operations. All which, without absurdity, may fall under the general notation of a Natural History, things of Art (as the Lord Bacon well observeth) not differing from those of nature in form and essence, but in the efficient only; Man having no power over Nature, but in her matter and motion, i.e. to put together, separate, or fashion natural Bodies, and sometimes to alter their ordinary course.

Yet neither shall I so strictly tie my self up to this method, but that I shall handle the two first, viz. The several Species of natural things, and the errors of Nature in those respective Species, together; and the things Artificial in the end apart: method equally begetting iterations and prolixity, where it is observed too much, as where not at all. And these I intend to deliver as succinctly as may be, in a plain, easie, unartificial Stile.[24]

This method, it will be observed, differed considerably from the short general description of a region by a chorographer, which usually prefaced a particular study of its hundreds of parishes, and which allowed the contrast between various kinds of land, or soils, to appear only by selection. The method Plot followed in *Oxfordshire* was the same he employed later in *The Natural History of Staffordshire*, because, as these two books were to form part of a series, he adopted a consistent treatment from the start. The ten chapter headings are 'Of the Heavens and Air,' 'Of the Waters,' 'Of the Earths,' 'Of Stones,' 'Of Formed Stones,' 'Of Plants,' 'Of Brutes,' 'Of Men and Women,' 'Of Arts,' and 'Of Antiquities.' (Because of the close similarity of the two books, only one, *Staffordshire*, will be examined in detail here.)

The publication of *Oxfordshire* was greeted enthusiastically by the learned gentlemen of the day, and facilitated Plot's entry into the Royal Society the same year. It was also among the fellows of the society that he circulated his inquiries. Even before, however, he had been actively involved in the society's intellectual orbit, participating in the discussion of problems concerning husbandry, occasionally meeting Robert Hooke in the coffee-houses of London, or contributing to the society scientific communications that were as often as not published in the *Philosophical Transactions*.[25] In May 1682 he presented *Oxfordshire* to the Royal Society, and by the end of

November had been elected its secretary.²⁶ He was involved in the Philosophical Society of Oxford, of which he had been a principal founder in 1683–4, and directed its experiments, while occupying a position roughly comparable to that of Hooke in the Royal Society. During the period of his secretaryship he made many donations to the repositories of the organizations in which he proudly served, and while adding to his own collection of minerals, with the view of securing a representative series for the Oxford museum, he also made certain that a parallel series be made available for study at Gresham College.²⁷ In 1683 he obtained the post of editor of the *Philosophical Transactions*, which he held from number 143 (1683) to number 166 (1684) inclusive. All of these activities, of course, kept him in direct contact with many of the leading British scientists.

Because Plot held the position of first 'Custos' of Elias Ashmole's new museum at Oxford (by 1683 he held the professorship in chemistry), it is not surprising that the pressures of his other duties forced him to relinquish his secretaryship of the Royal Society. This left him free to devote more time to setting up the Ashmolean Museum, as the new institution at Oxford came to be known, and where the Philosophical Society flourished, and to equipping a chemical laboratory in its basement.²⁸

As professor of chemistry, Plot prepared several works on the subject. Among them are many examples of Plot's interest in the speculative and philosophical side of science.²⁹ But he apparently kept up his interest in alchemy, manifested in an interest in the preparation of transcendental medicines and substances. In an article on Plot's alchemical concerns, we are told that he had devoted much attention to 'mysterious liquors, which he regarded as fundamental to transcendental medicine and alchemy,' and that there is evidence that in or about 1677 he set up in partnership with others to prepare and sell 'chymical medicines'; furthermore, Plot later came across a secret that involved his attempt 'to make an agreement with some [unknown] person ... whereby, in return for the knowledge of the secret ... he was to take the practical steps necessary for the preparation of the Elixir, the Alkahest, and the Grand Arcanum, and to share the proceeds with Plot.'³⁰ Scientist though he was, he nevertheless exhibited, as Gough explains: 'the frequent appearances of want of judgement [which] must be ascribed in great measure to the credulous temper of the age he lived in.'³¹

Throughout this period Plot acted as a major link between the two scientific societies in which he served.³² Meanwhile, in 1684 he began to visit Staffordshire, perhaps at the invitation of Walter Chetwynd of Ingestre, with the view of preparing a natural history of that county.³³ Plot began by issuing his second set of queries in 1679, but because of the burdensome work-load of his employments and the difficulty of the required field-work was not in any case an easy chore, *Staffordshire* did not appear until 1686. This was about the same time that natural history began to flourish elsewhere; in 1683, for example, the *Philosophical Transactions* advertised a regional natural history of Switzerland, compiled by Jacob Wagner, which was also intended 'to promote a true Experimental Philosophy.'[34] Plot's tour of Staffordshire seems to have begun in May 1680, and the material was collected within about a year, 'about which time the book will be put to the press.' However, at one point in *Staffordshire* (219), in reference to some 'deterrations, or falls of the Earth,' Plot mentions the current year to be 1684, thus indicating a delay in bringing the work to print.[35] Like *Oxfordshire*, it contains an elaborate map of the county executed in Plot's hand. Although Plot was not by profession a map-maker, this particular production merits due attention because it established the model for future maps down to the latter part of the eighteenth century, going further than earlier map-makers in using conventional signs to distinguish parishes, villages, houses, etc. Also, for the first time the relation of the county to the degrees of latitude is indicated, with the fifty-third degree drawn across the map and the margin divided into minutes.[36]

In *Staffordshire* Plot uses the same method as in his first regional study, except that here he became involved 'in the determination of more difficult Questions.' His first chapter, 'Of the Heavens and Air,' is concerned with natural phenomena, especially with unusual displays such as rainbows, solar haloes, winter lightning, strange echoes, etc, usually with the view of rationally explaining such phenomena in non-supernatural terms. Like many of the earlier chorographers, Plot had an interest in prodigious accounts of unusual objects seen falling with rain – accounts transmitted by the ancients, who are individually cited by the author. But unlike them he was quick to point out, for example, that frogs seen falling from the sky 'may be either blowne from the tops of Mountains, or drawn up with the vapours ... and be brought to perfection in the Clouds, and discharged thence in Showers.'[37] Plot was able, in other words, to separate fact from fancy, in most instances at least.

The second chapter, 'Of the Waters,' embodied a systematic discourse on the origin of spring-water, while utilizing particular local instances as a basis for several innovative general arguments.[38] In so doing Plot asked his readers:

Whether the Springs are supplyed with that great Expence of water, that we see they dayly vent, from Rains, Mists, Dews, Snows, Haile etc. received into the Spungy tops of Mountains and sent forth again at the feet of them, or somewhere in their declivities; or whether they are furnish't from the Sea through subterraneious passages, as from the great Treasury of the waters, and are return'd again thither by the Rindles, Brooks, and Rivers? Or in short, whether they have their Origine from the Sea by a superior Circulation through the Clouds; or by an inferior, through Channels in the bowells of the Earth? or from both?[39]

Then Plot set out a detailed classification of springs, which included a discussion of 'periodical waters,' such as those of major rivers like the Niger, Ganges, and Rio de la Plata. At first he presented in a deliberate manner the (correct) theory that the ultimate source of spring-water is rainfall. But as he continued, dissecting the hypotheses of other writers on this subject, he unfortunately discarded his original theory, and stated that most springs depended on the sea for their supply, on the basis of an 'inferior circulation.' He was able nevertheless to support his revised viewpoint with several persuasive scientific arguments.[40]

A good portion of chapter 3 is taken up with soils in relation to agriculture and the use of clays and marls.[41] From a general description of agriculture Plot embarked on an investigation of the constitution of the particular soil types, noticing the effects of denudation and deposition: 'It is also likely, if not certain, that all valleys rise by atterration i.e. by Earth continually brought down from the tops of mountains by rains and Snows, whence all Mountains are become lower than they were formerly, and the Valleys risen higher; So that in time all the Mountains (except the rocky, such as the Rockes in the Moorelands) will by great shoots of rain be quite washed away, and the whole earth levelled.'[42]

The poor condition of the roads had not apparently changed significantly since the chorographers first complained about it. Thomas Habington, for example, had described Worcester ways as singularly bad due to the character of the local soils and the flooding of the Avon. Plot attributed the deteriorating state of the roads around Sedgley, Wednesbury, and Dudley to the carriage of heavy loads of coal.[43] The remainder of the chapter dealt with coal.[44] First, Plot provided a list of the items that require consideration:

Whereof there being great plenty of diverse kinds found here, I shall first give an account of the severall Species of them. 2. of their dipping, basseting or cropping, and their Rows or Streeks, 3. of the measures or floores there are of them, their partings or Lamings, with the terms of Art for them in different places, 4. of the damps that attend them, by what means they seem to be occasioned, and how cured, 5. how the coal-pits come so many of them to take fire, and 6. of their several ways of finding and working them.[45]

Plot's stratigraphical account of the coal measures transcended Owen's geological outline or even Aubrey's work, and here we find some of the first explicit statements of fundamental conceptions, together with the terminology, of structural geology.[46] He described the 'profundity' (thickness) of beds, their succession, and gave examples of detailed sequences with measured thicknesses at different locations, thus presenting one of the first – if not the first – tables containing the core material of stratigraphical data. At Wednesbury, for example, he established the following divisions, with their respective depths and different denominations, for the layer of upper coal:

1. The top or roof floor, 4 foot thick.
2. The overflipper floor, 2 foot.
3. The gayfloor, 2 foot.
4. The Lain-floor, 2 foot.
5. The Kit floor, 1 foot thick.
6. The benchfloor, 2 foot and 1/2.
7. The springfloor, 1 foot.
8. The Lower flipper Floor, 2 foot and 1/2. ... [47]

The final few pages of this section centre on the practical search for coal.

The chapter that follows, 'Of Stones,' takes up a subject long inherent in regional study, it seems, namely the use of lime for fertilizer. But it also contains a rather amusing anecdote concerning Plot's experiment on the variation of the compass needle. While out in the field and finding that his compass reading was wide of the mark by six degrees, Plot 'could not imagine how this should come to pass otherwise than by the Magnet, unless by some old Armour that might be buryed hereabout in the late civil War': in truth, the problem was most likely the result of local deposits of magnetite, an ore that is now known to exist in abundance in Staffordshire.[48]

But it is for the content of chapter 5, 'Of Formed Stones,' that Plot's book is best known. (By 'Formed Stones' Plot meant, in effect, mineral crystals and genuine fossils.) He began with objects supposedly having some connection with the heavens, for example, 'selenites' and 'asteriae,' working his way downwards through the 'inferior heaven' (those objects generated in the air among the clouds) and the waters, to the earth below. This led him in some instances to indiscriminately disseminate descriptions of objects that obviously call for treatment as a class; but this method conforms to his handling of other topics throughout the book.

One of Plot's most important contributions lies in the field of palaeontology, specifically in his exact descriptions and illustrations of fossils. *Oxfordshire* is notable for its excellent illustrations of fossils from the Jurassic and Cretaceous periods. Similarly, in *Staffordshire* Plot described and illustrated, for the first time, some of the more familiar shells (brachiopods) taken from the Carboniferous and Silurian limestones.[49] Despite this, some of his views were of dubious value, at least as far as his central proposition regarding the origin of fossils is concerned.[50] 'The great Question now so much controverted in the world' had already been established in *Oxfordshire*:

Whether the stones we find in the forms of Shell-fish, be Lapides sui generis, naturally produced by some extraordinary plastic virtue latent in the Earth or Quarries where they are found? Or whether they rather owe their form and figuration to the shells of the Fishes they represent, brought the places where they are now found by a Deluge, Earth-quake, or some other such means, and there being filled with mud, clay, and petrifying juices, have in tract of time been turned into stones, as we now find them, still retaining the same shape in the whole, with the same lineations, sutures, eminencies, cavities, orifices, points that they had whil'st they were shells?[51]

Plot rejected the idea that fossils 'owed their forms and figure to the shells of the Fishes they represent' and took the former view, leaning 'rather to the opinion of Mr. Lister, that they are Lapides ... ' disagreeing therefore with Hooke, Ray, and others who maintained the opposite.[52] For Plot, fossils represented naturally created objects produced by some extraordinary plastic virtue latent in the earth where they were found.

In *Staffordshire* he elaborated on the argument that formed stones were not the actual remains of once living organisms:

But as for stones found, like Sea-fish, though in this Mediterranean County, I have met with many, and of many sorts; but chiefly resembling Shell-fish of the testaceous kinds, both univalves and bivalves; and of the former of these, some not turbinated, and others again of the turbinated kind. Of the first sort whereof, viz. Stones representing univalves not turbinated, I had two bestowed on me by the curious Observer the Worshipful Walter Chetwynd of Ingestre Esq; so altogether unlike any of the living Shell-fish, that alone they are sufficient to convince any unprejudiced person, that all these formed stones cannot be shaped in Animal moulds.[53]

Plot gave no fewer than seven reasons for adhering to this position. First he rejected opposition theories of a past flood, either the deluge of Noah or else a more localized flood that supposedly transported the shells inland. Second, he was unable to discover the kinds of shell bones that, he assumed, would have been deposited by flooding. He was able to locate only some testaceous shells. Among his other arguments he noted that many of the formed stones appear to have been created on the spot where they were found. Robert Hooke, however, believed that formed stones included several specimens that were so similar to living shells that they could have been nothing else but the remains of animal shells. He also reasoned that nature would not have wasted her time in the useless creation of such formed stones.[54] Plot could not agree to either argument. He thought that, in fact, there existed many things in nature that resembled living organisms, pointing to the auriculare and cardite stones that looked like those parts of human bodies from which they derived their names. As for Hooke's other hypothesis, Plot countered that formed stones were – like flowers – created by nature to beautify the world and for their medicinal properties. Furthermore, he noted that many former shells were found far inland, deposited there by different types of actions; for example, some were thrown up on the seashores, others were remnants of shellfish eaten and discarded by town dwellers, and so on. All of these shells had been permeated by 'petrifying juices' and thus, in time, became petrified.[55]

After describing the flora and fauna of Staffordshire, Plot incorporated into his work a study 'Of Men and Women.' In keeping with his love of displaying the unusual he treated the 'accidents' that have befallen mankind: first, those occurring 'at or before his birth, then in his course of life, and lastly at his death.' The entire chapter is riddled with examples, many concerning people with whom Plot had personal contact. Other examples are taken from the works of Erdeswicke, Stow, Dugdale, Wood, and others. His favourites include monstrous births, long periods of somnambulism, strange distempers and diseases, etc.

The final chapter, 'Of Antiquities,' is further proof of the continued and evolving interest in antiquarianism itself within the context of regional study, even if it usually remained a secondary concern when compared to natural history. As Plot explained:

For Satisfaction of the Reader, upon what terms I add this Chapter of Antiquities to my *Natural History*, it seeming to some altogether forraigne to the purpose: I take leave to acquaint him, before I advance any further, that I intend not to meddle with the pedigrees or descents either of families or lands, knowing a much abler pen ... nor of the antiquities or foundations of Religious houses, or any other pious or civil performances: it being indeed my designe in this Chapter, to omit, as much as may be, both persons and actions, and chiefly apply my self to things; and amongst these too, only of such as are very remote from the present Age, whether found under ground, or whereof there yet remain any footsteps above it; such as ancient Medalls, Ways, Lows, Pavements, Urns, Monuments of Stone, Fortifications, etc. whether of the ancient

Britans, Romans, Saxons, Danes, or Normans. Which being all made and fashioned out of Natural things, may as well be brought under a Natural History as any thing of Art: so that this seems little else but a continuation of the former Chapter [ie, 'Of Arts']; the subject of that, being the Novel Arts exercised here in this present age; and of this, the ancient ones.'[56]

Plot set out his program of scientific antiquarianism in his letter to John Fell. Plot endeavoured 'to make a full Collection of British, Roman, Saxon, and ancient English Money,' and also of urns, lamps, 'Lachrymatories,' ancient inscriptions, ruinous buildings, hill fortifications, barrows, and Roman roads.[57] Hunter, however, is correct in drawing attention to Plot's tendency to rather uncritically 'interpret antiquities piecemeal by received ideas.'[58] Thus, in *Oxfordshire*, Plot had already referred to such monuments as the Rollright stones and that famous stone circle located outside of Wiltshire. But he credulously repeated, at the same time, various wild claims as to their origins. In *Staffordshire*, Plot described not only monuments but also portable artefacts. He described in detail serrated points and spears, discussing their origins and use. He insisted that these were all man-made, and he compared the stone tools of Britain with those from America.[59] His illustrations of a stone projectile and of a spearhead are perhaps the first published British drawings of local stone artefacts. In regard to Stonehenge, Plot concluded that it was most likely a British forum or temple, and not one commemorating any Roman pagan deity, since the Romans were at one time skilled in architecture and, if they had been the builders, 'would have made a much more artificial structure,'[60] He found arguments, too, to counter claims that Stonehenge was built by the Danes.

Staffordshire crowned Plot's reputation, and one hundred years later it could still be said that in the compiling of regional natural histories 'he has not been excelled by any subsequent writer.'[61] *Staffordshire* also proved to be the only book on the natural history of the county until 1844.[62] Once *Staffordshire* was completed, Plot relinquished the chair of chemistry, married, and retired to the life of a country gentleman on his Kentish property. Not surprisingly, he could not resist compiling a natural history of that county. (He also intended to do the same for London and Middlesex.) In mid-August 1693, therefore, Plot engaged in a fact-finding excursion through Kent in company with a man named Thomas Browne.[63] By early September he was able to write that 'I have now finish't all the upper part of Kent, having travell'd as near as I can guess about 200 miles, whereof I believe not much above fifty on horseback, notwithstanding the weather here has been so bad.'[64] He then directed his attention to London and Middlesex, so that by

'Learning So Much Neglected' 205

November 1694, he had 'now actually enter'd upon my great work.'[65] However, the plan was to come to nought; Plot fell ill with the 'stone' and died in the spring of 1696.

That Plot's scholarship was held in high esteem is evident in the fact that a new post, that of 'Mowbray Herald Extraordinary,' was created specifically for him about one year before his death, at which time he was also appointed register to the court of honour. His name was kept alive among fossils by one of the sea urchins, *Clypeus plotii*. Partly thanks to Plot's two major works natural history became the dominant element in regional study, thus supporting the claim that Plot was 'one of the Oxford pioneers in the development of regional geography.'[66]

Plot, like Aubrey and the other natural historians, occasionally had precedents for their own work in that of the earlier chorographers, in some ways at least. One example is the Welsh chorographer George Owen of Henllys (1552–1613), author of a study of Pembrokeshire.[67] Owen is a rarity in the unchorographic interest he displayed, for example, in geology. He traced in detail the distribution of the carboniferous limestone and coal measures of southern Wales,[68] and he is credited with making 'a definite beginning to the delineation of the geology of Britain,' his work being 'the first attempt to map, if only in words, a British geological formation.'[69] The pioneering aspects of Owen's work resulted, not surprisingly, in inaccuracies in his findings; for example, at one point he confused the different types of limestone. But he was aware that rocks containing coal followed a course that closely parallels that of limestone, being sometimes found in close association with a limestone vein. Subsequent mining has proved him correct.

Owen's object in describing the distribution of the rocks was to indicate where they were known to occur in order to guide parties to seek limestone 'whereas yett it lyeth hid, and save labour to others in seakinge it where there is no possibilitie to finde it.' Thus Owen was stimulated by the same motive that in large part justifies the existence of many of today's geological surveys. His account was the first to map, if only in words, a British geological formation, though John Aubrey and Martin Lister were the first to propose the making of a geological map, in the period 1683–91.[70] It apparently did not occur to Owen to indicate the distribution of limestone and coal veins upon a topographical map like the one he prepared to accompany his *Penbrokshire*. Therefore, the value of Owen's contribution towards the founding of geology as a science tends to be exaggerated.[71] He did not demonstrate that he had a clear idea of geological structure, even though he established that bands of outcrop are traceable across the country; this latter development was to come only at a later time. However, because of the valuable work he did in chorographically displaying his native land he deserves to be remembered, in the words of Camden, as 'a singular lover of venerable antiquity.'[72]

Edward Lhuyd (or Lhwyd) (1660–1709) is known for his participation in the late-seventeenth-century revision of Camden's *Britannia*, edited by Edmund Gibson. Originally brought into Gibson's project by the antiquary William Nicolson to provide information on Wales, Lhuyd employed a network of correspondents to aid him in his study. His own commitment is more richly documented than are the efforts of his correspondents. Frank Emery demonstrates how Lhuyd took advantage of the strong connection between Jesus College, Oxford, and Wales, utilizing people at both locations in his task.[73] Lhuyd believed that the revised *Britannia* should contain those 'occurrences in Nature, as seem more especially remarkable,' and therefore he included in his contribution items such as those sent to him by one of his correspondents, Nicholas Roberts, including observations on migratory sea-birds, on the blow-hole at Bosherston on the south coast of Pembrokeshire, and on the making of black butter from seaweed.[74] Lhuyd also received specimens of fossilized plants from his informers, such as the 'Mock plants' found at a coal pit in Glamorganshire.[75]

Lhuyd is considered, in fact, to be one of the founders of the science of palaeontology.[76] His original curiosity in fossils was aroused by his early discovery of several varieties near Oxford. Once other investigators had learned of Lhuyd's interest in these, they began to send him specimens from the Bristol and Somerset coalfields. Like Lister, Plot, Hooke, and some of the other virtuosi, Lhuyd soon found himself embroiled in the heated controversy concerning the origin of fossils. He was of the opinion that such 'formed stones' grew out of the earth from seeds dispersed by vapours from the sea, although he was never fully satisfied with this hypothesis.

Frank Emery and other modern scholars have assessed Lhuyd's achievements as a natural historian, antiquary, and comparative philologist. But until relatively recently, archaeologists saw Lhuyd as having existed in a sort of limbo between the first stage of archaeology – the pioneer, preparatory, and speculative stage – and the second stage – the descriptive, or formative one. As Plot's successor, however, as keeper of the Ashmolean – 'a poor place, seeing there is no pay' – he was celebrated among his fellow scholars for his great learning and prodigious labours.

Lhuyd experienced a meteoric rise to prominence at a relatively young age. Although the exact place of his birth in Wales is unclear, he prided himself on his Welsh roots. In 1682 he entered Jesus College but did not follow the orthodox kind of academic career shared by John Ray or by his other friends.

'Learning So Much Neglected' 207

Although he never did graduate, he was elevated to the underkeepership of the Ashmolean in 1684, becoming keeper upon the retirement of Robert Plot. Plot furthered Lhuyd's scholarly career by putting him in communication with the chief authorities on the subjects in which he was interested.

The relationship between Lhuyd and Plot, his tutor, is a perplexing one. Plot appears to have taken pride in his young protégé, almost to the point where there developed a certain camaraderie between them, one of a professional nature at least. At one point he wrote to Lhuyd, requesting: 'Pray let me hear from you sometimes how the Musaeum and Natural History thrives, as you shall from me, for tho' the London and Oxford Societies sleep, yet let *us* be awake.'[77] And yet this attitude was not as much a reciprocal one on the part of Lhuyd, even though without Plot's help and the use of Plot's geological collections, Lhuyd in all probability would not have enjoyed the opportunity of engaging in high-quality research into British palaeontology. Lhuyd said of Plot: 'To give him his due, he is both curious and ingenious, but the most vainglorious and conceited man I ever met with; a fault which perhaps is chiefly owing to his education. His Book is garnished here and there with Divine Reflections and written in a very tedious and somewhat incongrous stile, but I hope it may contain some observations which may be acceptable to ye curious and in [some] measure make amends for the stile.'[78]

Lhuyd may have been slightly jealous of the success of his mentor's publishing activity. But his attitude was in part also due to his own professional system of collection and classification, which was thoroughly methodical. Lhuyd therefore, was able to fault Plot's overall methodology, finding Plot's approach to scientific investigation haphazard in many instances. Less willing to theorize than to observe, Lhuyd was also critical of the hasty assumptions he had found in some of Plot's antiquarian arguments. In the marginalia of page 164 of the Ashmolean copy of Plot's *Oxfordshire*, for example, Lhuyd indicated his scepticism of Plot's conclusion that all the pots listed are of Roman origin, not British.[79] Likewise, he viewed with disfavour John Aubrey's methodology, or rather his lack of one. Interested primarily in establishing the facts, Lhuyd stated that he would believe an illiterate shepherd sooner than a bishop in matters of 'mountanous and desert places,' which no doubt were better known 'to those of his profession than men of learning.'[80]

This ever-increasing insistence on original observation through field-work is one more illustration of the growing scientific awareness of the men engaged in regional studies in the second half of the seventeenth century. Lhuyd and others generally scorned the work of earlier writers, 'who, till this last century contented themselves with bare reading and scribling paper.'[81]

Thus, although he exhibited a certain amount of admiration for his predecessors, calling the chorographer Erdeswicke, for example, 'a very eminent man, who has nicely enquir'd into the venerable matters of Antiquity,' Lhuyd was reluctant to accept their authority outright on nearly every matter.[82] He had noted that Giraldus wrote 'in an age less cautious and accurate.' He also questioned the investigations of George Owen; he said of a site at Nevern, the Rocking-stone, 'having never seen it myself,' he could not entirely satisfy himself as to 'whether it be a Monument or, as Mr. Owen seems to suppose, purely accidental.'[83] Hence he rigorously insisted on undertaking extended perambulations that were devoted to field-study. He puzzled the publishers of the *Britannia* by insisting on a scientific expedition or 'Camden Tour' through Wales in 1693. Even before the terms of the contract were satisfactorily resolved, Lhuyd left Oxford for Wales in mid-August, returning by mid-October, so that by September 1694 the results of his travel were in the editor's hands.[84] Lhuyd was hopeful that his contribution to the *Britannia* would induce the Welsh gentry to 'encourage something more considerable,' and in this regard he was not disappointed. In May 1665 he wrote to John Lloyd:

Some gentlemen in Glamorganshire have invited me to undertake a 'Natural History of Wales'; with an offer of an annual pension from their County of about ten pounds for the space of seaven years; to enable me to travail etc.: but I know not how the gentry of other countrey's stand affected. If the like encouragement would be allow'd from each county, I could very well willingly spend the remainder of my days in that employment: and begin to travail next spring. Nor should I onely regard the Natural History of the countrey, but also the antiquities ... I must confesse the sallary may at first sight seem too much and the time of seaven years too long; but such as are acquainted with Natural History know there's no good to be done in't without repeated observations; and that a countrey of so large extent cannot be well survey'd, and the natural productions of it duely examin'd, under the space of four or five summers; after which time remaining will be short enough for methodizing the observations and publishing the History.[85]

Thus, in the summer of 1695, Lhuyd put forth a two-part proposal for 'A Design of a British Dictionary, Historical and Geographical; With an Essay entitl'd, "Archaeologia Britannica"; And a Natural History of Wales.' Presumably the plan had support at Jesus College, for the principal there, Jonathan Edwards, had suggested to Lhuyd that the work he had already accomplished towards the *Britannia* might serve as the nucleus of a greater study, that is, as a natural history of Wales after the style of Plot's studies. The idea appealed to Lhuyd, especially when couched in terms that he could, and most likely did, interpret as holding out hope of upstaging Plot in his own backyard, so to speak. He was aware, however, that the proposed task would be a difficult one and therefore decided to accept the following advice from Nicholas Roberts: 'It will be an insuperable Task to give so exact an account of each County in Wales, as Dr Plot has of Oxfordshire, unless you have many Correspondents, one for each 2 or 3 Counties at least,'[86] In all, his program of research was to last from 1697 to 1701, and to take Lhuyd not only through Wales but also to Ireland, Scotland and Brittany.

In his investigations Lhuyd and his helpers, travelling on foot with knapsacks, made extracts from manuscripts and collected diverse rarities in addition to examining the natural phenomena of the regions they visited. But Lhuyd supplemented this research by issuing in 1696 a number of *Parochial Queries* to be sent to each parish in Wales, Shropshire, and Herefordshire, often employing undergraduates as carriers to deliver these to the local gentry or clergy. The use of such queries had already been an accepted practice for at least thirty years. Lhuyd therefore had several models available on which to base his own and, as Emery points out: 'it is suggestive to find the placing together in one volume of his miscellaneous papers of the queries printed by Molyneux, Plot (both the 1674 and 1679 versions), and by Thomas Machell (1677).'[87] Machell's queries, which were the first to specify the study of antiquities, history, and geography at the parish level, and which were more elaborate than their predecessors, were of particular use to Lhuyd.

Lhuyd's own queries, however, were oriented more towards a wider range of natural history.[88] They were divided into two sections, relating to 'The Geography and Antiquities of the Country' (sixteen questions) and 'Queries towards the Natural History' (thirty-one questions). Following each query was a blank space for the reply. Lhuyd explained that their purpose was not to 'spare myself the least Labour of travelling the Country, but on the contrary be assured, I shall either come myself, or send one of my assistants into each Parish.'[89] He also admonished the recipients to confine themselves 'to that Parish only where they inhabit, and distinguishing always betwixt Matter of Fact, Conjecture, and Tradition.'[90] The first section of the queries contained a general question concerning the name of the parish and its derivation, and also more specific questions of a somewhat chorographic nature, regarding the seats of the gentry, old arms, inscriptions, customs, peculiar games and feasts, and so on. The second section, by contrast, delved deeply into topics relating to natural history. Query 26 even offered an inducement: 'If such places [eg, caves, mines, coalworks, quarries] afford any uncommon Oars, Earths, or other Minerals; Stones resembling Sea-Shells, Teeth, or other Bones of Fish;

or Crab-Claws, Corals, and Leaves of Plants; or in brief any Stones; or other Bodies whatever of a remarkable Figure; the Workmen are desired to preserve them, till they are called for by the Undertaker, or some of his friends; *in Consideration whereof, they shall receive some Reward suitable to their Care and Pains.*'[91] (Italics mine) Four thousand queries were printed for dispersion throughout Wales, Ireland, and Scotland, and were sent out previous to Lhuyd's journeys there. Many of the responses to these were published in the early twentieth century.[92]

Lhuyd devoted the first two years of his tour to combing Wales itself. He moved on to northern Ireland, then Scotland, where he spent the winter of 1699; after next visiting southern Ireland, he spent four more months in Wales and four in Cornwall. A short stint in Brittany, on the way back to Oxford, rounded out the fact-finding mission. Lhuyd kept no diary of the trip so that Emery has had to piece together the dates and details of his progress from letters and from the addresses and dates upon them.[93] While absent from the Ashmolean, Lhuyd was kept informed of museum news by letters from the librarian in charge, William Williams.

Unfortunately, the tour coincided with some of the severest climatic conditions of what is today known as the Little Ice Age. This posed a major impediment to his work. In Wales, for example, Lhuyd encountered freak hailstorms in May and June 1697, and there are replies to his queries that describe hail and snow at Easter 1698.[94] As if such experiences were not discouraging enough, Lhuyd and his expeditionary party encountered other difficulties, more typical of those experienced by other natural historians in the field. Everywhere he travelled Lhuyd aroused suspicion. In southern Wales he was mistaken for a Jacobite spy, while in Brittany he was arrested as an English spy.[95] In Ireland, he confessed, 'the tories [bandits] of Kil-Arni in Kerry obliged us to quit those mountains much sooner than we intended.'[96] The reception was no less inhospitable in Cornwall: 'Mr. Lhuyd [so the account goes] came into the country at a time when all the people were under a sort of panick, and in terrible apprehension of thieves and house-breakers; and travelling with his three companions ... prying into every hole and corner, [they] raised a strange jealousy in people already so much alarmed ... At Helston, as Mr. Lhuyd was poring up and down, and making many enquiries about Gentlemen's seats, etc., he (with his companions) was taken up for a thief, and carried before a Justice of the Peace.'[97]

It is regrettable that only one volume of Lhuyd's proposed design was to reach print, the 'Glossographical' section of the *Archaeologia Britannica: an Account of the Languages, Histories, and Customs of Great Britain*, and then not until 1707, four years after the manuscript was completed. In ten parts,

'Learning So Much Neglected' 211

this work dealt with etymology, philology, and grammar. It was unenthusiastically greeted by its subscribers and the Welsh gentry, who undoubtedly would have had a difficult time in understanding its Welsh language component.[98] The more scholarly types, the linguists, and the English and Celtic academics, however, acclaimed its appearance. This was enough to facilitate Lhuyd's entry into the Royal Society, where his work received the recognition it deserved. Lhuyd resolved to publish the 'Glossography' before the dictionary and the natural history because he believed, ironically, that it would meet with more buyers, and because he had 'a tolerable apparatus for it.' The intention was to issue the volume on natural history in five parts, dealing with the following topics: (1) 'A general description, soil, meteors, a comparative table of weather in general places, seas, rivers, lakes, etc'; (2) 'All the various sorts of earths, stones, minerals'; (3) 'Formed stones' (fossils); (4) 'Plants – those growing spontaneously'; (5) 'Animals.' However, this volume was never published. Emery claims that it had the potential, if it had ever reached the printer, of constituting 'a study of the first importance for the geography of Wales.'[99]

One cannot but be amazed at the many-sided genius of Lhuyd's scholarship. The 'Glossography' represents the first effort at a comparative study of the Celtic languages – something that was not attempted again until the mid-nineteenth century. Lhuyd had made basic discoveries in the study of linguistics, for example, phonetic analogies among the Celtic languages, and he made the earliest known collection of Manx vocabulary. But even before Celtic philology had found its place within his field of interest, Lhuyd had already occupied himself with the scientific study of botany and palaeontology. It appears that he was a born botanist, and had inherited his love of the plant kingdom from his father. As early as 1682 we find him climbing the peaks of Snowdonia, securing specimens. (His name is commemorated in his most notable discovery, *Lloydia serotina*, Mountain Spiderwort.) His interest in fossils has already been noted. One of his first duties in the Ashmolean under Plot involved the cataloguing of the fossil collection, which resulted in his presenting to the Philosophical Society in 1686 a catalogue of the shells. His major interest lay in 'formed stones'; he realized the need for a handbook of fossils, and he himself produced a systematized catalogue of a total of 1,766 fossils of Britain, entitled – in reference to fossil plants, which Lhuyd called 'Lythophyta,' or mineral leaves – the *Lithophylacii Britannici Ichnographia* (1699). This catalogue won acclaim for him as the foremost natural historian in Europe.[100]

Lhuyd's approach to scientific antiquarianism is in some respects comparable to that of Plot, if not to that of Aubrey. Both men observed the similarity between Indian stone tools brought from America and those discovered in Britain. In discussing the amulets of the Scottish Highlanders, which he believed to be relics of the Druids, Lhuyd observed: 'I doubt not but you have often seen of those Arrow-heads they ascribe to elfs or fairies: they are just the same chip'd flints the natives of New England head their arrows with at this day: and there are also several stone hatchets found in this kingdom, not unlike those of the Americans.'[101] Both Plot and Lhuyd stressed the relationship between European thunder stones and stone implements on the one hand, and weapons from other parts of the world on the other, demonstrating that these were used for hunting and wood trimming, not merely for purposes of defence. Lhuyd's comparative knowledge of archaeology exceeded that of Aubrey.[102] Overall, be it in the area of natural history or scientific antiquarianism, Lhuyd was advanced in his methodology, thus justifying the praise heaped on him by his contemporaries, just as his fellow countryman, George Owen, was counted among the leading chorographers of his day. Like Aubrey and Plot, therefore, Lhuyd was instrumental in promoting and publicizing regional natural historical study.

CHAPTER TWELVE

'Industrious Searchers into the History of Nature'

Men have Travelled far enough in the search of Foreign Plants and Animals, and yet continue Strangers to those produced in their own natural Climate.
  Martin Martin *A Late Voyage to St Kilda*

THIS CHAPTER examines the regional natural histories produced near the turn from the seventeenth to the eighteenth centuries, focusing first on the work of the Scottish participants in the late seventeenth-century *Britannia* project, led by the famous Edmund Gibson (1669–1748), which aimed at revising William Camden's monumental sixteenth-century historical-topographical tome of the same name. Gibson recruited many scholars from various regions of Britain to help in this endeavour. Collectively they attempted to correct the mistakes found in the original work, to supplement it with publications that had appeared since Camden's death, and to produce a satisfactory translation of the original, which was in Latin. The focus here is on the Scottish contributors: Sir Robert Sibbald and other lesser-known figures, whose work is given insufficient recognition by modern scholars, although their English and Welsh colleagues readily admitted its value both to the project and to natural history as a whole.

The 1695 *Britannia* was kept to one volume, despite the desire of the country gentry that their pedigrees and details of their estates be included. This allowed the inclusion of more natural history than otherwise might have been possible. Thus the work of the Scottish natural historian Robert Sibbald proved of immense use to the revised study, even though Sibbald at the time was not quite as well known to contemporary scholars outside of Scotland as he was to his fellow countrymen.

An exception to this statement was Edward Lhuyd, a close personal friend of Sibbald and, as noted in chapter 11, a participator in the *Britannia*. Sibbald was host to Lhuyd while the Welshman was travelling through Scotland in 1699, and proudly showed him the collection of coins and artefacts located in the museum, which had just been purchased by the 'College of Edenborrough.' Sibbald, for his part, benefited from this friendship by means of letters from Lhuyd, which contained the latest information on the subject of natural history.[1] The Lhuyd-Sibbald connection, in fact, is only one example of the larger link between the Scottish regional writers and those to the south. The Scottish scholars were also linked, by tradition, to those in the Low Countries.

Although Emery has taken brief note of the work of Sibbald, Scottish chorography and natural history have by and large received insufficient attention.[2] Despite Scotland's scholarly connections with England and the Low Countries, regional study there, especially in its early stages, to a considerable degree developed on its own. Sibbald was aware of this insular tradition and devoted a work to this subject.[3] In it he noted that the earliest accounts, written by foreigners, constituted 'only some short touches of it [Scotland], and these [were] obscure.'[4] The studies executed by early domestic observers, however, gradually increased in number until a climax in chorographic writing was reached with Timothy Pont's cartographical and chorographic efforts at the very end of the sixteenth century.[5] Pont and his fellow countrymen had to endure the same types of difficulties in their field-work that faced Lhuyd a hundred years later. The chorographer Robert Gordon, following in Pont's footsteps, relates how 'with small means and no favouring patron, he [Pont] ... travelled on foot right through the whole of this kingdom, as no one before him had done: he visited all the islands, occupied for the most part by inhabitants hostile ... and with a language different from our own; being often stripped, as he told me, by the fierce robbers, and suffering not seldom all the hardships of the dangerous journey, nevertheless at no time was he overcome by the difficulties, nor was he disheartened.'[6] Emery points out that Gordon's variety of chorography 'was more realistic and geographical than most landscape interpretation' that was being done in England at about the same time, and that it set the pattern for his son, James Gordon.[7] Thus, in his depiction of Fifeshire (1654), Gordon preferred not to list the numerous seats of the gentry of the shire, because to do so 'belongs ... to history.'[8] This is of importance because most seventeenth-century regional study in Scotland centred on the work of the two Gordons, at least until Sibbald enters the picture.

The link between the Gordons and Robert Plot was developed when, in 1686, Charles Gordon, the grandson of Robert, travelled to Oxford to attend some of the meetings of the Philosophical Society, ostensibly to 'improve himself in Natural Philosophy' and thereby take some of this new-found knowledge back to Scotland, where the 'industrious searchers into the history of Nature' were attempting to organize their efforts.[9] At about the same time James Gordon in Scotland had effected the merger of the earlier chorography with current natural history by donating all his material to Sibbald.[10]

Robert Sibbald (1641–1722) personifies both the Scotland-Holland and the Scotland-Oxford linkage. Born in Edinburgh, he began his education there, studying first theology and later medicine. In 1660, after academic stints in Paris and Angers, he studied medicine at Leyden for one and a half years and took the degree of MD. It seems only natural that Sibbald studied at Leyden, for in medicine, history, and law, if not also in theology, the Dutch university excelled those in Sibbald's homeland. By 1700, in fact, Leyden could boast of having 'about 20 Scotsmen al very studious so that ther was never a parcell of better students at Lyden.' While there, Sibbald studied under Christian Marcgraf, the brother of George Marcgraf, author of a renowned natural history, the *Historia Naturalis Brasiliae* (1648).[11] No doubt Sibbald had the opportunity of thoroughly familiarizing himself with this production, or at least, on a general level, with the budding new field of natural history; this opportunity perhaps first inculcated in him an interest in the description of natural phenomena.

Upon settling in Edinburgh to practice medicine in association with Sir Andrew Balfour, Sibbald established the botanical garden there. This was in keeping with his desire to investigate what 'materia medica' in the way of herbs Scotland was capable of producing: 'I came by conversation with him [Balfour] to know the best writers on natural history. I had ... a designe to informe myself of the naturall history this country could affoord.'[12] Sibbald was also chiefly instrumental in the founding of the Royal College of Physicians of Edinburgh (1681), and in 1682 he was appointed physician to Charles II as well as geographer of Scotland. This latter appointment he obtained through the generous assistance of the earl of Perth, whom he had met in 1678. Perth encouraged Sibbald to enlarge upon his study of the 'naturall products' of Scotland. In order to do so, in 1682 Sibbald decided to 'publish in our language one advertisement, and some generall queries, copies wherof were sent all over the kingdome; and from severall shyres and Isles, especially by the care of the reverend Mr. Murdo Mackenzie, Bishop of Orkney and Shetland, full informationes were sent to me by several learned men.'[13] The queries were typical of the day, covering general physical geography, flora and fauna, minerals, and also ancient inscriptions, customs, and so on.

Sibbald's *Nuncius Scoto-Britannus* (1683) explains the order of the planned work and laid out his sources, which included Cluverius and Varenius and the classical geographers. The *Nuncius Scoto-Britannus* was prefixed to his first volume, the *Scotia Illustrata, sive Prodromus Historiae Naturalis*, which may be translated as 'Scotland Illustrated, or an Introduction to its Natural History.' It was published in 1684. In the *Nuncius Scoto-Britannus*, Sibbald explained he was undertaking the project to update the account of Scotland on the express command of the king and also because of certain alterations in its physical appearance that had occurred since the ancient classical Greek and Roman writers had written of it. He also sought to improve upon the imperfect or 'partial' descriptions of the more recent regional writers. The *Scotia Illustrata* itself consisted of no fewer than twelve hundred copies. No other volumes were published because Sibbald failed to procure the required financing, although he had the material ready for the printer.[14] Like the contemporary works issues in England and elsewhere, the first volume contained a discourse on the quality of the air, and on the flora and fauna, waters, mountain ranges, forests, mines, arable lands, and other physical features. The inhabitants were described as being qualified for both arms and the arts by virtue of the roughness of their soil and the purity of the air. Interestingly enough, Sibbald also discussed diseases in general, relating information about extraordinary cases, for example those involving hysterical fits and dropsy. He also included a description of the types of medicine produced in Scotland plus their respective virtues. In an attempt to show how little need there was for exotic drugs, Sibbald cites a comparison of domestic and foreign medicines. Also included was a large section on the history of the fossils of Sibbald's native land. Although he held few original ideas about them and patterned his views on those of Martin Lister – who believed they were *lapides sui generis* – Sibbald did give a great number of accurate descriptions of fossils.[15]

The methodology exhibited in his work was not quite as scientifically oriented as that of Plot. For example, Sibbald relied to a much greater degree on literary sources for his information. Also, many of his distant correspondents throughout Scotland proved to be too credulous or ignorant to be able to provide him with the kind of accurate information that he required. For this, in fact, he has been resoundingly criticized by some of his more vocal detractors.

Sibbald's other works also gave evidence of both traditional and new modes of regional study. He is probably best known for *The History, Ancient and Modern, of the Sheriffdoms of Fife and Kinross* (1710), which was dedicated to the earl of Rothes, sheriff-principal of Fife, and to Sir William Bruce of Kinross. Sibbald had acquired an intimate knowledge of this region, as he was of a Fifeshire family. In his study he incorporated chapters on etymology, on the perceived ancient prospect of the countryside, and on its ancient inhabitants the Picts and the Caledonians, about whose languages, religions, and customs he speculated. He also recounted the Roman exploits in Scotland. Many of these topics are more common to chorographic literature, but they are found alongside the ones on natural history, flora and fauna, minerals, soils, beach erosion, and so on. Scientific antiquarianism also has its place in the work. Sibbald drew attention to the recumbent stone circles of northeastern Scotland.[16] The record of prehistoric copper artefacts also begins with his report of a Late Bronze Age hoard from Fife.[17]

Sibbald's interest in antiquarianism is evident in some of his other studies. In connection with his work on the Roman settlement he had this to say about the new science of archaeology:

Amongst the Sciences and Arts much improved in our time, the Archeologie, that is the Explication and Discovery of Ancient Monuments, is one of the greatest use: For the Ancients by Triumphal Arches, Temples, Altars, Pyramids, Obelisks, and Inscriptions upon them, and Medals, handed down to Posterity, the History, Religion and Policy of their Times, and an Account of the Sciences and Arts which then flourished. Certainly in these times, of which Records are not found, the only sure way to write History, is from the Proofs [which] may be collected from such Monuments. And accordingly the best Historians in the Age, lately elapsed, have followed that way in writing of such Ancient Times. In imitation of them I have written this Essay of Historical Inquiries concerning the Monuments [that] were left by the Romans in this North part of the Isle.[18]

Sibbald's interest in the Romans extended to *The History, Ancient and Modern, of the Sheriffdoms of Linlithgow and Stirling* (1710), a work similar to his *Fife and Kinross*. Utilizing the papers of Pont and others, Sibbald not only described the Roman walls, forts, garrisons, and monuments, but also attempted to reconstruct life as he envisioned it in those days. He even extended this retrospective reconstruction to natural history; for example, 'the cutting down of the Woods, and the rotting of the Timber, occasioned the great Mosses yet to be seen in several parts of the Shire [Stirling].'[19] With the reservations mentioned above, Sibbald's scholarship generally evoked a positive response from scientists across Britain and the rest of Europe. Gibson, in his comments regarding the 'Scotland' section of the 1695 *Britannia*, spoke for them all when he referred to the 'very learned Sir Robert Sibbald, Dr. of Physick,' as one 'who has given sufficient testimonies to the world, of his knowledge of Antiquities, and particularly those of his own Country.'²⁰ But the value of Sibbald's work was enhanced by the information he received from several of his more reliable informants.

For this reason a little more may be said of some of his distant collaborators. One of these, Andrew Symson (1638–1712) proved to be one of the more trustworthy correspondents. Curate of the Scottish Episcopal Church and minister of the parish of Kirkenner in Wigtownshire and for a short time a printer in Edinburgh, he compiled a 'Description of Galloway' in 1684 while he was still at Kirkenner.²¹ When he came upon Sibbald's *Nuncius-Scoto* and Sibbald's queries, he decided to 'comply something with my genius' and thus to contribute his efforts towards Sibbald's work. In his study of 'Galloway' he depicted the three natural units still seen in the present-day land-use pattern of Wigtown: the Moor area noted only for its fine wool, the coastal tract marked by sandy soil, and the Machairs, which were composed of 'white ground' with arable fields on thin gravel and coastal clay.²²

But Symson's interest was neither in pure natural history – 'I pretend no great skill in Ichthnology' – nor in pure antiquities. Using information gathered on the basis of his own observations or what he could obtain from others, he included in 'Galloway' a mishmash of miscellaneous topographical detail. Not surprisingly, he seems to have taken a special interest in accurately establishing the names of the patrons of each place and the extent, size, and boundaries of each and every parish. In a letter to Lhuyd, who was then at Oxford, he states a plan to investigate the parishes with regard to their names: 'But if it be only the names of places that you desire, I hope God willing to give you a large account thereof in a Book which I intend to publish under the Title of 'Villare Scoticum,' wherein I intend to give an account of all the parishes in Scotland, as spelled of old and as now, together with all the Severall titles and places of our Nobility Knight Baronets, etc. which shall furnish you abundantly with Pictish names.'²³ Some sound historical evidence on the point of ecclesiastical patronage might well have supported the claims of the Episcopal Church as against the Kirk in local ecclesiastical differences.

Sibbald received information from many other parts of Scotland, much of which is printed in Mitchell and Clark's *Geographical Collections*. But special mention should be made of Martin Martin and his regional studies, which were produced not only in direct response to Sibbald's queries, but also to those of Lhuyd. Martin (d 1719) is perhaps the most famous and the most important of all of Sibbald's collaborators. Born at Bealach and member of a prominent family of Skye, in 1681 he graduated MA from Edinburgh University. It is also known that in 1710, in his later years, he entered Leyden University, that hotbed of dissenting English and Scottish students, and there graduated MD. At some period Martin lived in the Western Isles of Scotland, holding the post of governor to the laird of Macdonald. After his stay at Leyden, however, he apparently resided in London until his death.

It was primarily out of curiosity that Martin undertook the survey of the Western Isles, in a manner 'more exactly than any other,' and so he enlisted as one of Sibbald's observers. On hearing of Martin's enthusiastic stance, Lhuyd also sent him queries, asking him to compile an account of the various dialects of the Irish language. Martin, like the two men whom he served, proved diligent in the cause of natural philosophy, informing his readership that his goal was 'to oblige the republic of learning with any thing that is useful.'[24] He was critical of the earlier chorographies that described the Western Isles, commenting on the 'great change in the humour of the world, and by consequence in the way of writing' that had taken place since the great age of chorographic writing. He arrived at the conclusion that the improvements in natural and experimental philosophy meant that 'descriptions of countries, without the natural history of them, are now justly reckoned to be defective.'[25]

One of Martin's studies was of St Kilda, to which he had sailed in an open boat in 1697. It was published in the following year as *A Late Voyage to St Kilda, The Remotest of all the Hebrides, or Western Isles of Scotland*. Martin dedicated the work to Charles Montague, then president of the Royal Society. He stated that he wrote for the benefit of 'the intelligent reader.' In the preface to this work we find evidence of the growing tendency among regional writers to resist the popular demands for accounts of little-known, distant lands, and to encourage instead knowledge of the regions of Britain itself. This tendency had already manifested itself in the work of Philip Falle, who discovered that as far as Englishmen were concerned, 'we were as great strangers as if these [Channel] Islands had been some degrees beyond the line'; as a result Falle tried to alleviate the situation through the publication of his study. Martin expressed a similar feeling:

Men are generally delighted with Novelty, and what is represented under that plausible invitation seldom fails of meeting with acceptance. If we hear at any time a Description of some remote Corner in the Indies Cried in our Streets, we presently conclude we may have some Divertisement in Reading of it; when in the mean time, there are a thousand things nearer us that may engage our thoughts to better purposes, and the knowledge of which may serve more to promote our interest; and the History of Nature. It is a piece of weakness and folly merely to value things because of their distance from the place where we are Born.[26]

A Late Voyage contains the now standard items of natural history, including descriptions of the inhabitants – a flourishing population of 180 – climate, springs, natural products of the earth, flora and fauna, an earthquake, 'Chrystal, how it grows,' etc. The favourable reception that this work received encouraged Martin to continue to travel about the rest of the isles, observing the natural phenomena, and to publish the results of his findings in his major production, *A Description of the Western Islands of Scotland* (1703).

In tune with the scientific spirit of the age Martin displayed an intense interest in topics such as those relating to archaeology. He determined, for example, that a particular stone circle that he had encountered was once used as a place of worship in heathen times. His interest in geomorphology led him to claim that in the Hebrides the islands of Tiree and Coll had at one time been severed from each other by the tempestuous Atlantic. He likewise claimed that some of the still more exposed Outer Isles had been separated by the same agent.[27] Martin was aware of the implications of such studies towards building a natural philosophy that had its foundations in the laws of the universe.[28] Furthermore, he realized the value of his observations to the field of religion: 'The land and the sea that encompasses it, produce many things useful and curious in their kind, several of which have not hitherto been mentioned by the learned. This may afford the theorist subject of contemplation, since every plant, of the field, every fibre of each plant, and the least particle of the smallest insect, carries with it the impress of its maker; and if rightly considered may read us lectures of divinity and morals.'[29] Martin's verdict was that man's salvation lay in a sort of primitive simplicity in his way of life. Thus, the inhabitants of the Western Isles appeared to him to be totally content, living a blissful, harmless existence, perfectly ignorant of most of those vices that afflict mankind elsewhere. He found these people to be governed by the dictates of both Christianity and reason. A similar theme is apparent in Martin's *Description of Skye*, written for Sibbald.[30] Although much of this work concentrates on natural history, the author takes the time to point out that most of the inhabitants of that island honour their ministers to a high degree, 'to whose care under God they owe their freedom from Idolatrie and many superstitious Customes.'[31]

Martin's work has been described as coming closest to the standards established by the Royal Society.[32] One of Martin's contemporaries, however, obviously did not express the same sentiment. John Toland criticized Martin's work on a multitude of counts. His critical comments are written in his hand in the marginalia to a copy of Martin's *Western Islands* (dated Putney, Sept 1720).[33] Here Toland summarized his view: 'The Subject of this book deserv'd a much better pen ... [These] Islands afford a great number of materials for exercising the talents of the ablest antiquaries, mathematicians, natural philosophers, and other men of Letters. But the author wontes almost every quality requisite in a Historian (especially in a Topographer) ... '³⁴

Notwithstanding Toland's remarks, the quality of the work of the Scots was generally quite high. A very brief look at the endeavours of one other Scottish regional writer, James Wallace, confirms this. Wallace (d 1688) was one of a number of Puritan natural historians. He graduated MA at Aberdeen in 1659 and became the Episcopalian minister of Kirkwall. In addition to several tracts on matters of religion and church discipline, Wallace, at the request of Sibbald, wrote *A Description of the Isles of Orkney*, based on personal residence there. His work was incorporated almost verbatim into the 'Additions to the Orcades' section of the 1695 *Britannia*, in which Wallace is described as 'a person very well vers'd in Antiquities.'³⁵ In some respects his interests are closest to those of Gerard Boate, who wrote half a century earlier, for example in Wallace's preoccupation with the coastline, the ebb and flow of the sea, sea storms, and other related oceanic features. This preoccupation with the Orcadian littoral might also have been based on a practical interest; by the end of the seventeenth century Hudson's Bay Company vessels were regularly putting in to recruit Orkney men for the factories along the bay; and therefore, a sound knowledge of these treacherous waters was of the utmost importance. Wallace and Boate also exhibited a similar interest in the furies of nature, as in Wallace's account of two whirlpools in the sea, 'occasioned, as it is thought, through some hiatus that is in the earth below.' They affected vessels with such violence, 'that if any boat or ship come within their reach, they will whirl it about till it be swallowed up and drowned.'³⁶ Beyond these similarities to Boate's work, *Orkney* bears considerable resemblance to the other natural histories of the late seventeenth century, exhibiting an interest in features such as climate and soil. Wallace's son, James, edited and reissued his father's study in 1700, pretending to be the original author.³⁷

The Scottish regional writers, it is clear, not only maintained a high quality in their work but also kept in close touch with workers in the field who were investigating other parts of the British Isles. The *Britannia* project made this kind of contact imperative. Thus in 1699 we find William Nicolson of Carlisle working on the northern counties of England, and visiting Sibbald at Edinburgh. Similarly, Sibbald requested Edward Lhuyd to take time to 'acquaint me what is publisht relating to Natural History.'³⁸

These men display a markedly increased awareness of process and change, a more vibrant feeling for the past. New techniques of investigation in the natural sciences were not only adopted but quite often initiated by the

Scottish activists in the *Britannia* project, making possible new geological and archaeological approaches to the past. Utilizing these new methodologies, they studied antiquities for their intrinsic value, not just as elements in the cultural landscape. Generally, regional study was no longer thought to be the realm of the rank amateur, whose major claim to such study lay in his legal training or in his general familiarity with the area he was investigating, as it had been in the first half of the seventeenth century. Much of the credit for this attitude must go to Sibbald and his Scottish co-workers.

But it should be noted that not every book met these standards. *The Natural History of Lancashire, Cheshire, and the Peak in Derbyshire*, published by Dr Charles Leigh in 1700, must have proved disappointing to its subscribers, even though it styled itself – as the title indicates – a natural history. At least, it was criticized later by Richard Rawlinson as being too selective, too vague, and as having no sense of locality.[39] In many respects this book is little more than a translation of earlier Latin treatises; some of it, however, is a summary of a much longer regional study by Dr R. Keurdan, to whose five manuscript volumes on Lancashire Leigh had access. In the manner of a learned dilettante, Leigh set forth an imposing welter of concepts, which he discussed in a pedantic rather than a critical manner.

Charles Leigh (1662–1701?) graduated BA in 1683 from Brasenose College, Oxford. Wood records that he left Oxford in debt and went to Jesus College in Cambridge. In May 1685 before he graduated MA and MD (1689) at Cambridge, he was elected a member of the Royal Society, where some of his papers were read. His connections with the universities and the Royal Society allowed him the opportunity to strike up acquaintance with some of the leading scholars of the day, including Plot, whose scientific lectures gained Leigh's attention, as his correspondence bears out.[40]

Lancashire is divided into three books: the first relating to natural philosophy, the second about 'chiefly Physick', and the third focusing on the British, Phoenician, Armenian, Greek, and Roman antiquities of the counties he covers. The first book, therefore, contains the type of material one normally associates with natural history: information on flora and fauna, river systems and springs, minerals, and so on. Leigh's observations on geology are particularly erudite, noting such geological structures as the basin type common in coalfields, rock-salt in Cheshire, and lead ore deposits found in Derbyshire.[41] He also included two plates of fossils, with descriptions. Leigh believed in a non-organic source of fossils, borrowing all his ideas on this topic from the writings of others. Endorsing the universality of the deluge of Noah, he cited as support for the occurrence of the deluge such evidence as the mounds of oyster shells in Virginia, and the skeleton of a buck discovered in England. His theory, though, rested primarily on biblical references.⁴² Plants and formed stones were the two exceptions to his theory that organic remains were depositions of the deluge – plants could be made by chemical processes, while formed stones were the result of the principle of 'ovism' – that is, the development and growth to maturity of animal eggs in stone cavities in the earth.⁴³

Leigh's second book largely follows Plot's example, detailing incidents in the lives of eminent persons, the various arts, professions, trades, 'accidents,' etc. The most interesting material, however, lies in the third book. First, referring to Camden at some points, Leigh describes the antiquities. Then, obviously influenced by the Phoenician theories of Aylett Sammes, he tries to prove that all of Britain was at one time colonized by Asian settlers, who arrived long before either the Greeks or Romans. The basis for this claim was Leigh's comparison and examination of the Armenian, British, and Phoenician languages, the 'Asiatick' manner of fighting, the Eastern and British way of computing time, 'the Reverse of a Coin and divers other Things.'⁴⁴ He develops his argument along elaborate lines, for example:

I shall now ... endeavour from another Relick of Antiquity to make it evident, that the Britains traded with the Eastern Nations, to wit, Phoenicia, and that is from a Torques lately found in Staffordshire, belonging, as may be supposed, to the British Queen Boadicia: Of which take the following Account:

In the County of Staffordshire, in the latter end of April, 1700, a poor Man in the Parish of Patingham, found a large and ponderous Torques of fine Gold ...

It seems to be a British Piece of Antiquity for the following Reasons:

First, Its being found near an old British City, as Dr. Plot in his *Natural History of Staffordshire*, makes that Place to be.

Secondly, Dion Cassius assures us, that Bonduca or Boadicia wore a Golden Torques.

The Romans, as well as the Britains, wore these Torques's, and probably both of them had it from the Asiaticks.⁴⁵

Nevertheless, Leigh's conjectures on the Phoenician origin met with ridicule, as did his book as a whole, even though as prominent a scholar as Edward Lhuyd had expressed some interest in *Lancashire*.⁴⁶

In fact, one of the best of all of the county natural histories was John Morton's *The Natural History of Northamptonshire*, published in 1712. Morton (1671–1726) was born at Whitton in Lincolnshire, where his curiosity about natural phenomena was stimulated by the abundant fossils found locally. He matriculated at Cambridge in 1688, graduated BA from Emmanuel

College in 1691, and received his MA in 1695, by which date he had already become curate of Great Oxendon in Northamptonshire. There he searched the nearby medieval quarries for fossils, having been further inspired to do so by reading Lister's works on the subject. By the time he was inducted into the Royal Society in 1703 he had forged friendships with Lister, Sloane, Lhuyd, Nicolson, and many of the leading virtuosi of the day.

Palaeontology was the cornerstone of Morton's friendship with Lhuyd, with whom he corresponded regularly from 1694 until 1709.[47] On occasion the two men went out into the field together, as when they followed the Jurassic limestones to their intersection by the Humber. When they did not journey together the one usually kept the other informed as to his activities. When Lhuyd was in Sligo, Ireland, for example, he wrote the following to Morton:

We have travaild above a hundred miles in the highlands of Scotland; and afterwards proceeded to ye Lowlands as far as Glasgow and Edenburgh ... In ye Winter Monts ... we could make few Observations as to Natural History ... Since our Return to this Kingdome we have been in Quest of their Form'd Stones; their Shells, and other exangusous Sea animals, and their Plants. We viewd ye Giants Causey in ye Country of Antrim; and found it to differ from a Basalte we had observ'd at Cader 'gon in Wales in that it is much more elegant and breaks off in joynts.[48]

It was on Lhuyd's advice that Morton decided in 1698 to compile a natural history of Northamptonshire. Lhuyd offered to help in any way possible, for example, with the subscriptions or by supplying information, and so on.[49] In 1704 Morton printed proposals for subscriptions in which he acknowledged that although the natural history of England would be an undertaking of too great weight and extent for any one single person, 'it may be happily accomplished successively and by Parts,' if only 'several Persons of due Skill, Application, and Abilities' were to engage in this work, and each 'take upon him the Search, and Survey of only a single County.'[50] Morton proposed to follow Plot's example, the latter having shown 'that such an Undertaking is not like to be unacceptable in any other part of England.'[51]

Northamptonshire did not appear until 1712. In a progress report contained in a letter to Sloane, Morton indicated in 1706 that completion of the study was delayed because of the growing accumulation of materials he had acquired during his 'hunts,' and because of the difficulty of assembling these in a useful manner.[52]

Northamptonshire opens with a general description of the county, in which Morton systematically divides it into several formal 'territories' and 'distinct

'Industrious Searchers into the History of Nature' 225

Tracts.' The former encompass large areas bounded by the main rivers; the latter are based on the different types of surface physical features, such as fen, heath, wood, and pasture. Although his information was largely derived from a personal acquaintance with the county, Morton does note what earlier writers said about Northamptonshire; Leland, Camden, Speed, and Plot all have their place here. Often, however, these writers are used by the author as vehicles for exhibiting his own pride in his native region. He notes, for example, that Speed thought no other county was better stored with grain, and that Camden observed the soil was exceedingly fertile both for tillage and for pasture. He even quotes Drayton's verse on the fruitfulness of the soil.[54] At another point he comments disparagingly on Camden's not having taken note of 'the Prospects,' that is, hills.[54]

Morton proves himself to be a keen observer of the changes in the county that have taken place over the years. He delves into the developments in agriculture, addressing the problem of enclosure, the principal reason for which appears to be

the great Progress the Woodlanders had made in Agriculture, when the Ground was clear'd of the Wood which was cut down in Plenty for the use of the many IronWorks we had here in the ancient Woodland Part. Hereupon the County began to stand in need of Pasture to balance the Tillage. They had no Wants of Bread for the many People employ'd about the Fields: But there needed a Proportion of Cheese, Butter, and Flesh. This is suggested [by Camden] as the Cause of the Warwickshire Enclosures, and was probably the great Occasion of them here.[55]

In a similar vein, in his descripton of the markets and towns of the county, Morton exhibits curiosity about their character in the past. Thus, he is quick to point out that Leland in his *Itinerary* found Oundle to have 'a very good Market, and was all builded of Stone, and that the Paroch [Parish] Church was very fair.'[56] In the main body of the book he describes in detail the composition of the soil, the stone in strata, etc, and the mineral deposits. He derived much of his information on these through field-work, which included his rummaging through some stone pits, while investigating the 'Interiour Parts of the Earth expos'd to view in the Quarries.'[57]

Much of the rest of the work is reminiscent of those by Plot. Morton arranges his material similarly, examining – chapter by chapter – the 'Waters,' 'Air and Heavens,' 'Plants,' 'Brute Animals,' 'Human Bodies,' 'Arts,' and 'Antiquities.' He is perhaps at his best when it comes to fossils, to which a large amount of the text is given. Of the fourteen plates in *Northamptonshire*, nine are exclusively figures of fossils. They show representatives of the leading Jurassic groups: lamellibranchs, brachiopods, gasteropods, ammonites, echinoids, and ichythyosaurians. The treatment of these follows Plot; the figures do not attain quite the same standard of excellence, however, although Morton had more information to go on. He had access to other newly published works on the field, as well as the standard ones issued by Lister and Woodward. The book includes the customary accounts of unusual or amusing incidents. Thus, Plot's 'echoes' receive attention, as do 'monstrous births,' and other 'accidents' and diseases. Morton seems especially intrigued by cases involving the effects of lightning, for example:

But of all the ill-Accidents that have happened here by Lightning, that at Everdon, July 27. 1691. was the most dismal. Of Ten Persons who upon the Approach of a Storm removed from their Harvest-Work to a Hedge in the Field for Shelter, Four were kill'd and Six wounded. One of the poor men that was kill'd had a little Dog on his Lap: His Hand upon his Head, and with Bread and Cheese, or one of them in his Hand, as if just ready to give the Dog a Bit. And thus in that very posture he continued as Stiff as if the Lightning had suck'd up all his Moisture, or had coagulated and fix'd it.[58]

More than thirty pages are devoted to the Roman antiquities of the county; Roman ways, coins, fortifications and Danish and Saxon features are included in the chapter on 'Antiquities.' A considerable proportion of Morton's source material on these comes from previous authors such as Kennett. On the whole, Morton adds new knowledge to the field, often offering his own assessment of a particular site or artefact, based on personal observation:

My Opinion of the Hunsborough Encampment, in brief is this: that it was a Summer Camp of the Danes, and particularly of one of their many Parties which sustain'd themselves by Plunder and Rapine. I take it to have been pitched at the time when Towcester was built by King Edward the Elder Anno 921, or however in that Interval betwix 913 and 921. They chose to Post themselves rather there than in the Town in the Summer Quarter; for by the Advantage of this Situation, they could the more conveniently command and overawe the adjacent Countrey.[59]

Northamptonshire was an improvement over all the natural histories that came before it. Yet Morton's aim, including the practical one espoused by all the scientists of the day, was to glorify God by presenting to the world His works of nature. Morton's attitude to regional study is also revealed in his reply to a suggestion by Lhuyd that he might describe some other county; Morton decided to confine himself to a study of the county with which he was most familiar, and where he could find patronage, thus circumventing many of the problems encountered by former regional writers who may have 'overextended' themselves. As far as Morton was concerned he had been, in Northamptonshire, 'a Searcher after Nature already many years, and the Gentlemen of the County are pleased to Naturalize me. The field is sufficiently rich and large in that extent I take it. In one Word, one county is enough for me.'[60]

When it came to natural history, states Roy Porter, in some ways Morton 'out-Plotted Plot,' as, for example, in visiting all the villages in his county, and also in his strict methods of inquiry:

Most of the Observations that this Second Chapter consists of have been made upon the Interiour Parts of the Earth expos'd to view in the Quarries. My Method in taking them had been usually this. First I noted the Species, Thickness, Order and other Circumstances of all the Strata, beginning with the uppermost and so proceeding downwards to the Bottom of the Pit; particularly the several Kinds and Varieties of Stone, the Uses of Each, and how each is affected or wrought upon by Water, Frost or Heat; the Number of the Strata into which by means of Horizontal and Parallel Fissures it is divided; and the Eavenness or Inequality of the Surface of the Stratum; As also the Foreign Matter and Heterogeneous Bodies, especially the Sea shells, enclosed in the Stone.[61]

By the time Morton conducted his research regional or county studies had become increasingly natural-historical in their orientation, superseding, for instance, those pure compilations that merely listed seriatim all that an investigator knew about his region – eg, Childrey's *Britannia Baconica* – and those works that still contained considerable elements of civil history. Plot and others had tended to organize their material after the universal classificatory scheme suggested by Bacon, but rarely had anyone offered a general introduction to the physical geography of the region covered. Morton, however, did just this, by introducing his study with a comprehensive discussion of the position of Northamptonshire within the physical structure of Britain; he also went further than anyone before in presenting integrated accounts of the relief of the region, relating this to the strata and topography.[62]

Morton's study, in a sense, capped a century of vibrant developments in regional study, the last half of which, in particular, marked an important stage in the description, identification, and classification of natural historical material as a whole.[63] Through empirical research, men were pioneering new methods to investigate and describe the natural history of the earth. Lhuyd's insistence that 'with Natural History know there's no good to be don in't without repeated observations,' observations that in his view could only be accomplished by means of extensive travel, is indicative of their commitment to Baconian philosophy.[64] In attempting to discover essential characteristics and differences within nature at a local level, and to classify these, they opened the door for the future reconceptualization of the entire country, and indeed the entire earth. Men like Sibbald, in Scotland, who proposed a 'Royal Society' or 'Academy' for Scotland, utilitarian and encompassing the study of all natural knowledge like its English model, were now actively combining the study of regions with that of antiquarianism and science.[65]

CHAPTER THIRTEEN

Conclusion: 'Rust of Old Monuments'

... all the main Heads of Natural History have receiv'd aids and increase from the famous Verulam, who led the way to substantial Wisdom, and hath given most excellent Directions for the Method of such an History of Nature.
Joseph Glanvill *Plus ultra*[1]

THE STUDY of localities or regions made considerable progress in the seventeenth century. It was not, however, an entirely new subject, for the works of ancient geographers such as Strabo and Ptolemy had originally established similar study long ago. The sixteenth-century voyages of discovery and trading, a revitalized patriotism, and an interest in antiquities helped to stimulate interest in the subject during the late sixteenth and early seventeenth centuries.[2] Even before the stimuli of these activities Britain had a long tradition of regional study, but it had not been firmly established and the descriptions of regions were perfunctory for the most part. The antiquary John Leland, who sought to glorify England by the presentation of topographic and antiquarian information, inaugurated a steady stream of such chorographic literature. Most subsequent chorographers recognized the importance of his work and sought to imitate it. From their point of view, chorography constituted a distinct branch of antiquarian study. Writers who came much later still acknowledged their debt to their sixteenth-century predecessors. Nathaniel Salmon, the eighteenth-century surveyor of Surrey, regarded his own work as falling into the Lelandian tradition in regional study. He expressed his 'Acknowledgment to all those Gentlemen who have gone before me,' while accepting the extensive use of documentary sources such as Leland. Therefore, both literary sources and field studies – the latter being a feature of regional study or geography that matured in the seventeenth century – proved of equal value to Salmon's research.[3]

Although F. Smith Fussner and others have described the late sixteenth- and early seventeenth-century emergence of empirical theories of history in general, and have outlined the development of antiquarian researches, such approaches are often confusing and lack unity. While modern authors may show clear evidence of the wide range of historiography during this period, they do not always demonstrate how one specific body of historical, geographical, or antiquarian writings developed in form and function as time progressed, and as newer ideas replaced old. Such is the case, for example, with Fussner's *Historical Revolution*, where one finds short analyses of historical writing – universal, theoretical, territorial, local, and problematic; these analyses, however, do not show the continuity of each as the century went by.[4] Thus, although the new techniques, attitudes, and facilities developed for geographic, historical, and antiquarian research have been examined rather closely, there is plenty of room to study these in the light of a particular context, or body of literature, and over an extended period of time.[5]

The chorographers were in the forefront in making history, 'perhaps the most prized learning in Tudor England because it seemed the most immediately useful.'[6] Just as the natural historians of the latter part of the seventeenth century promoted the practical usefulness of knowledge of natural phenomena, so, too, are there statements of the utility of history in the chorographic literature. This is not surprising, for many chorographers emphasized the historical elements of their work over the topographical. However, the 'usefulness' of the historical content in chorographies was never intended to be compared with the 'usefulness' of histories such as the tragedies in *The Mirror for Magistrates*, which were designed in part to teach moral lessons by example. Often in chorographies, even history played a second fiddle to antiquarianism, which was held in high esteem by the practitioners of the art for its reliance on non-literary evidence. The kind of 'history' contained in chorography sometimes amounted to little more than the dry setting out of dates, names of people, and facts – as in genealogy – with little or no explanation of causes. Nevertheless, even this process was considered as having utility, especially in legal disputes over land inheritance or related issues. This traditional treatment of history was maintained in a sense by the natural historians, many of whom avoided explicit explanations of causes and theorizing. There are exceptions, of course, as in the case of Camden's work, where theorizing is implicit, at least, in his reconstruction of Romano-British society, or in the statements of Aubrey or Plot.

The authors of these works were usually country gentlemen, and details of their existence may be construed in some cases from their participation in other phases of Elizabethan or Stuart life, and from their other works. Their

Conclusion: 'Rust of Old Monuments' 231

chorographic productions constituted more or less comprehensive descriptions of counties and regions, often combining geographical and historical information with notes on the local families and antiquities. As Fussner states: 'Such works probably served as guidebooks and introductions to local history, but primarily they conveyed information about the history and topography of the area and about contemporary life'; that is why they are 'of considerable interest to present-day historians who want to know what the Tudors thought important in their own society and in their own past.'[7] Not all these works were free from defect; occasionally authors were haphazard in their use of evidence, or did not display 'a sufficient regard for accuracy [nor] ... a nice discrimination in ... choice of materials.'[8] But, in most instances, men such as Camden were careful to base their narrative on documentary evidence rather than guesswork or preconception.[9] When documentary proof was not accessible, these men rarely resorted to conjecture; when they did, the general practice was to inform the reader that guesswork was involved. At other times all that was required of the authors, besides their training and practical experience, was a keen eye and a reliable horse, in order to examine visually the places they describe. Nevertheless, this offered no safeguard from the scorn of critics such as John Earle, who, in his *Microcosmographie* of 1628, ridiculed the antiquary:

Hee is a man strangely thrifty of Time past ... Hee is one that hath that unnaturall disease to bee enamour'd of old age, and wrinckles, and loves all things (as Dutchmen doe cheese) the better for being mouldy and worme-eaten. He is of our Religion, because wee say it is most ancient; and yet a broken Statue would almost make him an Idolater. A great admirer hee is of the rust of old Monuments, and reads only those Characters, whose time hath eaten out the letters. Hee will goe you forty miles to see a Saint's Well, or a ruin'd Abbey: and if there be but a Crosse or stone footstoole in the way, hee'll be considering it so long, till he forget his journey. His estate consists much in shekels, and Roman Coynes, and he hath more Pictures of Caesar then James or Elizabeth. Beggars coozen him with musty things which they have rak't from dunghills, and he preserves their rags for precious Reliques. He loves no Library, but where there are more Spiders' volums than Authors, and lookes with great admiration on the Antique works of Cob-webs ... His chamber is hung commonly with strange Beasts' skins, and is a kind of Charnel-house of bones extraordinary.[10]

Chorography itself did not remain static. While in nearly all cases utilizing history and/or antiquarianism as a base, the chorographers shifted the focus of their studies from geography – or, more specifically, topography – if it had ever been there, to both genealogy and heraldry. By the time of the English civil war and the Protectorate chorography appeared to have lost a sense of purpose. Even as Hartlib and others were actively setting the stage for a new type of regional study, a number of workers in the field were viewing such occupation primarily as a solace in days of danger. But antiquarian studies continued to flourish in the later seventeenth century in Britain, with the regional writers once again coming to the forefront. It is somewhat ironic that the renewed interest in antiquities and in regional study was in part facilitated by the waning of chorography as an attractive *scholarly* pursuit. The new course took a slightly different direction, charted by the scientific ideas being promulgated by scholars such as Bacon and Hartlib, by the members of the Royal Society, and by the new standard in scholarship established by erudite men. As Barbara Shapiro indicates, humanistic and scientific learning – fields conventionally seen as belonging to different and perhaps mutually exclusive traditions – were now combined in the person of the virtuoso, who used historical and natural materials and scientific methods to make accurate statements of fact about both past and present, and who applied these to regional description.[11] Before the ideas on natural philosophy found full expression, therefore, 'it was only in the face of abnormalities or calamities that writers paused to consider phenomena belonging to the realm of Nature.'[12]

Herbert Butterfield has pointed out that the history of science 'could never be adequately reconstructed by a student who confined his attention to the few men of supreme genius,' and the work of the contemporary natural historians or natural philosophers is the work of a group of men who did not necessarily possess great wisdom, but who played an important role in spreading the ethos of the scientific revolution.[13] Many of the regional natural historians/geographers were leading exponents of the Baconian philosophy, which stressed the detailed study of natural phenomena to the detriment of the reliance on human events that was characteristic of chorographic works. The English civil war, by restricting the ability of researchers to seek out and examine literary material, contributed for at least half a century to the downgrading of chorography as the main conduit for regional study. So, too, did Hartlib's example. Soon the regional writers began to rely heavily on the advice of the Royal Society, of which some of them were members, while some of their most important observations were published in the society's *Philosophical Transactions*, founded by Henry Oldenburg.

Bacon's philosophy in many respects was unsystematic; perhaps his principal importance was in generating the psychological motivation for natural philosophy. He believed that science should be advanced so as to establish 'the power and dominion of the human race itself over the universe,' and that the only way of commanding nature was by 'obeying her.'[14] Like much of the British scientific community generally, a considerable segment of regional writers of the second half of the seventeenth century adhered uncritically to the aims of Bacon, believing that the natural world must be studied because discoveries bear tangible results.

Regional study after Bacon was characterized by at least two major features. First, like history itself, regional study required a critical use of authorities, which had to be carefully compared with one another. The natural historians took this requirement one step further, prescribing that authorities be checked against natural phenomena known by means of experiment, and especially observation. This attitude emphasized the importance of observation and often necessitated the rejection of classical authority. Second, there was a growing faith in the ability of human reason to classify data and to arrive at conclusions from facts in order to formulate 'laws' to which natural phenomena seemed to conform.

These regional writers thus collected most of their evidence from outside the traditional literary sources. Minerals or accounts of changes in river courses and in other topographical features were sources generally overlooked by the chorographers. In the same fashion, the natural historians/ antiquaries made precise drawings and plans of barrows and ruins, and recorded the archaic speech of Wales, Ireland, and other regions of Britain. There was, then, a markedly increased awareness of process and change, a more vibrant feeling for the past.[15] New developments in the techniques of investigation in the natural sciences were not only adopted, but quite often initiated and developed by the regional writers, making possible new geological and archaeological approaches to the past. In any modern studies of the origins of such approaches, therefore, we find that the regional writers are usually – and quite justifiably – considered to be among the founders or at least precursors of these branches of scientific study. The concern of the regional writers for studying intensively the natural history of a place was fostered primarily by the ideas of Bacon, who conceived of knowledge as an hierarchical pyramid, with natural history as the essential basis, and metaphysics, as generalized physics, forming the apex.[16] This alone would have encouraged them to devalue the importance of civil history in their study of a region. Yet, these men were more 'natural historians' than 'natural philosophers,' when one considers the issue closely.

The enthusiasm for the inclusion of natural history in regional study, generated by the early workers in the field, is still evident in the following suggestion, extrapolated from John Morton's 'Proposals for Subscriptions' to his natural history (1704):

> Indeed the Natural History of England would be an Undertaking of too great Weight and Extent for any one single Person; but it may be happily accomplished successively and by Parts. Would several Persons of due Skill, Application, and Abilities, engage in this work, and each take upon him the Search, and Survey of only a single County, we might hope to see the whole finished in good time. Dr. Plot began in this Method; and the great Encouragement he met with from the Nobility and Gentry of Oxfordshire, and Staffordshire, the Nat-History of which two Counties he wrote, shew that such an undertaking is not like to be unacceptable in any other part of England. And indeed partly this, and partly the Commands of some Persons of great Note and Judgment, have engaged me in the Composure of the *Natural History of Northamptonshire*.[17]

It is perhaps significant that it was only as an afterthought that Morton added the following words to the above: 'with some Remarks on the Antiquities thereof.'[18] Regional study had become centred on the study of natural history, with the study of antiquities, even through scientific inquiry, of less importance.

The majority of the regional writers of the second half of the seventeenth century concretely expressed their faith in the harmony of science and religion, as did many of the other virtuosi, concluding that 'the discoveries of natural philosophy would not contradict the teachings of Christianity which they believed in and practiced.'[19] Again, John Morton is representative of this attitude, as applied specifically to the study of natural history:

> Whereas Exact Descriptions of Things, however small and seemingly contemptible: and faithful Accounts of what is observable in them, will always be of Use to those who study Nature, to what End soever that be: Whether to take a clearer View of the Infinite Wisdom of the Great Creator, in the Artful Contrivance of so vast a Variety of Organiz'd Bodies, which appear as remarkable in the smallest as in the largest Animals and Vegetables: or to enquire into the Structure of the Terrestial Globe, and the Changes it has undergone: Or lastly, to improve or apply Natural Products of any Kind, to the Uses of Human Life.[20]

Morton's position is also revealed in his handwritten note found in a copy of *Northamptonshire*, which reads: 'I will give Thanks unto Thee, O Lord, with my Whole Heart: I will speak of all Thy marvellous Works [of Nature].'[21]

Natural theology informed the natural philosophy of these men; that God's benevolence was everywhere revealed in the universe, was perhaps given its

Conclusion: 'Rust of Old Monuments'   235

best-known topographical expression in John Ray's designation of mountains as serving eight specific and divinely ordained ends, which ranged from forming convenient political boundaries and creating habitats for upland-dwelling flora and fauna, to being the seat from which rivers flow. In so mundane a thing as denudation, Ray could thus observe the workings of the Almighty: 'the Rain brings down from the Mountains and higher grounds a great quantity of Earth, and in times of Floods spreads it upon the Meadows and Levels, rendering them thereby so fruitful as to stand in need of no culture or manuring.'[22] Many others were equally convinced of the benevolence and power of God as displayed in nature, but this is not always clear in all of their writings. When Ray's friend Edward Lhuyd, for example, discussed the power of denudation, he hit upon the modern concept of morphoclimatic regions. Having reached the conclusion that the steepest slopes in Wales were to be found on the highest mountains, he reasoned: 'This I can ascribe to nothing else but the rains and snow which fall on those great mountains, I think, in ten times the quantity they do on the lower hills and valleys ... I affirm That by this means not only such Mountains as consist of much Earth and small stones, or of softer Rocks, and such as are more easily dissoluble, are thus wasted, but also the hardest Rockes in Wales.'[23]

Despite their juxtaposition of religion and nature, few of the regional writers were deeply original thinkers in the philosophy of science. Their major contribution to the new philosophy, as a group, was in their leadership in the areas of empirical research and observation. Some of them, such as Robert Plot and Edward Lhuyd, were men of genius who contributed substantially to scientific knowledge. Others remained relatively obscure, tending to be followers of current fashions in experiments and scientific disciplines. But then, a similar observation might be made of the membership of the Royal Society as a whole. Similarly, the regional writers who claimed to be exponents of the new philosophy varied widely in outlook and – as in the case of John Aubrey and Robert Plot – were often ready to abandon rational scientific beliefs when faced with disturbing religious, supernatural, or mystical phenomena.

The conflict between Baconian and pre-Baconian philosophy was especially pronounced in the early years of the Royal Society; we can see this in the thought and work of John Aubrey, who was as interested in astrology as in serious field-studies. Displaying varying degrees of credulity, he thus demonstrated far fewer scientific scruples than did some of the later regional writers. Occasionally his means of verification was faulty. A typical example is his belief in the tradition that when an oak is being cut down, 'it gives a kind of shreikes or groanes, that may be heard a mile off, as if it were the genius of the oake lamenting.' By way of verifying this tradition, Aubrey merely adds that his friend 'E. Wyld, Esq. hath heard it severall times.'[24] The fascination of these early virtuosi with the freakish or monstrous seems today to have been antithetical to sober scientific thought – unless, perhaps, one views such activity as having been based on a desire to rationally explain the mysterious. As for their continued interest in astrology, Keith Thomas believes that it was only because of the pretensions of astrology to be a genuinely scientific system that it retained its credibility for some time to come.[25] Influential figures such as Thomas Hobbes began condemning it as a mere 'human device for mulcting the stupid populace,' and yet it still thrived.[26]

In most respects the regional writers were no different from many other scientific thinkers of the age, who, although not necessarily protectors of traditional beliefs, still thought that the limits of the possible were wide and therefore warranted investigation. Although some men did not achieve greatness in scientific inquiry, their failure cannot be blamed on a backward-looking mentality. Stuart Piggott, a renowned archaeologist in his own right, called Aubrey an 'engaging and whimsical dillettante,' but in the next breath he acknowledged Aubrey's contribution to science: 'when he [Aubrey] turned to prehistoric antiquities his collecting of notes came as near to system and method as his grasshopper mind could ever achieve, and his observations on such Wessex monuments as Stonehenge and Avebury are still of value, and show the beginning of what has become a great British tradition in archaeology, the study of field monuments by observation and survey without excavation.'[27]

In the work of the regional writers we see the birth of serious field-studies. Utilizing new methods, they studied the antiquities of a region for their intrinsic values and not merely as one element in the cultural landscape. Regional study was no longer considered the realm of the rank amateur, whose major claim to such study lay in his legal training or in his familiarity with the area he investigated.

Aside from their scientific interests, the regional writers had a normal curiosity about the prehistoric sites and pre-Roman monuments that had long been prominent features in many parts of Britain. John Aubrey's interest in them is not surprising, for he came originally from Wiltshire, which had more than its share of such sites. Through their studies these men helped to popularize these places, to the extent that they soon became the tourist attractions they remain today. Of Stonehenge, the eighteenth-century archaeologist William Stukeley wrote: 'The almighty carcase of Stonehenge draws great numbers of people, out of their way every day, as to see a sight: and it has exercis'd the pens of the learned.'[28]

Conclusion: 'Rust of Old Monuments' 237

In this same period, measurement was slowly becoming part of archaeological field method, and at times the researchers consciously set down field procedures, perhaps as a guide for others. There was a growing tendency, also, to think in terms of plans, maps, and drawings rather than merely written descriptions, as can be seen in Aubrey's conceptualization of a land-use map and in Plot's drawings of local stone artefacts – the first British ones published. As Piggott states, the appearance of archaeological illustrations in the seventeenth century was intimately joined to the development of technical draughtsmanship in the natural sciences.[29]

Few of the artefacts and structures known to belong to ancient times could be directly related to written records, a fact that widened the distinction between archaeology and history. An article was rejected for publication in the *Philosophical Transactions* because it was regarded as 'being chiefly Historical, [therefore] it seems not so properly the subject of these *Transactions*.'[30]

As antiquaries, the regional writers were faced with such problems as determining the date and method of construction, the function, and the value within society of ancient artefacts, manuscripts, and so on. Some progress was made towards solving these. Plot, for example, speculated on the flake scars on flint points while trying to recognize some of the processes involved in their manufacture, and arrived at the correct conclusion that they had been worked with a tool. Both he and Lhuyd were perhaps the first to attempt to determine the possible uses for stone tools, largely through analogy with Indian and Oceanic uses for similar artefacts.[31]

What may be stated with certainty is that seventeenth-century field archaeology was partly a by-product of the great range of antiquarian work motivated, as Douglas has emphasized, by the search for precedent in ecclesiastical and political matters, and evident in the work of the chorographers.[32] But now new questions were being asked based on information gained in the field, and the answers to these were eventually to provide fresh insights into the origins of the pre-Roman peoples of Britain. This meant, however, that the regional writers, as pioneers, invariably implanted some tenacious misconceptions in the public mind. Aubrey, for example, gave substance to the false notion that stone circles were the work of Druids.[33] But such false explanations are perhaps unavoidable where any new approaches to human inquiry are concerned.

Closely related to the interest of the regional writers in archaeology was their involvement in geological and palaeontological pursuits. In cataloguing the strata and fossils of Britain, at least some of the regional writers understood that they were founding a new science, yet even they remained unaware of its full implications. Fossils especially were used as important elements in any account of the history of the earth. They were utilized to support the theories both of their inorganic and of their organic origins. Lhuyd was particularly active in the collection and classification of fossils, but in general this is one field where non-regional writers led the way; like Lhuyd, they rightfully regarded Martin Lister as 'the first great discoverer of our country fossils.'[34]

It seems logical to conclude that, overlooking natural history, many of the most important scientific achievements of the regional writers resulted from their antiquarian pursuits, and that science and antiquarianism were mutually reinforcing areas of human thought. In this respect, a scientist searching for the antiquities of a region was not as retrogressive as one might at first think. Over the course of the seventeenth century antiquaries undertook actual measurement and investigation of the 'shipwrecks of time,' as Bacon called them. Thus, history was studied more methodically and natural history was approached more scientifically. As the scientists acquired an increasingly sophisticated perception of the past, many of them chose to apply this perception to their study of the regions of Britain. The linkage of science and history was eventually to produce a historical science of the earth, and a large share of the credit for this has to go to the men engaged in regional study in late seventeenth-century Britain.[35]

Whether involved in antiquarianism or in natural history, men began to recognize that they could not always depend on written reports and the reports of witnesses, but that firsthand observations were also required; that stories of fabulous and mythical beasts, or even of men such as Brutus, needed dispelling, and that impartiality was the key to accurate investigation. It was thought that even conjecture, which was sometimes unavoidable in the absence of hard data, would prove beneficial if it was accompanied by meticulous experiment and accurate observation. Truth, therefore, might be reached through the use of induction and experiment. Nevertheless, the men involved in regional study, the natural historians, appeared to have been considerably less inclined to theorize than were many of the other virtuosi, on whom Bacon may have had less of an influence. A few of the regional writers mentioned in the present study explored, in a systematic way, such methodological and epistemological problems. But they often left them to men such as Hooke or Newton. Perhaps it is for this very reason that when one speaks of early-modern scholars, the natural historians usually take a back seat to mathematicians, physicists, and other theoretical scientists where histories of science and philosophy are concerned.[36]

On certain occasions the natural historians themselves found it impossible

Conclusion: 'Rust of Old Monuments' 239

to avoid totally the use of documentary evidence. Also, their explanations of the physical changes they observed in geology were often fitted into the scriptural account of the earth's past, for example, the Creation and the Great Deluge or Flood. Like many of the other scientists of the late seventeenth century, they were not averse to becoming involved in chronological studies or in scriptural commentary, although many of the regional writers preferred to leave such studies to the ecclesiastical historians, or 'world makers,' who were increasingly treating scripture as historical documents.

With the arrival of the eighteenth century, a problem arises in choosing a satisfactory description of the variety of sciences which Kuhn and others have attempted to solve by separating the 'mathematical' tradition from the 'experimental' tradition.[37] Natural historical study in the eighteenth century is sometimes associated by modern historians of science with 'experimental' natural philosophy, and studied in relationship to its audience. The latter, at least, is not totally unacceptable, for there is now recognition on a wider scale that science did not entirely sink 'into torpor between the age of Newton and the nineteenth century triumphs of Davy and Dalton.'[38] On one scale, then, science in this period was on the rise in the provinces as well as in London, a rise that owes more than is usually recognized to the regional historians of the previous century. Porter presents a convincing argument that the growth of eighteenth-century provincial science is directly attributable to the prominence given to its cultural value:

The popularization of provincial science had more to do with cultural status than with factories, more to do with adjusting social relations than with refining engines. This point is confirmed into the nineteenth century by the indifference of many lit and phils to questions of direct utility and their extensive coverage of antiquarian and literary topics; their bans on investigations which might interfere with matters of trade; and their frequent failure in directly utilitarian enterprise. It is further underlined by the continuation of similar traditions of polite learning well into the Victorian period, even in industrial areas – for example, the Yorkshire Philosophical Society; or by the vogue amongst the middle classes of Victorian England for establishing naturalists' clubs, and antiquarian and archaeological societies.[39]

Taken in this context, science is seen more as a rational amusement for the leisured classes, 'much more widely diffused through Georgian society via the commercialization of leisure, the entrepreneurship of knowledge, and the rise of professional communications popularisers.'[40] This makes it easy to identify the 'science' of the period before the Enlightenment solely with 'physics and mathematics, correlatively privileging categories and procedures of an abstract, systematizing, deductive, and nomological type.'[41] In this scenario natural history can degenerate into a 'non scientific' natural philosophic or literary tradition – when not ambiguously lumped together with the experimental tradition – useful as a vehicle for satisfying a popular appetite for the marvellous. As such, it can then be exploited by charlatans, quacks, and other itinerant public lecturers. In fact, by the end of the eighteenth century natural historical systematics, which appealed 'to a contrasting conception of the management of nature,' could offer an alternative model of knowledge to that of the charlatans, or 'radicals' – those who exploited natural philosophy in order to undermine established authority, often by appealing to those excluded from political decision-making.[42]

The process of science as spectacle had already begun by the late seventeenth century, at least where the 'humble' non-mathematical sciences are concerned.[43] By 1697 William Wotton could decry 'the public ridiculing of all those who spend their Time and Fortunes in seeking after what some call useless Natural Rarities.'[44] Inadvertently, the regional writers, in their exuberance for collecting and examining natural specimens, were somewhat to blame for this. The prestige of some of these men, and their success in their activities, influenced many who were much less scientifically knowledgeable or capable of undertaking similar pursuits. The 'public,' though taking an interest in these lesser lights, found the quality of such activity deteriorating and even laughable, and began to distinguish between the 'pure' mathematical (deductive) sciences and the inductive, 'applied' ones. The social prestige of the latter was on the decline, to be pursued by pleasure seekers rather than the utilitarians, and applied science became a genteel part of natural philosophy, as distinguished from 'real' science. As John Locke stated in 1693, 'a gentleman must "look into" natural philosophy, [mainly] "to fit himself for conversation."'[45] Depending on how one looks at it, the decline of the Royal Society was either a contributor to, or symptomatic of, these end-of-century developments: 'By the 1690's most of the early Fellows were dead, and although there were a few gifted younger men besides the great Halley, yet the Society as a whole was in a languid state; under an administration which was often mediocre and seldom much interested in science.'[46]

Even the search for evidence of the glory and wisdom of God had already been scorned – in this case by Meric Casaubon – as a pretence by 'sick brains' of 'hunting after Novelties.'[47] All in all, though becoming much more popular and widespread, natural historical and philosophical interests were now being lumped together and regarded by contemporaries as rather trivial pursuits, unworthy of the attention of the true scientist, although of common interest to the rank and file virtuoso.[48] The replication of spectacles led to such things

Conclusion: 'Rust of Old Monuments' 241

as the fashion for grottoes, as displays of nature, which in turn has been linked to geological theory and to Gothic architecture; hence an interest in grottoes has been said to have speeded up the 'Gothic revival' of the 1740s.[49]

By the mid-eighteenth century and certainly after its first few decades, elegant language and literary values seemed to be of primary concern to anyone involved in regional study. And so, few workers in the field retained the same spirit of scientific inquiry that Edward Lhuyd did, a man who was aware of the passing of an age-old culture, especially in Wales, and who considered himself fortunate to meet anyone who still possessed knowledge of the old traditions and languages.[50]

By about 1750 the term *natural history* encompassed the study of flora, fauna, and minerals.[51] Even if the general public did not generally consider it a serious scientific pursuit, at least it has been argued more recently that 'during the eighteenth century no branch of science developed so rapidly as natural history.'[52] But the case should not be overstated, either, at least at the regional level. As this study has attempted to show, that development in the eighteenth century owes much to the seventeenth. With the passage of time, more and more 'hack writers' and 'scholars who are not at all recognized as scientists' took up the cause, many inspired as it were by Linnaeus, but rarely attaining his level of scholarship.[53] The design and development of a sculptured hedge or ornate flower garden, for example, enticed more than a few men to label themselves 'botanists.' As Allen points out, by the 1720s, as the 'exceptional generation' of seventeenth-century naturalists passed away, causing a drastic reduction in the number of naturalists in the ranks of the Royal Society, a marked narrowing resulted in the society's outlook as a whole.[54] By then Wotton had already sought to explain why 'Natural and [even] Mathematical Knowledge ... begin to be neglected by the generality of those who would set up for Scholars.'[55] Men came to 'paint the landscape' in terms that were much more figurative than those of their predecessors. The interest in the natural history of a region was primarily rekindled not only by the activities of Linnaeus but also by those of the Board of Agriculture, which was set up near the end of the century to stimulate land-owners to follow the newly developed agricultural techniques.

Turning to antiquarianism, we notice a similar decline, although the writing of regional works continued unabated during the eighteenth century, and was invigorated near its end by the romantic movement. Although the classical emphasis began before the growth of romanticism – it was classicism that urged the antiquaries back to Roman times – it was left up to the Romantics to pick up the slack by focusing on British and Anglo-Saxon times. Until then, many regional writers, as antiquaries, continued to be 'distracted by the more ephemeral elements of classical learning and by the Grand Tour,' unable to appreciate the fundamental issues of prehistory.[56]

It could be argued that in some respects regional study had gone the full circle between the mid seventeenth and the eighteenth centuries, with the pioneering scientific researches of the natural historians providing the main break between the two periods.[57] The most popular eighteenth-century works were occasionally closer in appearance to the early chorographic literature than to the natural histories. This group included the following studies: John Harris's *The History of Kent* (1719); Nathaniel Salmon's works; John Bridge's *The History and Antiquities of the County of Dorset* (1774); Edward Hasted's *The History and Topographical Survey of the County of Kent* (1778–99); and Samuel Rudder's *A New History of Gloucestershire* (1779).

This is not to say that there were no exceptions. Some eighteenth-century regional writers inherited, even if unconsciously in some cases, certain traditions developed by their late seventeenth-century predecessors. The basis of a gentleman's education continued to be scripture as well as the classics, and in a few quarters the trend toward quantification in natural science continued to increase. That antiquarianism and natural history on the whole did not simply fade away is reflected in the continued establishment and growth of scientific societies, such as the Linnaean Society and the Society of Antiquaries, the latter formed in London on much the same lines as had been the earlier Elizabethan society of the same name. But at least one writer has argued that 'the world of [scientifically based] natural history,' however, 'came to exist almost solely on paper.'[58] The odd regional writer maintained an abiding interest in science; William Borlase, for example, in his *The Natural History of Cornwall* (1758) followed Plot in the framework for a natural historical study, while devoting one volume of this work to antiquities.

Generally, however, during much of the eighteenth century, field-work and the exact study of natural phenomena faded to a considerable degree from regional study, even if topography still had its place. It seems that the public eventually came to accept this loss without much displeasure. This is in contrast to the late seventeenth and early eighteenth centuries when, as noted above, a critical public had soon learned to discriminate against regional work that neglected a scientific study of natural history and scientific antiquarianism. As Hunter pointed out, for this very reason, Sir Henry Chauncey's *Historical Antiquities of Hertfordshire* (1700) had been 'not much commended,' and William Sacheverell's *Account of the Isle of Man* (1702) was reported by one astute observer to make 'but a small figure here' (that is, in London).[59]

Conclusion: 'Rust of Old Monuments'   243

Martin Martin perhaps stated the situation most succinctly: 'Descriptions of Countries, without the Natural History of them, are now justly reckoned to be defective.'[60]

It was not until near the end of the eighteenth century, then, that there occurred a significant leap in interest in natural history, partly due to the activities of the Board of Agriculture, partly Linnaean in inspiration and, admittedly, partly the result of the continued interest of rich collectors in acquiring cabinets of natural 'curiosities' in their pursuit of self-aggrandizement. (Unlike Cotton's cabinet, however, theirs were not based upon a useful consideration of the value of scientific endeavour.) These factors, combined with the influential interest generated within the court circles of George III for natural history, went hand in hand with the discovery of the picturesque, and with the general aesthetic advance associated with the romantic outlook. But by then the lead in the study of natural history had long fallen from the hands of the regional writers to others – and even then not always into the hands of 'scientists.'

The results of this for natural history were not altogether happy. In a few, rare cases the power of observation, and even more the later recording remained ... unclouded and undistorted ... Far more often the eye misted over, the pen trembled, Sense gave out as Sensibility came in. The accepted approach to nature had become no longer to set down what one saw plainly and accurately; the aim now was to record one's reactions – and the livelier these reactions appeared, the more beneficial, the more exalting, the more 'tasteful' the contact with nature was assumed to have been.[61]

It is also significant that 'For fifty years after 1705 not one British scholar made a contribution of any significance to the literature of geomorphology,' a period that saw the production only of 'scraps of material [in this field] culled from the publications of untutored travellers, second-rate topographers, and long-forgotten clerics.'[62] Except for mineralogy, natural history and the earth sciences at this time provided little scope for the type of serious taxonomic work that had become favoured by the majority of researchers – researchers who were strongly imbued with the Linnaean desire for classification.[63] It was only in the last decade of the eighteenth century that British interest in geomorphology was totally restored. When one considers scientific antiquarianism in general, one discovers, too, that by the eighteenth century it was conducted by men with few pretensions to true archaeological learning. Its eclipse lasted even longer; by the middle of the nineteenth century any interest in prehistoric archaeology that remained came in the form of 'a companion to the all-pervading interest in Gothic.'[64]

In conclusion, one cannot take anything away from the flowering of the scientific spirit in regional study that occurred in the second half of the seventeenth century. This flowering 'evolved' from, and did not revolt against, a long British tradition in both chorography and scientific endeavour, although many of the changes with which it is associated seem on the surface to have been sudden and unprecedented. Even if this spirit faded during the next epoch, its legacy remained, to be revived and continued in some forms to this day. Even where the eighteenth century is concerned, who is to say that 'collecting, classifying, comparing, re-evaluating in the light of continually appearing new evidence rather than bursting forward with the grand hypothesis' is not an important element in the advance of science?[65] Perhaps science does develop by the accumulation of individual discoveries and inventions. Certainly the social-psychological background or world view, or 'paradigm,' of seventeenth-century virtuosi was fundamental to their scientific – and other – investigations. But in the case of regional study, at least, the new direction taken in the second half of the century was based to a considerable degree on a cumulative process derived initially from the older established world view, and not entirely from new fundamentals – important as they were. In effect, two such mentalities were able to exist side by side, to a degree commensurate with each other. Eventually the more 'novel' paradigm prevailed, in some quarters. The process owed less to information gained accidentally than to the systematic seeking out of data. If the observation and evaluation of such data was not completely 'pure' – that is, if this sort of activity was partly determined by the perceptual mind-frame already established in the information gatherer – this was generally recognized as the case and therefore was partly responsible for the reluctance of the observer to arrive at conclusions, and especially at what some modern historians of science still call 'scientific truth.' In the long run, however, new methods of verification and falsification, which were being developed in the period under study here, were to gradually and eventually establish a new world view, or 'paradigm.' In our example, then, one can accept the idea of paradigms, but question – or add to – the way one paradigm replaces an existing one, if not speculate further on their very nature; for example, perhaps they are more compatible than Kuhn allows. But this is treading on shaky ground, for if the latter is the case, then can one legitimately speak of distinct 'paradigms'?

To be fair, more recently Kuhn has deviated somewhat from the imprecise notion of what paradigms are, or at least he uses in place of, or in addition to this term, catch-words such as 'disciplinary matrix' and 'exemplar.' Where seventeenth-century science is concerned, there is more leeway now, in his scheme of things, for more than Copernican astronomy and Newtonian physics, and an admonition that science advances not merely through

Conclusion: 'Rust of Old Monuments' 245

revolutionary upheaval, but also through 'normal science.' Still, not enough credit yet is given to the Baconian branch of science – assuming one accepts Kuhn's two-tiered system – as opposed to that given over to the sciences inherited from antiquity. For Kuhn the latter still seems to form the major driving force behind the intellectual transformation of Western thought in early-modern times, because of the result of major changes taking place within them at this period; but again, these changes were based on the rethinking of old concepts and problems rather than on Baconian principles of observation and experimentation. By comparison the Baconian sciences, if 'new,' remained undeveloped throughout the period in question, growing 'more systematically experimental then, but their theories remained vague and qualitative.'[66] In this light the classical sciences are still seen as more essential to the scientific revolution than the Baconian, even if the latter made some contribution to it.

The best criticism of Kuhn – and for that matter of Hall and Butterfield – is that of Joseph Levine, who defends the value of Baconian science. Using John Ray as an exemplar, Levine points out that a major revolution did take place within the body of the non-classical sciences in the early-modern period, especially when one considers those branches linked most closely with natural history: 'botany, zoology, palaeontology, and the like,' and that Ray was correct in thinking that a good number of the sciences had made great progress through 'what appeared to be the mere accumulation of new information, much of it during his own lifetime.'[67] In this respect, then, Aubrey, Sibbald, Plot, and the other regional natural historians, like the more famous Ray, Malpighi, and Leeuwenhoek mentioned by Levine, must be seen in the forefront in adding, through systematic investigation, much new knowledge to science as a whole, a development that in itself might be considered 'revolutionary,' and one that, as Wotton stated, raised new hypotheses, in the long run, at least.[68] Wotton, like many contemporaries, felt that the business at hand lay in natural historical research, and that the responsibility of the Royal Society was 'to collect a perfect history of nature in order to establish thereupon a body of Physics.'[69] Through natural history, based on collection, observation, experimentation, the ground was being prepared not only for Bacon's natural philosophy, but also in a way for mathematics and physics. If mathematics and physics mark the high point of the scientific revolution, then, as Levine has intimated in his article, natural history may in fact constitute its take-off point. This being the case – which many modern historians of science would deny – here are further grounds for diminishing the segregation of the classical and the Baconian 'sciences.' As Levine remarks:

246 'Speculum Britanniae'

If, then, we have any desire to understand the history of science in the seventeenth century and to evaluate it fairly, we must first recall this primacy of natural history, both in the scientific labor and the scientific theory of the time. And if that raises some questions for the present philosophy of science, so much the better. What science is, after all, is what science has been, and philosophy must always look to history for its data – a notion which, besides its modern proponents, must certainly have appealed to Francis Bacon, who wrote history like a philosopher and philosophy (as William Harvey said) like a Lord Chancellor, and who understood better than most the intimate and necessary relationship between the two ancient and rival disciplines.[70]

Regional study – 'evolutionary' in the transformation of emphasis from chorography to natural history, 'revolutionary' in its contribution to knowledge, and to the scientific revolution – is a difficult matter to fit into a paradigmatic schema.[71] (This is not necessarily to discourage such an effort.) Francis Bacon in particular brought in much that was new, but in some ways he also represented a reworking of the old. Similarly, both chorography and natural history owed a debt to the ancients and to the humanists, at least, but at the same time constituted a newer formulation of the efforts of their precursors. As for antiquarianism in particular, Levine has the last word:

Looking back across the two centuries that separated [George] Hickes [the Saxonist] and his friends from John Leland and his successors, one can see a single great antiquarian dream being slowly realized. Leland had never passed beyond announcing it and accumulating notes for it, but he understood virtually every one of the great antiquarian tasks of the next generations ... Was it surprising that Leland's notes were read with consuming interest by the antiquaries of the later seventeenth century ... and that various plans for their publication were soon realized? ... The works of Hickes and Lhwyd and the new *Britannia* were in a real sense the climactic achievements of Renaissance antiquarianism rather than the harbingers of anything new.[72]

But whether old or new, ancient or modern, chorography or science, one thing is clear; the total effect of these works was one of displaying to the whole world the human and natural bounties of Britain. Thus, to paraphrase Strauss's praise of the Germans, the men who executed these works of merit, and the works themselves, are entitled to a lofty place in both the intellectual and the literary canon of Britain – works that cast a spellbinding reflection of a real wonderland, Britain, that would no doubt have caught the attention of Alice.

Notes

PREFACE

1 A. Rupert Hall *The Revolution in Science 1500–1750* (London and New York 1983) 2
2 Michael Hunter, review of *'Concerning Natural Experimental Philosophie': Meric Casaubon and the Royal Society* by Michael R.G. Spiller in *Annals of Science* 39 (1982) 187–92
3 Hall *Revolution* 16
4 Ibid 19
5 Ibid 16–17
6 Thomas Kuhn 'Mathematical vs Experimental Traditions in the Development of Physical Science' *Journal of Interdisciplinary History* 7 (1976) 15

CHAPTER ONE

1 Greater attention will be paid to terminology later, but for now it should be stated that, strictly speaking, 'local history' in the narrow sense is not the focal point of this book; the narrow sense focuses more on the village or town level than on county or regional history. Also, few readers would probably associate the term with natural historical study. It is hoped that the regional studies described here are viewed as more than merely forming an expedient methodological tool for threading the writers and their works together, but as a distinct genre in English language literature, or at least in historiography, in the sense, perhaps, that the Elizabethan voyage literature is so called.
2 John Aubrey on Malmesbury Abbey, quoted in Margaret Aston 'English Ruins and English History: The Dissolution and the Sense of the Past' *Journal of the Warburg and Courtauld Institutes* 36 (1973) 231

3 That is, of course, if one managed to avoid the fate of John Leland, who went mad in the end, the result of antiquarian study's 'overtaxing' his brain (according to W.G. Hoskins *Local History in England* 3rd ed (London 1984) 19.)
4 Richard Gough *British Topography* 2 vols (London 1780) 1:iii
5 Ibid 1:1
6 Ibid 1:ii. Gough is referring specifically to the work of Giraldus Cambrensis. By the same token, an earlier observer, Flavio Biondo, dismissed Geoffrey of Monmouth's *History of the Kings of Britain* as 'stuffed with lies and frivolities'; see E.B. Fryde *Humanism and Renaissance Historiography* (London 1983) 22.
7 J. Strzelcyk, quoted in L.S. Chekin, 'Elements of the Rational Method in Gervase of Tilbury's Cosmology and Geography' *Centaurus* 28 (1985) 211
8 Sir Francis Bacon *Sylva Sylvarum or A Naturall History* 2d ed (London 1628) 235
9 A.N. Whitehead *Science and the Modern World* (London 1927) 49
10 David C. Douglas *English Scholars, 1660–1730* (London 1939; 2d ed rev, London 1951)
11 Ibid (1951 ed) 32
12 On sixteenth- and seventeenth-century historical scholarship see F. Smith Fussner *The Historical Revolution* (New York 1962), *English Historical Scholarship in the Sixteenth and Seventeenth Centuries* ed Levi Fox (London 1956), F.J. Levy *Tudor Historical Thought* (San Marino 1967), and W.R. Trimble 'Early Tudor Historiography, 1485–1548' *Journal of the History of Ideas* 11 (1950) 30–41.
13 E.G.R. Taylor *Late Tudor and Early Stuart Geography, 1583–1650* (London 1934, reprinted New York 1968) chap 4
14 Levy *Historical Thought* 159
15 Ibid 160
16 Here I am paraphrasing Gerald Strauss 'Germania Illustrata: Topographical-Historical Descriptions of Germany in the Sixteenth Century' (PH D diss Columbia University 1957) ii. Strauss's reference is to the German topographical-historical writings, which constitute a 'distinct genre in German humanist literature.' Interest in German antiquity was intense, 'awakened by the view of tribal life found in Tacitus's recently discovered *Germania* and rooted in the general humanist orientation toward the classical age' (ibid 1).
17 See Victor Skipp 'Local History: A New Definition' *Local Historian* 14 (1981) 325–31. Skipp attempts a definition of 'local history,' commenting that 'the chance of producing a verbal definition that would satisfy all the different types of local historian must be very remote indeed' (330). Quoting D.D. Marshall (330–1), Skipp advocates the placing of more emphasis on regional study within the context of local history.

18 Biondo is discussed in Strauss 'Germania Illustrata' passim and in Denys Hay 'Flavio Biondo and the Middle Ages' *Proceedings of the British Academy* 45 (1959) 97-128. Other European figures, besides Leland, who displayed an interest in antiquarian study include Konrad Celtis and Beatus Rhenanus in Germany, Guillaume Budé in France, and John Bale in England. Stuart Piggott *Ruins in a Landscape* (Edinburgh 1976) 6 says that medieval antiquarianism in England 'is no more than sporadic and exceptional,' growing by the time of the Renaissance. Roberto Weiss covers many of the workers in this field in *The Renaissance Discovery of Classical Antiquity* (Oxford 1969).
19 This point is made by J.R. Hale, ed *The Evolution of British Historiography: From Bacon to Namier* (Cleveland 1964) 11.
20 Margaret Aston 'English Ruins' 231, 234-6, 250, 252. In *The Gothic Revival* (London 1928) 19, Kenneth Clark has written that as the 'monasteries [were] destroyed and libraries dispersed,' the antiquaries 'were moved to perpetuate their vanishing glories,' while boasting about 'their country's treasures as poets did her wars.'
21 The intellectual background of the Elizabethan antiquaries is the subject of T.D. Kendrick *British Antiquity* (London 1950).
22 Arnoldo Momigliano 'Ancient History and the Antiquarian' *Journal of the Warburg and Courtauld Institutes* 13 (1950) 313; the article is reproduced in *Studies in Historiography* (London 1966) 1-39, and it is to that publication that later citations refer.
23 Joan Evans *A History of the Society of Antiquaries* (Oxford 1956) 9
24 See Antonia Gransden 'Antiquarian Studies in Fifteenth-Century England' *Antiquaries Journal* 60 (1980) 75.
25 Ibid
26 Ibid
27 R.W. Southern 'Aspects of the European Tradition of Historical Writing: The Sense of the Past' *Transactions of the Royal Historical Society* 5th ser, 23 (1973) 246-56 (see Gransden 'Antiquarian Studies' 75)
28 Gransden 'Antiquarian Studies' 75
29 See Robert W. Hanning *The Vision of History in Early Britain* (New York and London 1966) chap 1.
30 For Momigliano, 'original sources' include eyewitness accounts or documents and other contemporary material remains. 'Derivative' sources or authorities include historians and chroniclers, who did not witness the events they describe. Momigliano bemoaned the lack of a 'History of Antiquarian Studies.'
31 Momigliano 'Ancient History and the Antiquarian' 3
32 Ibid 4
33 Ibid 7

34 Ibid
35 Of course the men of the scientific revolution were assisted in such work by the development of new scientific instruments. And, even if the standards of the latter-day scientific antiquaries were more exacting than those of their predecessors, both groups were fond of collecting, attentive to irrelevant detail, 'lived to classify,' and avoided the a priori attitude inherent in the generalizing approach of some historians (see ibid 25).
36 Ibid 10–18. Even by the early sixteenth century historical research was indebted to antiquarian study. The collections of Cotton, the work of Spelman and Lambarde, and 'the new use of archaeological evidence and etymological analysis produced some sophisticated historical interpretation in the discourses of the Society of Antiquaries during the 1590s,' though 'they never reached a really high standard of historical scholarship'; see Kevin Sharpe *Sir Robert Cotton 1586–1631* (London 1979) 24. Fussner *Historical Revolution* 115 also states that the career of Sir Henry Spelman illustrates 'the transition from the study of antiquities in isolation to the study and interpretation of the historical context of English feudalism,' adding that 'the conventional distinction between research and interpretation continued to be made, however, largely on the basis of the difference between antiquarian collecting habits, and historical ideas of explanation.'
37 Momigliano 'Ancient History and the Antiquarian' 24. By the nineteenth century philosophic history and antiquarian research, however, came closer together (25).
38 Ibid 26–7
39 Gransden 'Antiquarian Studies' 81
40 Ibid 83
41 Ibid 85
42 For Rous see ibid 85.
43 Ibid 89
44 Sharpe *Cotton* 28
45 Ibid 33–4. William Burton, for example, the chorographer of Leicestershire, obtained material for the history of the parish of Theddingworth from Cotton (35).
46 Fussner *Historical Revolution* 69
47 Gransden 'Antiquarian Studies' 75
48 Fussner *Historical Revolution* 7
49 Wallace K. Ferguson *The Renaissance in Historical Thought* (Boston 1948) 17–18, quoted in Fussner *Historical Revolution* 9
50 Ibid 16
51 Ibid 46

52 See F. Smith Fussner's review essay in *History and Theory* 8 (1969) 386–7.
53 Quoted in Felix Gilbert 'The Renaissance Interest in History' in *Art, Science, and History in the Renaissance* ed Charles S. Singleton (Baltimore 1967) 377
54 Ibid 384
55 Zachary Sayre Schiffman 'Renaissance Historicism Reconsidered' *History and Theory* 24 (1985) 170
56 Ibid
57 John Pocock, quoted in ibid 174
58 To this Schiffman (ibid 175) begs the question: '... what keeps us from saying that the roots of historicism lie in the antecedents to these developments [in legal scholarship], and so forth?'
59 Ibid 182
60 Donald R. Kelley *Foundations of Modern Historical Scholarship* (New York and London 1970) 271
61 Ibid. Kelley (272) states that although Pasquier's aim was to write 'a modern history,' by this he did not mean conventional narrative but the investigation of specific problems on the basis of firsthand evidence, and he used philologists – not historians such as Livy – as his models.
62 Ibid 291
63 Schiffman 'Renaissance Historicism' 181
64 Ibid 180–1
65 Ibid 181
66 Ibid
67 Ibid 176; here Schiffman is alluding to Kelley's comments.
68 Isaac Kramnick 'Augustan Politics and English Historiography' *History and Theory* 6 (1967) 36
69 Sharpe *Cotton* 43
70 See Donald R. Kelley 'The Rise of Legal History in the Renaissance,' *History and Theory* 9 (1970) 176, 178. The ruins in the case of the legal humanists-cum-historians were 'the defaced and fragmentary monument of Roman jurisprudence, the *Digest*' (178). The chorographers usually extended 'ruins' to include ancient monumenta, inscriptions, etc.
71 Fussner *Historical Revolution* 27
72 Ibid 32
73 Louis A. Knafla 'The Matriculation Revolution and Education at the Inns of Court in Renaissance England' in *Tudor Men and Institutions: Studies in English Law and Government* ed Arthur J. Slavin (Baton Rouge 1972) 250. According to Knafla, these works could only be put to effective use by contemporary lawyers through massive cross-references to the 'modern' reports of prominent judges.

74 Ibid 251
75 Ibid 244–5
76 Ibid 246
77 Ibid 248
78 Ibid
79 See A.W.B. Simpson *An Introduction to the History of the Land Law* (Oxford 1961) 219. Sir Orlando Bridgman was a notable conveyancer who established considerable order by developing model conveyances that satisfied the desires of land-owners and kept within reasonably well-settled doctrines of the law (ibid). This gave rise to a conveyancing tradition 'which has continued to this day.' Simpson details the most notable achievement of the conveyancers, the classical 'strict settlement' (220–4).
80 Retha M. Warnicke *William Lambarde, Elizabethan Antiquary, 1536–1601* (London and Chichester 1973) 23
81 Ibid 24–5
82 Ibid 25
83 Ibid 27
84 Ibid 35; here Warnicke is quoting Fussner.
85 Ibid 58–9
86 Ibid 60–1. Lambarde printed the charge to the grand jury in his handbook for justices of the peace, the *Eirenarcha*, which became the basis for other works on this position.
87 Warnicke *Lambarde* 84; the title of 'prince' is attributed to Faith Thompson.
88 Claudius Ptolemy *The Geography of Claudius Ptolemy* ed and trans Edward Luther Stevenson (New York 1932) 1. It should be noted that by the seventeenth century regional study also drew upon sixteenth-century native and Continental models as well as classical ones; see Sharpe *Cotton* introduction and the present study, passim.
89 Robert Stafforde *A Geographicall and Anthologicall description of all the Empires and Kingdomes* (London 1607) 1
90 John Dee 'The Mathematical Preface' to *The Elements of Geometrie of the most Auncient Philosopher Euclid of Megara*, London, BL, Sloane ms 888 f A4; see also the first published English edition of Euclid, trans H. Billingsley (London 1570). Dee is quoted in William Bourne *A Booke called the Treasure for travellers* (London 1578) preface.
91 Arthur Hopton *Speculum topographicum: or the Topographicall glasse* (London 1611) sig B
92 William Pemble, *A Briefe Introduction to Geography* (Oxford 1630) 1; Peter Heylyn *Microcosmus; a little description of the great world* (London 1621) 10; see also Nathanael Carpenter *Geographie Delineated Forth in Two Bookes* 2d ed

(Oxford 1635) bk 2, 2. I am here indebted to Jack Oruch 'Topography in the Prose and Poetry of the English Renaissance, 1540–1640' (PH D diss, University of Indiana 1964) 4–9 for such definitions.
93 Dee 'Mathematical Preface'
94 The main authority on sixteenth-century German cosmography appears to be Gerald Strauss *Sixteenth-Century Germany: Its Topography and Topographers* (Madison 1959), which is based on his PH D dissertation (see n 16 above).
95 Apparently cosmography was introduced to England by Richard Eden's *A Treatyse of the newe India ... after the description of Sebastian Munster in his boke of universall Cosmographie* (London 1553), followed by William Cunningham's *The Cosmographical Glasse* (London 1559).
96 Thomas Blundeville *M.Blundeville, His Exercises* (London 1594) 134
97 See William Camden *Britain [Britannia]* ed and trans Philemon Holland (London 1610) 363
98 Ibid 343, quoted in Oruch 'Topography' 6
99 William Lambarde *A Perambulation of Kent* 2d ed rev (London 1596) C1
100 John Stow *Survey of London* ed C.L. Kingsford 2 vols (Oxford 1908) 1:xcvii; 2:253
101 John Throsby Supplementary Volume to the Leicester Views (London 1790) 5
102 *The Life and Times of Anthony Wood* ed Andrew Clark, 5 vols (Oxford 1891–1900) 1(1891) 209
103 These ties are discussed at various points in the present study.
104 Barbara Shapiro has made some intresting points on this and related issues. For example, see her 'History and Natural History in Sixteenth- and Seventeenth-Century England: An Essay on the Relationship between Humanism and Science' in Barbara Shapiro and Robert G. Frank, Jr *English Scientific Virtuosi in the 16th and 17th Centuries* (Los Angeles 1979).
105 Lytton Strachey *Portraits in Miniature* (London 1931) 19
106 Herbert Butterfield *The Origins of Modern Science* rev ed (London 1957). Margery Purver *The Royal Society: Concept and Creation* (London 1967) also contains some useful information on this topic.
107 See Sir Francis Bacon *The Works of Francis Bacon* ed James Spedding, Robert L. Ellis, and Douglas D. Heath, 14 vols (London 1857–74; reprint ed Stuttgart-Bad Cannstatt 1963) 4:68.
108 Thomas Hill *A Contemplation of Mysteries* (London 1571) 9
109 Bourne *Treasure* bk 5 chap 3 for example
110 The Royal Society obtained its charter in 1662.
111 White Kennett *Parochial Antiquities Attempted in the History of Ambrosden, Burcester, and other Adjacent Parts in the Counties of Oxford and Bucks* (Oxford 1695)

254 Notes to pages 26–9

112 Evans *Society of Antiquaries* 29
113 Ibid; Stuart Piggott *William Stukeley: An Eighteenth-Century Antiquary* (Oxford 1950) 1–17 passim. Charles Webster *The Great Instauration: Science, Medicine and Reform 1626–1660* (New York 1975) passim discusses the interest generated by Interregnum groups in the exploration of the earth, an interest taken up by the Royal Society, which tended to act as both a promoter and a diffuser of such knowledge; that is, it provided an institutional focus for science.
114 Sydney Ross 'Scientist: The Story of a Word' *Annals of Science* 18 (1962) 67. Ross says that the word *scientificus* as pertaining to demonstrable knowledge or science came into English as late as 1600. According to him (71), the name *scientist* was first propounded in the *Quarterly Review* for March 1834 by an anonymous reviewer.
115 Dorothy Stimson 'Amateurs of Science in 17th Century England' *Isis* 31 (1939) 40
116 Paolo Rossi *The Dark Abyss of Time: The History of the Earth and the History of Nations from Hooke to Vico* trans Lydia G. Cochrane (Chicago and London 1984) xiii. According to Rossi, Kuhn's most important point is that hermeticism, especially through the Faustian image of the magus as controller of nature, was a major contribution to Baconism and perhaps even to the whole scientific revolution. (According to Kuhn, Bacon should be seen as a transitional figure between the magus Paracelsus and Robert Boyle, the experimental philosopher.)
117 See Simon Schaffer's 'Essay Review: Mapping the Edge of the Abyss,' *Isis* 77 (1986) 321.
118 Nicoletta Morello '*De glossopetris dissertatio*: The Demonstration By Fabio Colonna of the True Nature of Fossils' *Archives internationales d'histoire des sciences* 31 (1981) 63–4
119 Rossi *Dark Abyss* 6–7
120 Ibid 7
121 Ibid
122 Ibid 8
123 Roy Porter, review of *The Meaning of Fossils* by J.S. Rudwick, in *History of Science* 11 (1973) 133
124 Rhoda Rappaport 'Hooke on Earthquakes: Lectures, Strategy and Audience' *British Journal for the History of Science* 19 (1986) 131. For Hooke, then, fossils revealed episodes in earth history, just as coins and other 'lasting monuments' (ibid) testified to events in civil history.
125 See Rossi *Dark Abyss* 13.
126 Rappaport 'Hooke on Earthquakes' 139
127 Ibid 134

128 Morello 'Fabio Colonna' 64
129 Ibid 66
130 Ibid 69–70
131 Ibid 70–1
132 Robert Hooke, quoted in Rossi *Dark Abyss* 13
133 Ibid 16
134 Ibid 16–17
135 In regard to climatic change, Hooke thought that those 'better versed in ancient Historians' than he search the 'ancient documents' for evidence of it; see Rappaport 'Hooke on Earthquakes' 134.
136 Ibid 140–1
137 Rossi *Dark Abyss* 17
138 Eric Cochrane 'Science and Humanism in the Italian Renaissance' *American Historical Association Review* 81 (1976) 1040–1
139 Ibid 1042
140 Ibid 1056–7
141 Karen Meier Reeds 'Renaissance Humanism and Botany' *Annals of Science* 33 (1976) passim. Of course Bacon had less use than the humanists for classical precepts and models for the study of nature, and in the recovery, editing, and translation of Theophrastus, Dioscorides, Pliny, etc. Dioscorides's *Materia medica* became the most important textbook of botany in the sixteenth century, and contained details of medicinal herbs.
142 Reeds 'Renaissance Humanism' 534
143 Ibid 533–4, quoting Rabelais
144 Ibid 540–1
145 W.R. Albury and D.R. Oldroyd 'From Renaissance Mineral Studies to Historical Geology, in the Light of Michel Foucault's *The Order of Things*' *British Journal for the History of Science* 10 (1977) 188
146 Ibid
147 Ibid 188–9
148 Ibid 189–90. The reference is to Agricola's *De natura fossilium* (1546).
149 Albury and Oldroyd 'Mineral Studies' 190. For this and for other reasons, the authors state that 'it is only by anachronism that this work [*De natura fossilium*] can be considered as a "natural history of the mineral kingdom."'
150 Ibid 190–1
151 Ibid 191–2
152 Thomas L. Hankins *Science and the Enlightenment* (Cambridge 1985) 117
153 Ibid 115
154 Bacon *Sylva Sylvarum* 234. Bacon admonished men 'that they doe not withdraw

256 Notes to pages 34–6

Credit, from the Operations by Transmissions of Spirits, and Force of Imagination, because the Effects faile sometimes'
155 Charles Webster *From Paracelsus to Newton: Magic and the Making of Modern Science* (Cambridge 1982) 41
156 Fussner *Historical Revolution* provides a typical example. He allows only about ten pages to what he describes as 'Territorial History.' In a footnote in his essay 'Ancients, Moderns, and History: The Continuity of English Historical Writing in the later Seventeenth Century' *Studies in Change and Revolution* ed Paul J. Korshin (Menston 1972), Joseph M. Levine has noted that 'unfortunately, we still lack [full-length] particular studies [even] of some of the major figures like Leland and Camden' (70n1).
157 Sir Maurice Powicke 'The Value of Sixteenth- and Seventeenth-Century Scholarship to Modern Historical Research' *English Historical Scholarship* ed Fox 116 states that the study of the subject of historical investigation in England between the time of Camden and the days of Queen Anne has 'hardly begun.'
158 J.W. Thompson *A History of Historical Writing* 2 vols (New York 1942); see also Kendrick *British Antiquity*.
159 A.L. Rowse *The England of Elizabeth: The Structure of Society* (New York 1950) 31–65
160 Douglas *English Scholars* passim
161 Taylor *Stuart Geography*; Margarita Bowen *Empiricism and Geographical Thought* (Cambridge 1981)
162 Michael Hunter *John Aubrey and the Realm of Learning* (London 1975)
163 David Elliston Allen *The Naturalist in Britain* (London 1976); F.D. Hoeniger and J.F.M. Hoeniger *The Growth of Natural History in Stuart England from Gerard to the Royal Society* (Charlottesville 1969)
164 Several scholars, including H.G. Fordham *Some Notable Surveyors and Mapmakers of the Sixteenth, Seventeenth, and Eighteenth Centuries and Their Work* (Cambridge 1929) and Edward Lynam *The Mapmaker's Art* (London 1953) have made valuable studies of map-making techniques and of mapmakers. Works on the history of cartography contain useful references, but on the whole they do not make a great impact on the evolution in regional study itself. The early history of anthropology is discussed in John Howard Rowe 'The Renaissance Foundations of Anthropology' *American Anthropologist* 67 (1965) 1–20 and in Margaret T. Hodgen *Early Anthropology in the Sixteenth and Seventeenth Centuries* (Philadelphia 1964). The anthropological aspects are also covered to a degree in the work of Keith Thomas (see bibliography).
165 Anthony à Wood *Athenae Oxonienses* 2 vols (London 1691–2). As for the *Dictionary of National Biography*, for many entries its completeness and accuracy are not always unassailable.

166 Some of the works that touch on the topic include H.F. Paget, ed *The Early Maps of Scotland* 3rd ed rev (Edinburgh 1973); Sir Arthur Mitchell and James Tosach Clark, eds *Geographical Collections Relating to Scotland, made by Walter Macfarlane* 3 vols (Edinburgh 1907–9); Frank Emery *Edward Lhuyd* (Cardiff 1971); and the other works by Emery.
167 Gough *British Topography*, for example
168 Glyn Daniel *A Hundred and Fifty Years of Archaeology* (London 1950) 20
169 Thomas Warton, as quoted in J.E. Jackson *The History of the Parish of Grittleton, in the County of Wiltshire* (London 1843) xii. The reference here is specifically to the county histories. Some of these 'works of entertainment … ' have been reprinted by the Scolar Press, where details of their authors' lives may be located (*Classical County Histories* ed Jack Simmons), and by other presses.
170 See David N. Livingstone 'The History of Science and the History of Geography: Interactions and Implications' *History of Science* 22 (1984) 271.
171 See Joseph M. Levine 'Natural History and the History of the Scientific Revolution' *Clio* 13 (1983) 73 n 51. Levine comments that 'the history of natural history during the scientific revolution still remains a desideratum.'

CHAPTER TWO

1 Joseph Hunter *Hallamshire* (London 1819)
2 J.O. Thompson *History of Ancient Geography* (New York 1965) 224–5; see Leonard Guelke, 'Regional Geography' *Professional Geographer* 24 (1977) 1.
3 Quoted in Strauss 'Germania Illustrata' 165; also ibid 166
4 Strabo, *Geography* ed and trans Horace Leonard Jones 8 vols (London 1917–32) 1 (1917) 3–5
5 This point is made by Evans *Society of Antiquaries* 2. Ptolemy is discussed in G.R. Crone *Maps and their Makers* (London 1962) chap 5 and in Strauss *Sixteenth-Century Germany*. Christiaan Van Paassen in *The Classical Tradition of Geography* (Groningen 1957) 3 states that 'Strabo's influence was … far less important than that of Ptolemy' until the nineteenth century, because Strabo worked 'very unsystematically and inaccurately when it comes to the location of places.' Paassen goes on to say that 'while Strabo gives geography a wider field of activity than does Ptolemy, the accent is still placed upon the quantitative aspects and upon the map' (5). Most important for our purposes, however, is Strabo's view of geography not merely as a natural science, but as a 'political' one where historical, cultural, and social aspects – and the relationships among them – all have a vital role to play. 'To Strabo the *oecumene* means a human entity, an anthroposphere, in particular a social unity in the adversity of space' (ibid 19). This holistic approach to geography, character-

ized by 'an appreciative or deprecatory appraisal of the region in question on the basis of social and political values' (29), certainly mixed, in the minds of many of the chorographers, with Ptolemy's tenets on the subject, to comprise an understanding of what 'chorography' should be or entail. Generally it is clear that geography and history were distinct disciplines but could be closely related, depending on the kind of geography. Geography was understood to be either a mathematical discipline in the Ptolemaic tradition, a part of applied geometry and connected to surveying, astronomy, and navigation, or a historical discipline, in the footsteps of Strabo, complementing the chronological aspect of the record of human actions and the fate of nations. Geography was also vital in histories of nations because of the astrological determinants of national characteristics. Munster's 1540 edition of Ptolemy's *Geographia* has been reissued (Amsterdam 1966).

6 Evans *Society of Antiquaries* 2
7 Antonia Gransden *Historical Writing in England* 2 vols (vol 1 Ithaca 1974, vol 2 London and Henley 1982) 1 *c 550–c1307*, 10–11
8 Ibid 1:23–4
9 Bede *A History of the English Church and People* foreword and trans Leo Sherley-Price, rev ed (Harmondsworth 1981), 37, 39. Bede, in fact, was one of the outstanding European figures in the history of 'scientific endeavour' during the Dark Ages. His other important scientific works include *De temporum ratione* (c 730) and *De natura rerum liber* (c 703); see *The Complete Works of Venerable Bede* ed J.A. Giles 12 vols (London 1843–4). In some instances Bede followed Pliny, whose natural history was the main authority on science available to Western European scholars until the translation of Greek scientific writings into Latin in the twelfth century; see Pliny *The 'Natural History' of Pliny* trans J. Bostock and H.T. Riley 6 vols (London 1855–7) and William H. Stahl 'Dominant Traditions in Early Medieval Latin Science' *Isis* 50 (1959) 95–124.
10 Gransden *Historical Writing* 1:68
11 Ibid 1:107
12 Southern 'Aspects of the European Tradition' 255–6. An interesting recent book on the topic is Rodney Thomson *William of Malmesbury* (Bury St Edmunds 1987).
13 Gransden *Historical Writing* 1:244–6. Also see Giraldus Cambrensis *The Historical Works of Giraldus Cambrensis* trans Thomas Forester and Sir Richard Colt Hoare, ed Thomas Wright (London 1887) and David Walker 'Gerald of Wales: A Review of Recent Work' *Journal of the Historical Society of the Church in Wales* 24 (1974) 13–26.
14 Charles Homer Haskins *The Renaissance of the Twelfth Century* (Cleveland 1961) 315. It is interesting to note, as does Gabriele M. Spiegel 'Genealogy:

Form and Function in Medieval Historical Narrative' *History and Theory* 22 (1983) 44 that the historian of the Middle Ages – almost like the ancient chorographical 'painters' – 'apprehended history itself as a perceptual field, to be seen and represented instead of constructed and analysed, an object more of perception than of cognition'; thus, even chronicles were 'images of histories, likenesses, semblances, pictures.' Giraldus and other contemporary historians are treated in Nancy F. Partner *Serious Entertainments: The Writing of History in Twelfth-Century England* (Chicago and London 1977).

15 Several 'urban surveys' were made during this period. Descriptions of King's Lynn and Great Yarmouth, apparently produced in response to governmental inquiries during the reign of Edward I, provide examples of a type that fits more neatly into the *Domesday Book* rather than the chorographic category. See Elizabeth Rutledge and Paul Rutledge 'King's Lynn and Great Yarmouth, Two Thirteenth-Century Surveys' *Norfolk Archaeology* 37 (1978) 92–114.

16 Gransden *Historical Writing* 2:45

17 Ibid 2:50–1

18 This attitude is made clear towards the end of the first chapter of the first book; Ranulph Higden *Polychronicon* ed C. Babington and J.R. Lumby, Rolls Series 41, 9 vols (London 1865–6) 1 (1865) 1.

19 Gransden *Historical Writing* 2:52

20 In his translation Trevisa added some stimulating comments concerning the original text, and upheld, on the whole, Geoffrey's authority against Higden; see David Fowler 'John Trevisa: Scholar and Translator' *Bristol and Gloucestershire Archaeological Society Transactions* 89 (1971) 99–108.

21 John Taylor *The 'Universal Chronicle' of Ranulph Higden* (Oxford 1966) passim. One of the first writers to imitate Higden is John of Fordun. Chapter 10 of the second book of Fordun's *Chronica gentis Scotorum*, a description of the islands off the west coast of Scotland, is the earliest such account in Scottish historical sources; see William W. Scott 'John of Fordun's Description of the Western Isles' *Scottish Studies* 23 (1979) 1–13. The Da Capo Press has recently reissued Higden's 1480 work under the title *The Description of Britain* (Amsterdam 1980).

22 Arthur B. Ferguson 'Circumstances and the Sense of History in Tudor England: The Coming of the Historical Revolution' *Medieval and Renaissance Studies* 3 (1968) 193

23 Francis A. Gasquet 'The Note Books of William Worcester' in *The 'Old English Bible,' and Other Essays* (London 1897) 293–4. Worcester's haphazard notes were first edited by James Nasmith in *'Itineraria' Symonis Simeonis et*

Willelmi de Worcester (Cambridge 1778). William Worcester *Itineraries* ed and trans John Harvey (Oxford 1969) contains a useful index. Worcester is briefly mentioned in V.H. Galbraith *Historical Research in Medieval England* (London 1951) 44–5.

24 Roberto Weiss *Humanism in England during the Fifteenth Century* (Oxford 1941) 178
25 Oruch 'Topography' 32
26 H.B. Walters *The English Antiquaries of the Sixteenth, Seventeenth, and Eighteenth Centuries* (London 1934) 2. Walter's treatment of the antiquaries is meagre at best. T.S. Dorsch in 'Two English Antiquaries: John Leland and John Stow' *Essays and Studies*, new ser 12 (1959) 29 states that Leland was 'the founder of a great school of antiquarian studies,' without whose pioneer work 'Camden and Stow and others of his successors could scarcely have given us works of so complete and detailed a nature as they did.' Leland, Camden, and some other sixteenth-century chorographers are treated in Joseph M. Levine 'From Caxton to Camden: The Quest for Historical Truth in Sixteenth-Century England' (PH D diss, Columbia University 1965), which I have used.
27 Kendrick *British Antiquity* 48–9
28 Dorsch 'Two English Antiquaries' 21
29 For Leland see John Leland *The Itinerary* ed Thomas Hearne 9 vols (Oxford 1710–12); William Huddesford *The Lives of Those Eminent Antiquaries John Leland, Thomas Hearne and Anthony Wood* 2 vols (Oxford 1772), and John Leland *The Itinerary* ed Lucy Toulmin Smith 5 vols (London 1907–10, reprint ed Carbondale 1964).
30 Momigliano 'Ancient History and the Antiquarian' 27–8
31 Leland *Itinerary* 1: xxxvii–xxxviii ('New Year's Gift'); also quoted in Robin Flower 'Laurence Nowell and the Discovery of England in Tudor Times' *Proceedings of the British Academy* 21 (1935) 47–8
32 Reprinted in Leland *Itinerary*, 1 (1907) xxxvii–xliii. Thomas Hearne was first to publish the manuscript of *The Laboryouse Journey* in his 1710–12 study.
33 Leland *Itinerary* 1:xlii–xliii ('New Year's Gift'). Leland is discussed in Gransden *Historical Writing* 2: 472–8 and in Joseph M. Levine *Humanism and History: Origins of Modern English Historiography* (Ithaca and London 1987). Levine (79–80) outlines Leland's debt to the humanists, including William Lyly, the first headmaster of St Paul's, stating that Leland realized that Italian humanism could be used 'to resuscitate classical antiquity' and 'recover the whole of the British past' (82).
34 Leland *Itinerary* 1:xlii ('New Year's Gift')
35 Levy *Tudor Historical Thought* 129–30
36 Leland *Itinerary* 1:xviii–xix
37 Levine *Humanism and History* 82

38 Leland *Itinerary* 1:xli ('New Year's Gift')
39 Kendrick *British Antiquity* 56
40 All of the post-1596 (2d) edition of the *Perambulation* are in essence reprints of the text of that issue.
41 Flower 'Laurence Nowell' 47–73
42 Wilbur Dunkel *William Lambarde, Elizabethan Jurist, 1536–1601* (New Brunswick NJ 1965) 38
43 The previously mentioned *Archaionomia, sive De priscis Anglorum legibus libri* (London 1568), is the first published collection of Anglo-Saxon laws. Notable works on Lambarde include *William Lambarde and Local Government* ed Conyer Read (Ithaca 1962) and *Archeion* ed Charles H. McIlwain and Paul L. Ward (Cambridge, Mass 1957).
44 The *Topographical Dictionary* was probably composed between 1566 and 1570, and published in London in 1730.
45 Lambarde *Topographical Dictionary* i
46 William Lambarde *A Perambulation of Kent* 1596 ed 'Dedication to Sir Wotton' (dated 1570)
47 See Levy *Historical Thought* 138.
48 Southern 'Aspects of the European Tradition' 259–60
49 Thomas Philpot *Villare Cantianum: or Kent Surveyed and Illustrated* (London 1659) 163
50 See Oruch 'Topography' 59–60.
51 Lambarde *Perambulation* 1576 ed 388–415; Oruch 'Topography' 61
52 Lambarde *Perambulation* 1596 ed AA6. Frederick P. Hard 'William Lambarde's *Perambulation of Kent*' (PH D diss, Indiana University 1966) lxi–lxii. Hard provides a valuable edition of this work with a critical commentary, which I have used, especially with reference to the 1596 edition.
53 Lambarde *Perambulation* 1596 ed U1
54 Lambarde *Perambulation* 1576 ed 317–20; see Levine 'Caxton to Camden' 399.
55 Lambarde *Perambulation*, 1596 ed x6, DD5
56 See John Nichols ed *Bibliotheca topographica Britannica* 8 vols (London 1780–90) 1 (1780) 525–6. James D. Alsop and Wesley M. Stevens 'William Lambarde and the Elizabethan Polity' *Studies in Medieval and Renaissance History* 8 (1987) 233–4 discuss Lambarde's attention to critical areas of the Tudor polity.
57 That the respect was mutual is evident in the fact that Lambarde called Camden 'the most lightsome Antiquarie of this age'; Lambarde *Perambulation* 1596 ed B2.
58 Nichols *Britannica* 1, 506
59 Lambarde *Perambulation* 1576 ed 387

60 F.J. Levy 'The Making of Camden's *Britannia*' *Bibliothèque d'Humanisme et Renaissance* 26 (1964) 78
61 R.J. Schoeck 'Early Anglo-Saxon and Legal Scholarship in the Renaissance,' *Studies in the Renaissance* 5 (1958) 104
62 Levy 'Camden's *Britannia*' 70. Denys Hay *Annalists and Historians* (London 1977) 150 says that Camden was 'far and away the most important figure in England before the seventeenth century.'
63 Levy 'Camden's *Britannia*' 76
64 Camden *Britain* 1610 ed, preface 'The Author to the Reader' quoted in Oruch 'Topography' 77
65 Camden used Leland's notes as transcribed by Stow. A jealous herald, Ralph Brooke, unjustly accused Camden of plagiarizing Leland in *A Discouerie of Certaine Errours ... in the Much-commended 'Britannia'* (London 1596). Camden refuted the charge in the preface to the 1600 edition of the *Britannia*.
66 Denys Hay 'History and Historians in France and England during the Fifteenth Century' *Bulletin of the Institute of Historical Research* 35 (1962) 127
67 Levy 'Camden's *Britannia*' 89. The Elizabethan Society of Antiquaries prospered into the first decade of the seventeenth century. While chorographical topics were included in the society's discussions, increasingly the attention shifted to the office of herald in England and on the antiquity of the Inns of Court, reflecting the legal background of many of the society's members; see R.J. Schoeck 'The Elizabethan Society of Antiquaries and Men of Law' *Notes and Queries* new ser 1 (1954) 417–21.
68 Kendrick *British Antiquity* 144–8
69 Camden *Britain* 1610 ed 28–34
70 Ibid 371
71 *Camden's Britannia* ed and trans Edmund Gibson 2 vols, 2d ed rev (London 1722) 1:215
72 Oruch 'Topography' 74
73 Kendrick *British Antiquity* 108–9 says that Camden's attack caused the most damage to the belief in the Brute fable. It was Polydore Vergil's *Anglica historia* (London 1534) that first attacked this legend; see Denys Hay *Polydore Vergil: Renaissance Historian and Man of Letters* (Oxford 1952).
74 Camden *Britain* 1610 ed 10–22
75 Ibid 65–6, 99–106. The long-standing tradition of collecting Roman coins was favoured by Cotton, from whose collection Camden obtained the coins, and by fifteenth-century princes and humanists such as the Medici at Florence and Niccolò Niccoli; see Roberto Weiss 'The Study of Ancient Numismatics during the Renaissance' *Numismatic Chronicle* 6th ser, 8 (1968) 179.
76 Camden *Britain* 1610 ed 349

77 Ibid 182
78 Ibid 393; Oruch 'Topography' 77
79 Levy *Historical Thought* 159
80 Quoted in F. Haverfield '*Cotton Iulius Fvi.* Notes on Reginald Bainbrigg of Appleby, on William Camden and on some Roman Inscriptions,' *Transactions of the Cumberland and Westmorland Antiquarian and Archaeological Society* new ser 11 (1911) 364
81 Hay *Annalists and Historians* 151
82 Stuart Piggott 'William Camden and the *Britannia*' *Proceedings of the British Academy* 37 (1951) 208–9
83 John Stow *Survay of London* 2d ed (London 1603) preface says that since the publication of Lambarde's *Perambulation*, he, Stow, had been told of similar projects for other counties. It was partly for this reason, and partly to 'giue occasion and courage to M. Camden to increase and beautify his singular work of the whole,' that he had 'attempted the discovery of London, my native Soyle and Countrey.'
84 Southern 'Aspects of the European Tradition' 262–3
85 On this topic of an earlier 'Enlightenment' see Roy Porter 'The Enlightenment in England' in *The Enlightenment in National Context* ed Roy Porter and Mikuláš Teich (Cambridge 1981) 1–18. It may be argued that virtuosi such as Morton, Molyneux, Sibbald, and others, believing in the practical uses of natural historical knowledge as well as the intrinsic value of such pursuits, not only began to develop new methods of examining nature, but also established, with other men of learning, a type of intellectual mind set that enabled eighteenth-century figures to realize the potential for improving the human condition. If this sort of atmosphere was on the wane by about 1720 or 1730, it did not totally fade away, and was revived in somewhat different and more sophisticated form by the philosophers of the Enlightenment.

CHAPTER THREE

1 Taylor *Stuart Geography* 45
2 May McKisack *Medieval History in the Tudor Age* (Oxford 1971) 143
3 Norden tells us of his employment by Lady Ann in the dedication of *A Pensive Mans Practise* (London 1584), which is addressed to Lady Ann's eldest son, Sir Henry Knyvet, and his wife, Lady Elizabeth.
4 John Norden *Speculi Britanniae pars: a Topographical and Historical Description of Cornwall* ed Christopher Bateman (London 1728) 'Dedication to Queen Elizabeth'
5 Lynam *Mapmaker's Art* 66
6 Ibid. In his 'English Maps and Map-makers of the Sixteenth Century,'

Geographical Journal 116 (1950) 15 Lynam has suggested that Norden had noted that 'Camden's *Britannia*, being in Latin, was not for the general public; that Saxton's maps showed no roads, had no index by which places could be easily found, often included three or more counties on one sheet and that both works were large and heavy tomes'; and that this made him 'determined to write a series of brief county chorographies illustrated by small but practical maps, to be published as duo-decimo books easily carried in a pocket.'

7 A.W. Pollard 'The Unity of John Norden: Surveyor and Religious Writer' *Library* 7 ser 2 (1926) 233–52. It is interesting to note that, in a similar vein, Lambarde's colleague, Laurence Nowell the antiquary, may not have been Laurence Nowell the dean of Lichfield, as previously thought; see Pamela M. Black 'Some New Light on the Career of Laurence Nowell the Antiquary' *Antiquaries Journal* 62 (1982) 116–23.

8 John Norden *An Eye to Heaven in Earth* (London 1619) dedication

9 John Norden *A Good Companion for a Christian* (London 1632) dedication; quoted in Harry Gordon Rusche 'John Norden's *Vicissitudo rerum*' (PHD diss, University of Rochester 1962) 17n. I am also indebted to Rusche for his interpretation of Norden's melancholic state.

10 Norden published some seventeen recorded books and pamphlets between 1585 and 1625.

11 In dedicating his *A Sinfull Mans Solace* (London 1585) to Sir Edmund Anderson, the lord chief justice of the common pleas, Norden attempted to find an influential patron to support his various activities; see Rusche '*Rerum*' 5–6.

12 Trying out his poetical skills in *The Labyrinth of Mans Life* (London 1614) B5, Norden wrote that 'melancholy, the mother of best artes, Hath greatest power, (grace absent) in men's hearts.'

13 Lynam 'Maps and Map-makers' 15. Although Lynam and Pollard consider Norden's law career to have been profitable, the opposite was closer to the truth. Even Pollard 'Unity' 241 cannot conceal doubt about this point.

14 In the letter to Elizabeth that accompanied the presentation copy of his *Speculi Britanniae pars: The Description of Hartfordshire* (London 1598), Norden says that he undertook his surveying at the advice of her 'honrable conccellors,' probably Burghley.

15 Norden's original holograph manuscript of his *Northamptonshire* was located by Y.M. Goblet in the Bibliothèque nationale in Paris; it had been considered 'lost' until then. *Northamptonshire* was published as *Speculi Britanniae pars altera; or, a delineation of Northamptonshire* (London 1720).

16 On this topic see Norden's *Speculum Britanniae, The First part; an historicall and chorographicall discription of Middlesex* (London 1593) A3.

17 See London, BL, Egerton ms 2644 f 49.

18 Pollard 'Unity' 243 suggests that Norden printed *Middlesex* at his own expense, and made his profit by the gifts he received in return for presentation copies.
19 Norden *Middlesex* 5
20 Ibid 9
21 Ibid 11
22 Ibid 27
23 Ibid
24 Ibid
25 Ibid 36
26 Ibid 37
27 London, BL, Harleian ms 570
28 Ibid
29 Ibid. This part is quoted by W.B. Rye, *England as seen by Foreigners* (London 1865) 100.
30 John Norden *Speculi Britanniae pars; An Historical and Chorographical Description of the county of Essex* ed Sir Henry Ellis (London 1840)
31 Ibid 7
32 John Norden, *The Mirror of Honor* (London 1597) 30–1. This particular work is one of Norden's most profound devotional tracts; see Rusche *'Rerum'* 8.
33 This work remains extant as London, BL, Additional ms 31853.
34 R.A. Skelton 'John Norden's Map of Surrey' *British Museum Quarterly* 16 (1951–2) 61–2. There is a complete copy of the text of the description of Sussex in the Northamptonshire Record Office (Finch-Hatton ms 113), apparently made for the antiquary Sir Christopher Hatton. Norden's place-name derivations here have a curiosity value as the first attempt at analysis of Sussex names; see John H. Farrant 'John Norden's Description of Sussex 1595' *Sussex Archaeological Collections* 116 (1978) 269–75 and R.V. Turley 'Printed County Maps of Wight 1590–1870' *Proceedings of the Hants Field Club and Archaeological Society* 31 (1976) 53–64.
35 John Norden *Nordens Preparatiue to his 'Speculum Britanniae'* (London 1596) 1
36 Ibid 3
37 Ibid 3, 17. Taylor discusses the *Preparatiue* in greater detail in *Stuart Geography* 46–7.
38 In Great Britain, Historical Manuscripts Commission, 7th Report, app 537, 540, there is a reference to the letters from the council directing Norden to map the countryside.
39 Pollard 'Unity' 245
40 Norden asked Elizabeth for her 'neuer fayling bountie for relief' (Norden *Essex* xxxiv).
41 The manuscript is in Lambeth Library, London (codex 521).

42 A British Library copy of Norden's *Hartfordshire* contains manuscript notes by John Stow.
43 Norden's use of symbols is discussed in E.M.J. Campbell 'The Beginnings of the Characteristic Sheet to English Maps' *Geographical Journal* 128 (1969) 411–15.
44 Norden, BL Additional ms 9062 f 237
45 Norden *Hartfordshire* 24–5
46 Lynam 'Maps and Map-makers' 19–20
47 *Cornwall*, as noted, was published in 1728 and *Northamptonshire* in 1720; *Norfolk* as *The Chorography of Norfolk: an Historicall and Chorographicall Description of Norfolk* ed Christobel M. Hood (Norwick 1938); the references to the survey of Kent are found in R. Rawlinson *The English Topographer* (London 1720) 79. Also see A. Hassell-Smith and D. MacCulloch 'The Authorship of the Chorographies of Norfolk and Suffolk' *Norfolk Archaeology* 36 (1977) 327–41 and *The Chorography of Suffolk* ed D. MacCulloch (Ipswich 1976); see also my note 34 above.
48 As quoted in Taylor *Stuart Geography* 48
49 Norden *Cornwall* 7. The 1728 edition, cited here, was reprinted by the publisher Frank Graham (Newcastle-upon-Tyne 1966). Bateman's 1728 edition remains true to the original manuscript (London, BL, Harleian ms 6252).
50 Norden *Cornwall* (1728 ed) 9–10
51 Ibid 22, 17, 19
52 Kendrick *British Antiquity* 163
53 See William Ravenhill 'The Missing Maps from John Norden's Survey of Cornwall' in *Exeter Essays in Geography in Honour of Arthur Davies* ed K.J. Gregory and William Ravenhill (Exeter 1971); see also Lynam 'Maps and Map-makers' 15–23.
54 Norden *Cornwall* (1728 ed) section on Kirrier Hundred
55 Quoted in Katherine Koller 'Two Elizabethan Expressions of the Idea of Mutability' *Studies in Philology* 35 (1938) 231 and (partly) in Rusche '*Rerum*' 10
56 Marjorie Nicholson *The Breaking of the Circle* rev ed (New York 1960) 100
57 Dorothy Stimson *The Gradual Acceptance of the Copernican Theory of the Universe* (Hanover NH 1917) 72, 73, 87. Bacon became more opposed to the Copernican hypothesis as the years passed. His objections are relatively mild in *The Advancement of Learning* (bk 2 sec 8, 1), but in *De Augmentis Scientiarum* (London 1623) bk 3 he describes the theory as absurd.
58 Norden *Vicissitudo rerum* (London 1600) stanza 6. The Arabian astronomers who preceded Norden, and who based their work on Ptolemy, were the first to add on this sphere.
59 Norden *Eye to Heaven* dedication

60 London, BL, Additional ms 5752 f 306
61 See Fordham *Notable Surveyors* 13 and J.A. Steers 'Orford Ness: A Study in Coastal Physiography' *Proceedings of the Geologists' Association* 37 (1926) pt 3.
62 The Survey of Windsor, done in 1603, is entitled *A Description of the Honor of Winsor, namely of the Castle, etc.*
63 John Norden *The Surveyors Dialogue* (London 1607) 21
64 Ibid 214–15
65 Ibid 204 contains an example of this.
66 Great Britain, *Calendar of State Papers, Domestic Series*. (James I): vol 45 (1609) 518; vol 50 (1609) 566; vol 48 (1609) 544, 553; vol 55 (1610) 616; vol 58 (1610) 642; vol 67 (1611) 97, 108; vol 68 (1612) 121; vol 71 (1612) 158; vol 84 (1615) 340
67 An example of this is Norden's *Abstract of divers manors landes and tenements latelie graunted unto Prince Charles* (1617) London, BL, Additional ms 6027.
68 London BL, Harleian ms 6288
69 Gough *British Topography* 1: 266. See also McKisack *Medieval History* 144–5.
70 Quoted in John Norden's *Survey of Barley, Hertfordshire 1593–1603* ed J.C. Wilkerson (Cambridge 1974) 32
71 Levy *Tudor Historical Thought* 162
72 *Cornwall* 1728 edition, introduction
73 Sixteenth- and early seventeenth-century English travellers abroad are discussed in Boies Penrose *Urbane Travellers, 1591–1635* (Philadelphia 1942); Samuel Chew *The Crescent and the Rose* (New York 1937); Lillian Gottesman 'English Voyages and Accounts: Impact on Renaissance Dramatic Presentation of the African' *Studies in the Humanities* 2 (1971) 26–32. There is also, of course, the mass of material by and about Richard Hakluyt.
74 China, for example, had its tradition; see Joseph Needham *Science and Civilization in China* 5 vols (Cambridge 1954–) 1 (1954) 517.
75 For the Dutch, see W. Redmond Cross 'Dutch Cartographers of the Seventeenth Century' *Geographical Review* 6 (1918) 66–70.
76 Taylor *Stuart Geography* 136. For Taylor, apparently, 'chorography' and 'regional geography' are interchangeable, and here the latter is used.
77 J.N.L. Baker 'Academic Geography in the Seventeenth and Eighteenth Centuries' *Scottish Geographical Magazine* 51 (1935) 129
78 J.N.L. Baker 'Nathanael Carpenter and English Geography in the Seventeenth Century' *Geographical Journal* 71 (1928) 261; also Taylor *Stuart Geography* 136 and Bowen *Empiricism* 72–5. Bowen (75) states that Carpenter was able to pay considerable attention to the concept of nature in conjunction with his empiricism, his model in the empirical approach being Gilbert rather than Bacon. Carpenter devoted four out of twenty-eight chapters to discussions of

plains, rivers, mountains, etc, and provided perhaps the first adequate discussion of the principles of British geomorphology, in a manner similar to, though more 'scientific' than, that of the chorographer George Owen (see chap 11).

79 Ibid. Carpenter's book leaned heavily on foreign sources; see Gordon L. Davies 'Early British Geomorphology, 1578–1705' *Geographical Journal* 132 (1966) 253.

80 E.G.R. Taylor in *The Mathematical Practitioners of Tudor and Stuart England* (Cambridge 1954) 312 claims that the author of this anonymously published work is a man by the name of John Fitzherbert. E.R. Kiely in *Surveying Instruments* (New York 1947) 104 also attributes the authorship to a man by that name, although he seems less certain of this than Taylor. See also Reginald H.C. Fitzherbert 'The Authorship of the *Book of Husbandry* and the *Book of Surveying*' *English Historical Review* 12 (1897) 225–36 and DNB under Anthony Fitzherbert.

81 Some of the early works on navigation or aspects of it include William Borough *Discourse of the Variation of the Compas* (London 1581), Edward Wright *Certaine Errors in Nauigation* (London 1599) and his *The Description and Use of the Sphaere* (London 1631), and William Gilbert *De Magnete* (London 1600).

82 These included Leonard Digges *A Boke named Tectonicon* (London 1562) and his *A geometrical practise, named Pantometria* (London 1571), Edward Worsop *A discouerie of Sundrie Errours* (London 1582), and William Folkingham *Feudigraphia* (London 1610).

83 Even so, surveying did not undergo great development in the seventeenth century, a time when the best methods of earlier periods were systematized. Most of the advances awaited the next century, when machine-divided circles and telescopic methods were used on a large scale. In general, earlier instruments, such as the compass, chain, plane table, and circumferentor, had their limitations, which resulted in considerable errors in the traverses of large areas. The many admirable property plans dating from this earlier period, however (the platts of parishes or small properties), stand comparison with modern plans rather well. It should be noted that the sophistication of instruments was less important to surveying than the new activity of assessing by the surveyor.

84 Frank D. Adams *The Birth and Development of the Geological Sciences* (Baltimore 1938) chap 10

85 See chap 1. Robert Hooke was among the first to advocate the study of fossils for this reason (Robert Hooke *The Posthumous Works of Robert Hooke* ed Richard Waller (London 1705) 335, 441). On the 'decay' and related issues see

Gordon L. Davies *The Earth in Decay: A History of British Geomorphology 1678-1878* (London 1969), which is the source for much of my information here.
86 By 'great age' we mean – in the context of contemporary views on the subject – a period still in the range of thousands, not billions, of years.
87 Quoted in Davies *Earth in Decay* 17. See also E. Lankester ed *The Correspondence of John Ray* (London 1848) 243.
88 It is doubtful, for example, that potential readers would have appreciated being reminded that denudation was one sign of impending doom. The chorographer-cum-natural historian Joshua Childrey, in 1661, was perhaps one of the first to relate actual topographical features to denudation. He associated what he took to be the barrenness of the English Chalk Downs with the fact that the loose earth 'is continually washed away by great rains.' He also noticed that in Gloucestershire 'the hills, and sides of hills are the most wet and clayie. The cause doubtless is the same with this, to wit, That the rains that fall, wash by degrees the uppermost mould down into the Valleys, because it is more loose and light.' He continued: '[it] leaves the under-clay behind, because more stiff and fast, and so very hardly to be tempted away' Joshua Childrey *Britannia Baconica: or, The Natural Rarities of England, Scotland and Wales* (London 1661) 58; see Davies *Earth in Decay* 53, 56-7.
89 In this task the natural historians were aided by two influential foreign books on the topic: Bernhard Verenius's *Geographia generalis*, published in 1650, and Nicholaus Steno's *Prodromus*, published in 1669, a treatise on the geological history of the Tuscan landscapes.
90 One can see this change in the way topography was interpreted in the work of three men: Carpenter, Owen, and Hooke. Carpenter adopted the theory of the Creation, viewing the flood as consisting of waters too placid to uproot a tree, let alone reshape the land. Owen took the contrary view: 'the violence of the generall flood, which at the departinge thereof breake southward and tare the erthe in peeces and seperated the Ilands from the Contynent,' also making 'the hilles and valleies as we now finde them' (quoted in Davies *Earth in Decay* 38). Although the flood theory was to find its fullest expression in the late seventeenth-century writings of Burnet and others, by that time it faced stiff competition from the geomorphologically based one. Seismic activity was widely thought to periodically shatter the earth's crust, raising some blocks of earth while depressing others, an idea even admitted by some proponents of the flood theory. Hooke, though believing in the flood, went further than most in postulating a post-diluvial shifting of large land and sea masses (39-40).
91 One of the few exceptions to this was Sampson Erdeswicke (d 1603); see his *A Survey of Staffordshire* ed Sir Simon Degge (London 1717). *Staffordshire* was

also edited and published by Thomas Harwood, on two separate occasions (London 1820 and 1844). Erdeswicke provided an extensive record of the changes in the social order effected by the dissolution of the religious houses. Though it went unpublished at the time, his survey was praised by Camden, Burton, Gerard, and other notable chorographers.

92 Richard Carew *The Survey of Cornwall* (London 1602), found in F.E. Halliday ed *Richard Carew of Antony, 'The Survey of Cornwall'* (London 1953)
93 Camden *Britannia* 1586 ed 79 contains Camden's assessment of Carew.
94 In the 1607 edition of *Britannia* Camden acknowledges Carew as his chief guide for Cornwall; see Francis Lord de Dunstanville *Carew's 'Survey of Cornwall'* (London 1811) xix. The *Survey* (1602 ed) has been reprinted by the Da Capo Press (Amsterdam and New York 1969).
95 Halliday *Carew* 82
96 Ibid 105
97 Ibid 57
98 Brief biographical notices of Speed are found in Taylor *Stuart Geography* 49–51 and in John Arlott's introduction to *John Speed's England; A Coloured Fascimile of the Maps and Text from the 'Theatre of the Empire of Great Britaine,' First Edition, 1611* 4 vols (London 1953–4) 1:7–8. Assessments of Speed's work are found in Fussner *Historical Revolution* 178–9 and Levy *Historical Thought* 196–9. More attention has been directed to his chorographical projects, as in John Speed *A Prospect of the Most Famous Parts of the World, London 1627* ed R.A. Skelton (Amsterdam 1966) v–xiii.
99 Greville, later Lord Brooke, was a statesman, scholar, and author, assassinated in 1628 by a servant.
100 John Speed *The Theatre of the Empire of Great Britaine* (London 1611) 53
101 John Speed *The History of Great Britaine ...* (London 1611). The paging of this work is virtually continuous to Speed's *Theatre*. The *Theatre* was reprinted in 1614, 1617 (as *A Prospect of the Most Famous Parts of the World*), 1646, 1662, 1668, 1676, and in Latin editions in 1616 and 1646. There have been many abridgements. The *History* came out in a new edition in 1623 (reissued in 1625 and 1627), and in further editions in 1632 and 1650.
102 Speed *Theatre* 1611 ed 4
103 This discussion is based on Skelton's introduction to the fascimile edition of the *Prospect* vii.
104 Speed *History* 1611 ed 'Summary Conclusion'
105 John Speed to Sir Robert Cotton, nd, quoted in Sir Henry Ellis ed *Original Letters of Eminent Literary Men* Camden Society no 23 (London 1843) 110
106 Kendrick *British Antiquity* 164 states that Speed 'was not prepared to make history of anything but the records of credible historians who were in a position to inform him of facts actually within their knowledge.'

107 Speed *Theatre* 1611 ed 'The Countie Palatine of Chester'
108 R.A. Skelton 'Tudor Town Plans in John Speed's *Theatre*' *Archaeological Journal* 58 (1952) 112
109 See ibid 109–10.
110 Agas quoted in ibid 115
111 Speed *History* 1611 ed 165–6
112 Ibid 166–70
113 Kendrick *British Antiquity* 124–5, 165
114 Speed *History* 1611 ed 424

CHAPTER FOUR

1 McKisack *Medieval History* 147–8. No doubt it was easier for less genteel members to belong to the reborn Society of Antiquaries, which was less august than its forerunner; that may be why Strangman was able to find a niche there.
2 John Guillim *A Display of Heraldrie* 2d ed (London 1632) 341. This book, with its mythical menagerie engraved as armorial bearings, was the type of work Sir Thomas Browne would later complain of as one perpetuating outworn creeds; see Robert R. Cawley and George Yost *Studies in Sir Thomas Browne* (Eugene 1965) 4.
3 Robert Reyce quoted in C.G. Harlow 'An Unnoticed Observation on the Expansion of Sixteenth-Century Standard English' *Review of English Studies* new ser 21 (1970) 170
4 Letters from Reyce to D'Ewes are contained in London, BL Harleian ms 376 f 149 and Harleian ms 380 ff 136–7.
5 Oxford, Bodleian Library, Rawlinson ms B f 1424
6 This work bears no title. It survives in a number of copies, including one found in a volume entitled 'Collections for Suffolk' in the College of Arms. It can be argued, as does Spiegel 'Genealogy: Form and Function' 51 that in medieval times – and evidence for this is found in many of the early-modern works described in the present study – genealogy was important in fashioning 'history as linear narrative.' Thus, 'beneath the apparent narrative disarray, the paratactic disjunction of episodic units ... lay a metaphor of procreative time and social affiliation which brought together into a connected historical matrix the essential core of the chronicler's material.'
7 Lord Francis Hervey ed *Suffolk in the xviith Century: 'The Breviary of Suffolk' by Robert Reyce* (London 1902) 32
8 C.G. Harlow 'Correspondence' *Review of English Studies* new ser 21 (1970) 471. Reyce's work probably also owed something to Erdeswicke's interest in pedigrees and land ownership.

9 Hervey *Suffolk* 25-6
10 Ibid 206
11 Ibid 210-11
12 Louis B. Wright 'The Elizabethan Middle-Class Taste for History' *Journal of Modern History* 3 (1931) 175-97 has more to say on this topic.
13 This was published by Daniel King, an engraver, along with the works of several other writers (William Smith, Samuel Lee, and James Chaloner) under the title *The Vale-Royale of England, or County Palatine of Chester* (London 1656). King intended to contribute to the honour of the county of Cheshire by 'restoring to light its glory.' The Dedication is by King, as are the engravings.
14 Ibid 1
15 Ibid 5. J.P. Earwaker *East Cheshire: Past and Present* (London 1877) 8-15 has more to say on the early surveys of that county.
16 By the time William Burton's *A Particular Description of Leicester Shire* (London 1622) was published, the chorographies of four English counties had already appeared in print.
17 Burton *Leicester Shire* preface
18 Ibid 221
19 John Nichols *The History and Antiquities of the County of Leicester* 4 vols (London 1795-1811) 3 (1800) xx. This work was reprinted by the Scolar Press in 1971.
20 The manuscript of this intended version, coming to light in 1798, fell into the hand of W. Chetwynd, a Staffordshire antiquary, and was utilized by John Nichols in his own history of Leicestershire.
21 Camden *Britain* 1610 ed 568; see also Sir William Dugdale *The Antiquities of Warwickshire* (London 1656) 711.
22 Oxford, Bodleian Library, English Letters mss 44, 53, 62
23 Kendrick *British Antiquity* 167; Oruch 'Topography' 105-6
24 W.G. Hoskins and H.P.R. Finberg *Devonshire Studies* (London 1952) 334; quoted in Oruch 'Topography' 106
25 Tristram Risdon *The Chorographical Description, or, Survey of the County of Devon* (London 1714, new ed London 1811). I refer here to the 1811 edition.
26 William Chapple *A Review of Part of Risdon's 'Survey of Devon'* (Exeter 1785) 3-4
27 Ibid 5
28 Risdon *Devon* 2
29 Ibid 4; see also W.G. Hoskins *New Survey of England: Devon* (2d ed London 1972) 148 and Oruch 'Topography' 106.
30 Risdon *Devon* v, 8
31 Ibid 7

32 Edward Cotton 'Of a considerable Load-stone digged out of the Ground in Devonshire' *Philosophical Transactions* 2 (1667) 423–4
33 Risdon *Devon* 11–12
34 Chapple *A Review* 35nt
35 Risdon *Devon* 14
36 Ibid 237. At one point (201), Risdon quotes Michael Drayton's verses on Plymouth in Drayton's *Poly-olbion* (London 1612).
37 Much of the information contained in *Devon* is also found in a separate manuscript (London BL, Additional ms 36748, written c 1633), which is attributed to Risdon. See also London, BL, Additional ms 33420 f 156, which concerns an eighteenth-century imperfect copy of Risdon's major work.
38 James Dallas and Henry G. Porter eds *The 'Note-Book' of Tristram Risdon* (London 1897)
39 Ibid xvi
40 Risdon *Devon* 29
41 John Prince *Danmonii orientales illustres; or, The Worthies of Devon* (Exeter 1701), quoted in Sir John William de la Pole ed *Collections toward a Description of the County of Devon* (London 1791) x
42 This letter is contained in London, BL, Harleian ms, 1195 f 37, and is quoted by de la Pole *Collections* iv–vi.
43 The 1616 volume was published in part by Sir Thomas Phillips as *Sir William Pole's Copies of, and Extracts from Old Evidences* (Middle Hill 1840?).
44 Pole *Collections* 121
45 Ibid 109–11
46 Thomas Westcote *A View of Devonshire in 1630* ed George Oliver and Pitman Jones (Exeter 1845)
47 Quoted in Risdon *Devon* xvi
48 Westcote *Devonshire* iv
49 Ibid
50 Ibid xvi
51 Ibid 34
52 Ibid 44
53 Ibid 43–4
54 Ibid 68
55 Ibid 67
56 Ibid 93
57 Ibid 251
58 Ibid 98, 348
59 Ibid 95
60 At his death in 1641, Sir Henry Spelman left behind an unfinished manuscript of

a county chorography, the 'Icenia; sive Norfolciae Descriptio topographica.' It is the only one written in Latin, and is primarily concerned with land ownership. However, Spelman also discusses etymologies, topography, and certain Roman antiquities. Spelman's major antiquarian production was the *Archaeologus* (London 1626). This work is a historical dictionary of legal and ecclesiastical terminology, remarkable for its historiographical significance. Spelman was familiar with all avenues of antiquarian research, not excluding chorography.

61 Thomas Gerard *A Survey of Dorsetshire* (London 1732) 'The Method Observed in this Survey of Dorset'; *Dorsetshire* was published under Gerard's pseudonym 'John Coker of Mapowder, Dorset.'
62 The similarity between the two works is discussed by John Batten 'Who Wrote Coker's *Survey of Dorsetshire?*' *Somerset and Dorset Notes and Queries* 5 (1890) 97–102.
63 Thomas Gerard *The Particular Description of the County of Somerset* ed E.H. Bates, Somerset Record Society no 15 (London 1900) xxii; quoted in Oruch 'Topography' 108
64 Gerard (John Coker) *Dorsetshire* 35
65 Gerard *Somerset* 12
66 Ibid xxii
67 For example, Gerard quotes a portion of Camden's description of Winburne, in *Dorsetshire* 112.
68 John Weever *Ancient Funerall Monuments* (London 1631) preface
69 He was one of the petitioners to the queen for the formation of an academy to house rare books and manuscripts; see McKisack *Medieval History* 145, and J. Hunter 'An Account of the Scheme for Erecting a Royal Academy in England in the Reign of King James I' *Archaeologia* 32 (1847) 132–49.
70 McKisack *Medieval History* 147
71 Sir John Dodridge *The History of the Ancient and Moderne Estate of the Principality of Wales, Dutchy of Cornwall, and Earldom of Chester* (London 1630) 77–8
72 Ibid 28
73 Ibid 92
74 Some works whose nature was more that of a diary or travel journal came close, at the Interregnum or earlier, to resembling the chorographic format or type. Studies based on the county unit temporarily disappeared. Among such quasi-chorographies there is Peter Heylyn's *A Full Relation of Two Journeys: The One into Main-Land of France, The Other, Into some of the Adjacent Islands* (London 1656), which included a religio-topographical account of the Channel Islands; John Smyth's description of the hundred of Berkeley, of which he

was steward, and his entertaining *Lives of the Berkeleys*, in three folio volumes containing nearly one thousand pages, completed in 1618 (on his work see Sir John Maclean ed *The Berkeley Manuscripts* Bristol and Gloucester Archaeological Society no 1, 3 vols (Gloucester 1883–5). William Lithgow *Present Surveigh of London* (London 1643) is largely a commentary on contemporary military and social affairs of war-weary London, whose populace, despite the hardships it endured, still delighted in public shows and 'frivole ostentations,' and enjoyed an abundance of various commodities, though 'wanting nothing except peace' (A4). As far as town surveys go, the period saw the issue of William Somner's *The Antiquities of Canterbury* (London 1640). Somner, the famous antiquary, compiled this work mainly from old manuscripts, ledger books, and other records found in the archives of Canterbury Cathedral, and focuses on the ecclesiastical government and on the histories of the religious establishments; ie, only vaguely can it be considered a chorography. The term is more appropriately applied to William Grey's *Chorographia: Or, a Survey of Newcastle-Upon-Tine* (published in 1649 and contained in *Harleian Miscellany* 3 (1745) 256–73), where it is evident that by then the Brute legend had totally fallen to the disaffection of Clio. In some respects it was similar to William Bedwell's much earlier *A Brief Description of the Towne of Tottenham Highcrosse in Middlesex* (London 1631).

CHAPTER FIVE

1 David C. Douglas 'William Dugdale: "The Grand Plagiary"' *History* 20 (1935) 203–4; see chap 2 of Douglas *English Scholars* (1951 ed) 42. As for the advances that are described in the text, see J.G.A. Pocock's *The Ancient Constitution and the Feudal Law* (Cambridge 1957).
2 See Stuart Piggott 'Archaeological Draughtsmanship: Principles and Practice' *Antiquity* 39 (1965) 171.
3 Walters *The English Antiquaries* 11
4 Kendrick *British Antiquity* 167
5 Dugdale *Warwickshire* 778
6 Ibid 3–5. Also William Hamper ed *The Life (written by himself and continued to his death), Diary, and Correspondence of Sir William Dugdale* (London 1827) 462; and Stuart Piggott *Ruins* 14
7 Dugdale is the subject of chap 2 in Douglas *English Scholars*. Much of the information we have concerning Dugdale's life and work was originally preserved for posterity by his contemporaries, such as Anthony à Wood, men who were considerably impressed by and influenced by Dugdale's scholarship.

8 Hamper *Dugdale* 5, 7–8
9 Douglas *English Scholars* 1951 ed 30–1
10 Hamper *Dugdale* 9
11 Ibid 14
12 Oxford, Bodleian Library, Aubrey ms 3 f 81
13 Dugdale's estate was sequestered by the Roundheads, requiring King Charles to give him a warrant for 13s 4d a day; Dugdale claims never to have received any of this. While in Oxford he lived, apparently, by conducting – in his capacity as herald – the funerals of prominent Cavaliers who were killed in action.
14 Royce MacGillivray *Restoration Historians and the English Civil War* (The Hague 1974) 56
15 Douglas *English Scholars* 1951 ed 36
16 Ibid 37–8
17 While collecting antiquarian information for his study of Leicestershire, Burton was also able to gather similar material for Warwickshire, which he was willing to provide to anyone 'that shall undertake the Illustration of the countye' (Oxford, Bodleian Library, English Letters mss 44, 53, 62, 76).
18 Hamper *Dugdale* 9; quoted in Douglas *English Scholars* 1951 ed 31
19 Some of the originals and drafts of their letters are found in Oxford, Bodleian Library, English Letters ms B1, collected by William Hamper.
20 Hamper *Dugdale* 273
21 For Archer see Phillip Styles 'Sir Simon Archer, 1581–1662,' an occasional paper published by the Dugdale Society in 1946.
22 Hamper *Dugdale* 185–6
23 Ibid 249
24 Dugdale *Warwickshire* in the dedication to Christopher Hatton
25 Ibid preface
26 Ibid 'To my Honoured Friends the Gentrie of Warwick-Shire'
27 Ibid preface. It is here that Dugdale states: 'As for the Work itself, it is an Illustration of the Antiquities with which my native country Warwickshire hath been honoured.' The book is opulently, almost gaudily, adorned with the arms and pictures of many prominent persons.
28 Ibid 'To my Honoured Friends the Gentrie of Warwick-Shire'
29 Ibid preface
30 Ibid
31 Ibid 297
32 For example, ibid 35 (in reference to Hyde), 60 (Cest-over), 219 (Radburne). These are all settlements noticed as having suffered a loss in population.
33 Ibid preface
34 For the entry on Hyde (ibid 35), for example, Dugdale refers the reader to Sir Henry Spelman's *Glossarii*, 'where may be seen the various acceptions thereof;

conceiving, that in this place it was first imposed, to impress a certain quantity of Land sufficient for one Plough to manage.'
35 Dugdale *Warwickshire* 92, 96. Aston 'English Ruins' 252n speculates that Dugdale's 'obsession with past losses' (eg, the dissolution and ruin of religious establishments) contributed to his 'hasty travels and work on the eve of the Civil War, which he rightly feared would entail more destruction.'
36 Dugdale *Warwickshire* 95
37 Similarly, with reference to the *Monasticon*, Roger Twysden, for example, a student of constitutional law and a friend of Dugdale's, did exactly this, by questioning the authenticity of certain documents used by Dugdale (Hamper *Dugdale* 335).
38 Ibid 182–3. A work by J. Ives *Select Papers Chiefly Relating to English Antiquities* (London 1773) contains Dugdale's 'Directions for the Search of Records, and Making Use of them, in order to an Historicall Discourse of the Antiquities of Staffordshire.' Ives claims that these 'Directions' were probably written for the benefit and use of Robert Plot, who was engaged in the collection of material towards his natural history of that county. In this work, Dugdale discusses the various types of source material and their locations. In so doing he sounds much like an experienced tutor, paternalistically educating a green student; eg, 'Next, the Charter Rolls (of these I can help you to an abstract, which will point out the Roll and number of what you shall finde for your purpose)' (37); or, 'Then the Patent Rolls, Clause Rolls, Fine Rolls, and all the Foreign Rolls, all these you must look over diligently, backsides as well as foresides' (ibid); Dugdale concludes by referring the reader – presumably Plot – to the example of his own work: 'It will be worth while to see my Collections for Warwickshire, as to the method and order which I used therein' (ibid). A valuable discussion of Dugdale's use of evidence is contained in H.A. Cronne 'The Study and Use of Charters by English Scholars in the Seventeenth Century: Sir Henry Spelman and Sir William Dugdale' in *English Historical Scholarship* ed Fox 88–91
39 Oxford, Bodleian Library, Ballard ms 14 f 6
40 Oxford, Bodleian Library, Wood ms E41 f 75. Stuart Piggott 'Antiquarian Thought in the Sixteenth and Seventeenth Centuries' in *English Historical Scholarship* ed Fox 110 discusses several other examples of Dugdale's objective interest in field archaeology. It seems that, on the whole, Dugdale's interest in this area became considerably more pronounced after the publication of *Warwickshire*, ie, after the natural historians were beginning to bring to light their own findings on the subject.
41 Hamper *Dugdale* 309
42 Richard Kilburne's *A Topographie; or Survey of the County of Kent* (London 1659) is enlarged from his *A Brief Survey of the County of Kent* (London

1657). White Kennett *Life of Mr. Somner* (Oxford 1693) 19 excuses Kilburne's sparse treatment of Canterbury on the grounds that 'Mr William Somner had so elaborately, judiciously, and fully wrote of the same, that there was left but little (if anything observable) which he had not set down.'

43 *Kent Surveyed* was published by and under the name of Philipot's son Thomas, who has been resoundingly criticized for dishonestly attempting to palm it off as his own production. Kennett *Somner* 37 and others have attributed the work to the elder Philipot; see H. Stanford London *John Philipot, M.P., Somerset Herald, 1624–1645* (Ashford and London 1948) 48.
44 Philipot *Kent Surveyed* 93
45 Ibid preface
46 London *John Philipot* 48
47 Philipot *Kent Surveyed* 182
48 Edward Leigh *England Described* (London 1657) 'To the Candid Reader.' Leigh listed the names of the authorities whose books he had perused. These included Burton, Carew, Drayton, King, Lambarde, Leland, Norden, Somner, and Stow.
49 Not only do the two books utilize the same types of sources, and therefore examine the same types of topics, but they also exhibit a similar language and the first editions (at least) of both works are remarkably close in their layout, typography, and use of illustrations. Gough *British Topography* 1:viii says that though he followed Dugdale's pattern, Thoroton's work shows 'all the marks of inferior abilities.'
50 John T. Godfrey *Robert Thoroton, Physician and Antiquary* (Nottingham? 1890) no pagination
51 Robert Thoroton *The Antiquities of Nottinghamshire* (London 1677) preface
52 It may simply have been an oversight on Thoroton's part, or else a reflection on Dugdale's nature, that Dugdale received no presentation copy of *Nottinghamshire* and had to pay a sum of 16 s to 18 s to obtain the book once it was published; see Great Britain, Historical Manuscripts Commission *12th Report* app vii.
53 The famous antiquary-historian John Throsby incorporated and enlarged upon Thoroton's study in his own work on Nottinghamshire, published in the 1790s; see the reprint edition of Throsby's volumes in the *Classical County Histories* series.
54 Among the chorographies produced after Dugdale, there is Sir Peter Leycester *Historical Antiquities in two books* (London 1673); James Wright *The History and the Antiquities of the County of Rutlandshire* (London 1684); and Sir Henry Chauncey *The Historical Antiquities of Hertfordshire* (London 1700). Silas Taylor, alias Domville, made extensive chorographic collections for Harwich,

and also focused on 'Natural Accidents and Productions.' However, 'altho' he hath here a large Field for Natural History, yet he is so very short, that in many Parts thereof there is little more than the Names of the things' (Samuel Dale *The History and Antiquities of Harwich and Dovercourt ... Collected by Silas Taylor ...* [London, 1730] iv).

CHAPTER SIX

1 Jerry Weinberger *Science, Faith, and Politics: Francis Bacon and the Utopian Roots of the Modern Age* (Ithaca and London 1985)
2 Baconism was used in more than a few instances by such apologists to portray science as an egalitarian co-operative venture, one removed from political antagonism and religious sectarianism, ie, a type of Baconism that is not identifiable with the 'vulgar Baconianism' of the Puritan reformers of the 1640s to 1650s. On the other hand, Baconism had its critics. For example, Robert Hooke's 'Discourse of Earthquakes' (1668) begins with a critical assessment of the 'naive' Baconian collection of data by some early members of the Royal Society; see D.R. Oldroyd 'Robert Hooke's Methodology of Science as Exemplified in His "Discourse of Earthquakes"' *British Journal for the History of Science* 6 (1972–3) 111. Hall *Revolution* 248 states his opinion that the utilitarianism of the Royal Society 'must be diminished by appreciation of the vast distinction between what men say, or even suppose themselves to be doing, and what they naturally achieve.' The desire to know, for the sake of knowing itself, it is said, was already institutionalized in the universities of Europe, and the fruits of scientific labour were applied only sporadically and usually not until long afterwards (ibid). Although probably true, much of this misses the point; that is, the utilitarian motive – whether it resulted in immediate practical benefits or not – did in fact spur many men into scientific endeavour.
3 Kuhn 'Mathematical vs Experimental' 1–31. This work is also reproduced in Kuhn's *The Essential Tension* (Chicago 1977) 31–65.
4 Kuhn 'Mathematical vs Experimental' 12
5 Ibid 15
6 Ibid 15–16
7 Francis Bacon 'Preparative toward Natural and Experimental Philosophy' in Fulton H. Anderson ed *The New Organon and Related Writings* (New York 1960) 272
8 Bacon *Works* 4:293–4. Barbara Shapiro *Probability and Certainty in Seventeenth-Century England* (Princeton 1983) 125 notes that most Englishmen understood civil history to mean 'conventional political and governmental

topics.' Fussner in *Historical Revolution* 153 notes that within the realm of civil history 'Bacon failed to discuss some titles which he ought to have recognized as historical. The survey, for example, Bacon ignored although most local history was written in survey form.' Fussner refers to, of course, local history written before Bacon's ideas began to take hold. It will be recalled, however, that chorographers should be regarded not merely as historians engaged in narrative history, but as antiquaries as well.

9 F.H. Anderson *The Philosophy of Francis Bacon* (Chicago 1948, reprint ed New York 1971) 152. See also Shapiro *Probability and Certainty* 130–5.

10 Bacon *Works* 4:303

11 Anderson *Francis Bacon* 152. According to Shapiro *Probability and Certainty* 126, 'Bacon used the term "history" in connection with astronomical observation, employing it to cover any kind of accurately reported data devoid of theoretical speculation.' Shapiro goes on ' ... much of the parallelism between history and natural science in the seventeenth century was based on the notion that history – any kind of history – consisted of an accurate report of past and present facts and events.'

12 See Stuart Clark 'Bacon's *Henry VII: A Case-Study in the Science of Man*' *History and Theory* 13 (1974) 102. Clark notes the gulf between historians 'proper' (those involved in literary or 'philosophical' history) and the antiquaries, who seldom claimed to be writing history proper, quoting (101) Thomas Madox's 1711 lament that 'Lovers of Antiquities are commonly looked upon to be men of Low unpolite genius, fit only for the Rough and Barbarick part of Learning.' It was difficult for historians proper to accept the use of non-literary sources, although the regional writers were among the first to have a foot in each camp.

13 Ibid 104

14 Ibid 104–6. Leonard F. Dean 'Sir Francis Bacon's Theory of Civil History-Writing" *ELH: A Journal of English Literary History* 8 (1941) 165 comments on Bacon's desire to eliminate the barriers between the departments of knowledge, quoting Bacon's assertion that the inductive method is applicable to parts of the subject matter of civil as well as natural history, but also pointing to Bacon's scepticism of employing scientific exactness in the treatment of civil affairs.

15 Thomas Sprat *The History of the Royal Society* (London 1667) 61. Roy Porter in *The Making of Geology* (Cambridge 1977) 12–13 points out that 'enumeration of the Earth was neglected by such leading naturalists as William Turner, John Gerard and Thomas Johnson, their passion being the animal and vegetable kingdoms,' and that 'general naturalists' service to later Earth science lay in transmitting and popularizing Classical and Renaissance knowledge in their books of natural history, wonders and secrets, lapidaries, pharmacopoeias and encyclopaedias' (13).

In this chapter in particular I am indebted to Barbara Shapiro for her views on the ties between civil and natural history.
16 Even though Carew inserted more material on natural history into his study than any other chorographer, as Charles E. Raven *English Naturalists from Neckham to Ray* (Cambridge 1947) 247 states: 'In the history of science Richard Carew may be a person of no importance: he made no great discoveries nor much contribution to knowledge,' even if 'in the history of civilisation he is a portent, a social type, representing a way of life, a quality of culture, a relation to his environment that are altogether novel.'
17 W.N. Edwards *The Early History of Palaeontology* (London 1967) 1; see also J.E. Hare 'Aristotle and the Definition of Natural Things,' *Phronesis* 24 (1979) 168–79.
18 They owed a large debt to the men who took the time to translate the classics, such as Philemon Holland, whose *The Historie of the World: commonly called, 'The Natural Historie of C. Plinius Secundus'* (London 1601) was read by many of the seventeenth-century natural historians; see also Charles G. Nauert, Jr 'Humanists, Scientists, and Pliny: Changing Approaches to a Classical Author' *American Historical Review* 85 (1979) 72–85.
19 See R.F. Jones *Ancients and Moderns: A Study of the Rise of the Scientific Movement in 17th-Century England* (St Louis 1936; 2nd ed 1961), which discusses how the learned men of the seventeenth century gradually came to realize that they were equal – if not superior – to the giants of antiquity; this view was thus shared by the natural historians.
20 Bacon *Works* 5:131. In a similar context (4:28), commenting on the 'Phenomena of the Universe,' Bacon describes it as 'experience of every kind, and such a natural history as may serve for a foundation to build philosophy [ie, sciences] upon.'
21 Bacon's plan was set out in *The Advancement of Learning*: see Bacon *Works* 3:330–3; and *Works* 4:294–9 (*De Augmentis Scientiarum*).
22 Ibid 4:294
23 Ibid
24 Ibid 4:295
25 Ibid
26 Ibid 4:295–6
27 Ibid 4:296
28 Ibid
29 Ibid
30 Ibid 4:294–5. Anderson *Francis Bacon* 151 summarizes this as 'whatever effects are produced by nature in her own workings, or by man in conjunction with nature by art, are all natural, the works of God's creatures.'
31 Bacon *Works* 4:297–8. Paolo Rossi *Francis Bacon: From Magic to Science* trans

Sacha Rabinovitch (London 1968) 218 says that Bacon's plan 'implies both a refutation of rhetorical culture and a new appreciation of the significance of the mechanical arts in intellectual spheres' and 'found its total fulfilment in the achievements of the Royal Society and the great Encyclopedia of the Enlightenment.'

32 Bacon *Works* 4:296
33 Anderson *Francis Bacon* 263. Purver *Royal Society* 35 describes it in these words: 'The practical skill and diligence which characterized the approach of the artisan, freed from an attachment to immediate ends, must be carried into the realm of intellect, itself freed from subservience to preconceived opinion.'
34 Bacon *Works* 4:299
35 Ibid
36 Ibid
37 Anderson *Francis Bacon* 261. It is noteworthy that Bacon's third directive specified that accounts of 'prodigies' in nature, under certain circumstances, are justifiable (ibid).
38 Ibid 263
39 Ibid 264–5
40 Ibid 266–9. J.A. May *Kant's Concept of Geography* (Toronto and Buffalo 1970) 37 notes that of the 130 titles Bacon recognized only one category of history as 'Natural,' that being the 'Natural History of Geography,' indicating that he is leaving out one aspect of the subject. According to May, 'Geography ... is intimately connected with "Nations, Provinces, Cities, and such like matters pertaining to Civil life." Thus, geography displays the unusual quality of cutting across the distinction between natural and civil.'
41 Rossi *Francis Bacon* 216
42 Ibid 217
43 Bacon *Works* 5:135–6
44 Purver *Royal Society* 43
45 Rossi *Francis Bacon* 218–19
46 Bacon *Works* 4:303–4
47 Shapiro *Probability and Certainty* 135
48 Ibid
49 It is true that antiquaries of all periods had to be familiar with non-documentary evidence such as coins, inscriptions, forts, etc, but it was only when the virtuosi of the post-Baconian period began applying some of Bacon's dicta regarding scientific method to the study of artefacts and other 'archaeological' evidence that we find the beginnings of recognizably 'modern' archaeological methods.
50 Shapiro *Probability and Certainty* 136
51 Ibid 136–7

52 Ibid 160. The rejection of some rhetorical features by history is attributed by Shapiro to the realignment of the natural sciences towards probabilistic knowledge (ibid 161).
53 Ibid 145–9
54 Bacon *Works* 4:299
55 It should be stated that Bacon is not taken here to be the only source of a 'quantitative' method, but as a major proponent of it.
56 Oxford, Bodleian Library, Aubrey ms 3 f 18. For a while Aubrey was involved in compiling a biography of Bacon; see Michael Hunter *John Aubrey* 44, 73n.
57 The influence peaked at about the mid-point of the seventeenth century.
58 Allen G. Debus 'The Medico-Chemical World of the Paracelsians' in *Changing Perspectives in the History of Science* ed Mikuláš Teich and Robert Young (London 1973) 90
59 Ibid passim
60 A. Rupert Hall *From Galileo to Newton* (New York 1981) 26
61 Ibid. Of course, not all the examples found in the work of the regional natural historians can be directly attributed to Paracelsus. The question of rationality of the science of the past is considered in Mary Hesse 'Reasons and Evaluations in the History of Science' in *Changing Perspectives* ed Teich and Young 127–47; Hesse rejects the significance of Renaissance hermeticism, at least, for the study of the history of science.
62 P.M. Rattansi 'Some Evaluations of Reason in Sixteenth- and Seventeenth-Century Natural Philosophy' in ibid 153
63 P.M. Rattansi 'The Social Interpretations of Science in the Seventeenth Century' in *Science and Society 1600–1900* ed Peter Mathias (Cambridge 1972) 10. More emphasis should be placed on the non-magical element in the thought of Paracelsus. For example, some recent historiography claims that his call for knowledge and manipulation of nature was part of his desire to denigrate the role of supernatural agents.
64 Ibid 12. Hesse 'Reasons and Evaluations' 133–4 downplays the role of sectarians such as Hartlib, Dury, and Comenius on the 'internal tradition' of the history of the mechanical philosophy, even if they contributed to the Baconism of the early Royal Society. But even if, as Hesse states (141), one cannot suggest a 'confluence of hermeticism and mechanism into the melting pot of the new science,' it may be argued that elements of hermeticism retained a fascination for at least some of the non-mechanically inclined virtuosi of the post-1660 period. Often the boundary between the 'rational scientist' and 'irrational Hermetic magician' is a hazy one at best.
65 Frances A. Yates 'The Hermetic Tradition in Renaissance Science' in Singleton

ed *Art, Science, and History* 262. Yates, in a rather convincing manner, states that this was 'the fatal blind spot in Bacon's outlook and the chief reason why his inductive method did not lead to scientifically valuable results.'
66 Hall *Galileo to Newton* 27
67 Ibid 26
68 K. Theodore Hoppen 'The Nature of the Early Royal Society' *British Journal for the History of Science* 9 (1976) 241, 248. Kircher himself could be classed either as a rational scientist one day or a 'credulous neo-medievalist' the next; see Joscelyn Godwin *Athanasius Kircher* (New York 1979) 85.
69 Allen G. Debus *The English Paracelsians* (London 1965) 80. Even earlier, alchemists such as Dee had to go to Europe for notice to be taken of their work (101).
70 Ibid 182; see also P.M. Rattansi 'Paracelsus and the Puritan Revolution' *Ambix* 11 (1963) 24–32.
71 See Walter Pagel 'The Spectre of Van Helmont and the Idea of Continuity in the History of Chemistry' in *Changing Perspectives* ed Teich and Young 104. Thanks to Boyle, chemistry became a science through the revival of the corpuscular hypothesis. Van Helmont, although given credit for his role in establishing chemistry as a science, in fact opposed the corpuscularians to some extent (105).
72 Ibid 104
73 Yates 'Hermetic Tradition' 268–9. Yates suggests that Bacon's rejection of Copernican heliocentricity and Gilbert's magnetic philosophy of nature was due to the fact that he associated both with Fician solar magic and/or animist philosophy, popularized by Giordano Bruno. In like manner Bacon may have downplayed the role of mathematics in his method because he associated it with a conjuring of numbers reminiscent of John Dee.
74 Yates 'Hermetic Tradition' 270
75 Bacon in the *Novum organum*, quoted in Rattansi 'Social Interpretation' 13
76 Ibid 17
77 W.H.G. Armytage 'The Early Utopists and Science in England' *Annals of Science* 12 (1956) 253–4
78 Mayling Stubbs 'John Beale, Philosophical Gardener of Hertfordshire, Part 1. Prelude to the Royal Society (1608–1663)' *Annals of Science* 39 (1982) 482–3. Bacon's philosophy is nicely summarized in Levine *Humanism and History* chap 5.
79 Hunter *John Aubrey* 118, 129, passim
80 Hoppen 'Early Royal Society' 259–61. Hoppen objects to the suggestion that Plot deliberately hid his alchemical concerns.
81 Rattansi 'Evaluations of Reason' 165
82 Francis Bacon, quoted in Rossi *Francis Bacon* 21
83 Ibid 23

84 Yates 'Hermetic Tradition' 267
85 Ibid 273
86 Webster *Paracelsus to Newton* 12 passim. In his discussion of the many interwoven strands of intellectual/political/scientific/pseudo-scientific thought of the age in Europe, Webster makes one wonder further about the appropriateness of compacting the situation into two neat phases; it is also incorrect to utilize the major categories of 'scientific' and 'occult,' as does Brian Vickers ed *Occult and Scientific Mentalities in the Renaissance* (Cambridge 1984). Both Vickers's and Webster's books are reviewed in Patrick Curry 'Revisions of Science and Magic' *History of Science* 23 (1985) 299–325.
87 Webster *Paracelsus to Newton* 69
88 Curry 'Revisions' 321

CHAPTER SEVEN

1 John Jay O'Brien 'Commonwealth Schemes for the Advancement of Learning' *British Journal of Educational Studies* 16 (1968) 42
2 Webster *Great Instauration* 511. It may also be noted that Hartlib himself was considered too radical in his affiliations to be installed as a member of the Royal Society.
3 More, quoted in Armytage 'The Early Utopists' 247. Armytage goes on to state that 'More's Utopians were pragmatists: the world of nature and man was a universal laboratory where operations had been going on ... to find out what was in harmony or in conflict with Nature' (248).
4 Ibid 249
5 Hoppen 'Early Royal Society' 3
6 Ibid 5
7 Webster's labelling of utilitarian – ie, scientific – traits as 'Puritan,' and his definition of what constituted a 'Puritan' itself, has come under scrutiny. Webster understands puritanism as reformist religion generally, ranging from certain sectaries to liberal Anglicans.
8 In this respect Webster is not alone. R.K. Merton, Christopher Hill, and Barbara Shapiro, among others, fall into this category; see Lotte Mulligan 'Civil War Politics, Religion and the Royal Society' *Past and Present* 59 (1973) 92–116.
9 Some recent work on the topic has suggested that, in fact, religion was viewed by Interregnum 'scientists' as a distraction from the day-to-day workings of science. See Lotte Mulligan 'Puritans and English Science: A Critique of Webster' *Isis* 71 (1980) 456–69.
10 A. Rupert Hall, review of Charles Webster's *The Great Instauration: Science, Medicine and Reform, 1626–1660* in *English Historical Review* 91 (1976) 857–8

11 Joan Thirsk 'Agricultural Innovations and their Diffusion' in *The Agrarian History of England and Wales* Joan Thirsk gen ed 8 vols (Cambridge 1967–) vol 5, II: *1640–1750: Agrarian Change* by Joan Thirsk ed (Cambridge 1985) 533. This statement, in fact, applies to most of the individual members of the Hartlib circle. Thirsk, however, also assesses the important role of individuals such as Sir Richard Weston, who, although known to Hartlib, who acted as his publisher, was strictly speaking outside the Puritan party.
12 Ibid 534–5
13 Ibid 552–3
14 Ibid 557–8
15 Mulligan 'Critique' 468–9 states that 'Certainly it is true that there was cross-fertilization between seventeenth-century theology and the explanations men gave of the natural world. But the emphasis should be on a general English providential view of the natural order rather than on the peculiar millenarian, pansophist element of the radical Puritans.' K. Theodore Hoppen in 'Early Royal Society' 9 points out that the Baconism of the Royal Society had 'a less immediately theological flavour than that which had influenced the concerns of Hartlib and his circle.'
16 Webster *Great Instauration* 479
17 Ibid. Despite this, Beale can, like most of his associates, be properly called 'A virtuoso rather than a scientific specialist'; see Stubbs 'John Beale' 489.
18 Ibid 478–9
19 Thirsk *Agrarian Change* 549. Like Boate, Beale also had his Irish connections.
20 James R. Jacob and Margaret C. Jacob 'The Anglican Origins of Modern Science: The Metaphysical Foundations of the Whig Constitution' *Isis* 71 (1980) 258, state that 'The initial Puritan reforming vision of the 1640s ... survived in a continuing belief in the material benefits of science,' one that was joined by the late 1650s to a new Anglican theology not essentially puritan but latitudinarian. The key words now became 'design, order, and harmony as the primary manifestations of God's role in the universe.' Predestination and radicalism came under the attack of this 'scientifically grounded latitudinarianism,' which was adopted as the 'public stance' of the Royal Society. Anglican science was, if one accepts the Jacobs' arguments, 'shaped by the dialectics of revolution: as a result scientific arguments gave vital ideological support to Protestant monarchy' (267). If this in fact was the case, then it could be argued that the Baconian inquiry of the Royal Society was based as much on ideological reasons as on scientific ones. The sectaries, with their belief that knowledge of God's works resulted from mystical experience or God's direct revelation to the saints, could thus be countered by the belief that this knowledge could be

gained instead through the reading of scripture and the study of nature, ie, scientific investigation (256). With reference to the post-1660 period Mulligan states in 'Civil War Politics' 111: 'science correlated less with Puritanism or latitudinarianism than with the waning role of religion.' In this scenario, then, the regional natural historians of the second half of the seventeenth century as a group, for example, cannot be neatly matched with any one particular religious affiliation. Mulligan, in fact, is one of those who argue that there is little substance to the thesis that Puritanism encouraged an interest in science (108). If anything, she believes, the distinction may be between Parliamentarians (pre-1660), who tended to applied, practical science, and Royalists, who tended to theoretical science.

21 Hall, review of *The Great Instauration* 858–9. Mulligan 'Critique' 468 draws attention to the non-Puritan origins of experimental science and the Royal Society.
22 Hartlib gave up his plan for the Office of Address in 1659, as the close association between Puritan patrons, politicians, and intellectuals began to dissolve with the end of the Protectorate (Webster *Great Instauration* 85).
23 Webster *Great Instauration* 64–5
24 Members of the Oxford circle were stigmatized as 'Vertuosi' by the upholders of the ancients; see Webster *Great Instauration* 172. The Boates, among other members of the Invisible College, were enlisted into participating in the Office of Address scheme.
25 Ibid 213
26 Ibid 377 points out that Aubrey believed it was this latter kind of attitude that united a broad group of intellectuals, even if there was some disagreement over precise social policies.
27 Margaret Espinasse 'The Decline and Fall of Restoration Science' *Past and Present* 14 (1958) 72–3 points out that although some headway was made toward a history of trades in the early Royal Society, this tended to decline, 'and the connexion between Royal Society science and industrial practice steadily dwindled during the rest of the century.' The practical applications of science retained some life in the development of statistics, called 'political arithmetic,' but these did not come into their own until after 1689 (ibid 74).
28 Joan Thirsk 'Plough and Pen: Agricultural Writers in the Seventeenth Century' *Social Relations and Ideas: Essays in Honour of R.H. Tawney* ed T.H. Aston, P.R. Coss, Christopher Dyer, and Joan Thirsk (Cambridge 1983) 295
29 Ibid 296
30 Ibid 300. The gentry, influenced by such writers, in turn 'spread the message by example' to those lower on the social scale (301). There are some exceptions

to this sequence, as in the case of the eminent agricultural writer Walter Blith, whose origins lie in the circle of yeomen and husbandmen. On some estates, the professional steward, often familiar with surveying and land law as well as estate management, acted as the improving husbandman. But since the main impact of this position was not felt until after 1750, it was usually up to middling gentry land-owners and, less frequently, the great landlord magnates to manage their estates themselves; see Christopher Clay 'Landlords and Estate Management in England' in Thirsk *Agrarian Change* 242-3.
31 Thirsk 'Agricultural Innovations' 533. Thirsk also points out (536-7) that writers of farming books were not the only ones paving the way: gardeners, herbalists, botanists, and others 'were continually on the lookout for better varieties of old and new plants.' One has to be wary, therefore, of overstating the influence solely of 'bookish' men in effecting change.
32 Ibid 537-8; Thirsk discusses reasons for this success.
33 Ibid 544. Again, in the following pages Thirsk does not fail to mention the deterrents to agricultural innovation, such as in the case of the husbandman, contented with a more meagre livelihood.
34 Ibid 560, 562, 566
35 See 'John Aubrey's Account of Sir William Petty' (transcribed from the Aubrey mss) in Geoffrey Keynes *A Bibliography of Sir William Petty* F.R.S. (Oxford 1971) 86
36 John Jay O'Brien 'Samuel Hartlib's Influence on Robert Boyle's Scientific Development, Part I. The Stalbridge Period' *Annals of Science* 21 (1965) 8
37 Ibid 10-13
38 Ibid 13
39 Richard S. Westfall 'Unpublished Boyle Papers Relating to Scientific Method I' *Annals of Science* 12 (1956) 63, 67. Leibniz, for one, believed that Boyle's espousal of experimental method was excellent, but that a better means of drawing consequences was required: see M.B. Hall 'Science in the Early Royal Society,' in *The Emergence of Science in Western Europe* ed Maurice Crosland (London 1975) 74.
40 Webster *Great Instauration* 425.
41 According to Webster (97), the Royal Society, though freed from state regulation, 'was also divested of a large element of the humanitarianism and utopianism of the [Hartlib's] Agency for Universal Learning,' which evolved from the Office of Address, and of which Boyle was a trustee. This lack of direct public responsibility probably diminished the Society's commitment to rigid Baconism, 'thus impairing the effectiveness of committees established to investigate particular problems of economic relevance'.
42 As evidenced by his diary, Bacon started contemplating a history of trades by

1608, expanding this idea in the *New Atantis* (1624) and especially in the *Parasceve, or Preparative towards a Natural and Experimental History*, added to his *Novum Organum* of 1620. Hartlib, as seen, tended to shift the emphasis of the history of trades away from Bacon's desire for a knowledge of causes and 'secret motions' of things to the immediate, practical reform of English society, or, at least, parts of it.
43 Keynes *Petty* 87
44 The Down Survey was so named because the topographical details were jotted down on maps.
45 Webster *Great Instauration* 434. Webster rightly makes the connection between these surveyors and the legacy of Hartlib's circle. Thus Petty's survey is viewed by Webster as vindicating the idea that a Baconian-like comprehensive natural history of a region was useful. However, one may argue that Boate had already clearly established this, even if Boate's work was not as cartographically oriented. Boate's combination of economic survey and natural history – or Bacon's 'history of Creatures' type – represents in one sense, even if not chronologically, a stepping-stone of a sort between Petty's work and that of future regional natural historians. The latter, like Boate, generally did not regard natural history as largely 'an extrapolation of conventional surveying'; their conceptual framework, rightly or wrongly, was never as rigorously organized or quantified along the same geometrical determinants as was that of Petty (ibid). Webster notices that, compared to the aspect of economic survey, natural history plays a secondary role in Boate's study. This is to be expected, perhaps, from an individual influenced so much by the Hartlibian emphasis on the history of trades, by the associated socioeconomic factors, and by the lack of a natural historical regional model upon which to build. But the two are so inextricably combined in Boate's book that it is often not possible to determine where one begins and the other ends.
46 Ibid 437–8. 'Common soldiers,' readily available in large numbers after the cessation of hostilities, fitted the bill perfectly. Like Lhuyd and others later, they had to endure hardships in the field. At the same time, many had a direct interest in the successful completion of the Down Survey, since they personally would benefit from the eventual distribution of land. John Andrews *Ireland in Maps* (Dublin 1961) 10 points out the remarkable accuracy of the resultant maps, even if they were not topographical in the modern sense: 'Petty was concerned with the boundaries of the forfeited town lands and the classification of forfeited land as cultivatable, bog, mountain, or wood,' with 'no attempt to make a complete record of roads, buildings, or relief features.' Petty has left us a history of the Down Survey, published by Sir Thomas Larcom in 1851.

47 William Petty *The Petty Papers* ed H. Lansdowne 2 vols (London 1927) 2:90–1
48 Webster *Great Instauration* 443–4

CHAPTER EIGHT

1 See Thomas Sprat *The History of the Royal Society* 2d rev ed (London 1702) 61–2 on the purpose of all prospective scientists.
2 Francis R. Johnson 'Gresham College: Precursor of the Royal Society' *Journal of the History of Ideas* 1 (1940) 413–38 was one of the first to argue that the informal circle of scientists who gravitated around the successive Gresham College professors, going back to the early seventeenth century, are to be regarded as the 'true precursors' of the Royal Society. Purver gives the credit to the Oxford group. Some others, on the contrary, view the Elizabethan Society of Antiquaries in the same light. To these groups – since Webster's work on Interregnum science has come out, at least – may be added the role played by Hartlib and the 'Puritan' reformers. Sprat is the source of argument for Wadham origins.
3 Hoppen 'Early Royal Society' 1
4 Ibid 5. The inference is that this view was promoted by Hartlib's circle.
5 Hoppen 'Early Royal Society' Pt II. Michael Hunter *Science and Society in Restoration England* (Cambridge 1981) examines the cross currents of thought at the time.
6 Hoppen 'Early Royal Society' 266. It is tempting to treat especially the 'earlier' workers in regional natural history, such as Aubrey or Childrey, along the lines of ancient as opposed to modern. Hall, 'Science in the Early Royal Society' 75 concludes that 'the English, on the whole, accepted the Royal Society's method as a model, even if they could not always emulate it,' and this is true of the growing number of provincial societies.
7 Hunter *Science and Society* 33–5
8 Ibid 49
9 Marie Boas Hall 'Oldenburg and the Art of Scientific Communication' *British Journal for the History of Science* 2 (1965) 285. On a more general note, it has been argued that it was principally the secretaries of the society in each period who were responsible both for who joined and for the kinds of science the Society sponsored; see Lotte Mulligan and Glenn Mulligan 'Reconstructing Restoration Science: Styles of Leadership and Social Composition of the Early Royal Society' *Social Studies of Science* 11 (1981) 327–64.
10 Hunter *Science and Society* 65. Again, Hunter's statement (70) that most of the members of the Royal Society came from the professional and landed classes is reflected within the ranks of the regional natural historians, many of whom

had the finances and the leisure time to indulge their scientific tastes, at least as dilettantes or virtuosi, if not in a more professional capacity. In this regard a parallel may be drawn with the chorographers of earlier decades, who generally were of gentry stock or otherwise had some direct interest in law, historical precedent, and the land.

11 Ibid 79
12 Ibid 89. Even so, members of the society continued to pursue technological applications of the new learning in scientific circles towards the end of the century.
13 The reference to Boyle is in ibid 90; Bacon's utilitarian aspect, according to Hunter, may have been especially useful to the society for publicity purposes, but as time passed, for a number of reasons it proved to be disappointing.
14 H.G. Koenigsberger *Politicians and Virtuosi: Essays in Early Modern History* (London and Ronceverte 1986) 212. Koenigsberger, in fact, sees this as a feature of the next two hundred years.
15 Ibid 223–4
16 Bacon left the scriptures themselves to be confirmed by the inner witness of the spirit.
17 Koenigsberger *Politicians and Virtuosi* 230–1
18 Harold Fisch 'The Scientist as Priest: A Note on Robert Boyle's Natural Theology' *Isis* 44 (1953) 255. Boyle, for one, was less prone than Bacon to separate the spheres of science and religion, believing that 'Reason, our primary instrument for dealing with natural phenomena, is limited in its powers and operates within a medium of Faith' (260). Ironically, Boyle's doctrine of nature as a machine – even if there is a master mechanic – may have added to the deistic trend of the next century (264). In Boyle's day atomism was under siege by those who associated it with atheism.
19 James G. Lennox 'Robert Boyle's Defense of Teleological Inference in Experimental Science' *Isis* 74 (1983) 38–52. Boyle's view is considered part of a 'Christianized' corpuscular and experimental philosophy; where Epicurean atomism is modified in that God, not random atomic collisions, is considered to be the force behind the formation of the world.
20 Hunter *Science and Society* 176–7
21 Keith Hutchison 'Supernaturalism and the Mechanical Philosophy' *History of Science* 21 (1983) 297
22 Piggott *Ruins* 17
23 Hall *Galileo to Newton* 28–9
24 Ibid 29. Figures such as Plot and Lhuyd come closer to the description of professional intellectual attitude than do many of the other regional writers.

25 Walter E. Houghton, Jr 'The English Virtuoso in the Seventeenth Century (II)' *Journal of the History of Ideas* 3 (1942) 193
26 Ibid 201
27 Hall *Galileo to Newton* 29 credits the virtuosi, who mingled a love of natural science with 'connoisseurship in art, a taste for history, or a desire to explore remote parts of the globe' with playing a large role in broadening the scope of the scientific movement in the second half of the seventeenth century, for example, by treating seriously subjects such as metallurgy, geology, botany, zoology, and so on.
28 Houghton 'Virtuoso' 194. This is not to say that their admiration of marvels was any less than that of Evelyn, but that they seemed to go beyond the marvellous alone, and beyond mere admiration in their attempts to analyse natural phenomena.
29 Ibid 202
30 C.S. Duncan 'The Scientist as a Comic Type' *Modern Philology* 14 (1916–17) 284–7
31 Hall *Galileo to Newton* 103
32 Ibid 104
33 Hall 'Science in the Early Royal Society' 58–61
34 Hall ibid 61 uses Joshua Childrey as one example of a dogmatically empirical figure, a collector of random facts about various areas of England.
35 Ibid 61–2
36 Ibid 62, 69
37 Ibid 71
38 Michael Hunter and Paul B. Wood 'Towards Solomon's House: Rival Strategies for Reforming the Early Royal Society' *History of Science* 24 (1986) 78
39 Ibid
40 Shapiro *Probability and Certainty* 10. The search for certitude was also typical of Descartes and Galileo, as well as Bacon, though along somewhat different avenues.
41 Ibid 17
42 Ibid 24. Metaphysics, in Bacon's pyramidal scheme, lay above natural history. The search for knowledge would conceivably lead one from natural history at the base to metaphysics, where one could find the much-sought-after 'Forms' of knowledge. Baconian 'science,' and the search for natural historical information, was ultimately intended to lead one to the certitude of physics and hence to such forms. Even if the search for the link between natural history and natural philosophy was in trouble, this did not prevent many regional writers from proclaiming themselves 'natural philosophers.'
   Bacon had hoped to unite the work of the mechanist or empiricist with that of

the philosopher, and thereby surpass that 'mechanic,' sometimes merely empirical and operative, that is dependent primarily on 'patience and the subtle and ruled motions of the hand and instruments'; true philosophy requires the coming together of the work of the empiricist and the rationalist, and the 'genuine' philosopher must go beyond obtaining his information from natural history and mechanical experience, 'and not take it unaltered into the memory,' but must 'digest and assimilate it for storing in the understanding' (Bacon as quoted in Paolo Rossi *Philosophy, Technology, and the Arts in the Early Modern Era* trans Salvator Attanasio, ed Benjamin Nelson (New York 1970) 86–7).

On a similar point it is interesting that Hall *Revolution* 243 states: 'The influence of the artisan, conceived as closer to the realities of nature than the abstracted philosopher, was an important element in many of the nascent sciences ... '

43 This may indicate, in some instances, a complete lack of thought on the subject. What is more likely is that they were well aware of opinions by Hooke, Boyle, and others on the subject, and either accepted or rejected these without always openly expressing their views, in their regional work at least.
44 Shapiro *Probability and Certainty* 36–7. Such ideas also depended upon a subjective view of the worth of the available evidence. In so doing, then, it is possible that consciously or not they were helping to free natural science 'from the unattainable goals of scientific demonstration ... [and to] shape a new mathematics more suited to scientific inquiry' (38).
45 See ibid 70–1. There were, of course, some individuals who adhered till the end to Bacon's dictum that through science certain knowledge of the works of nature was attainable.
46 See Hall 'Science in the Early Royal Society' 73–4.
47 Such hypothesizing is usually taken as indicative of the growing influence of Cartesian or Galilean ideas in England. The tendency to deviate from observation and data collection, already evident in some circles even before the foundation of the Royal Society, was much less pronounced in the realm of regional natural history. Hunter *Science and Society* 17 states that even 'Bacon's philosophy provided a general programme into which a whole range of particular ideas from other sources could fit, hypotheses being quite acceptable so long as they were seen as such and only espoused after rigorous testing.'
48 Shapiro *Probability and Certainty* 70. This is complicated by the fact that often opinions on various subject matters changed almost on a day-to-day basis.
49 Lately, the role of the universities in the development of natural philosophy has received greater attention. See, for example, John Gascoigne 'The Universities and the Scientific Revolution: The Case of Newton and Restoration Cambridge' *History of Science* 23 (1985) 391–434.

50 A. Rupert Hall 'Mechanics and the Royal Society, 1668–70' *British Journal for the History of Science* 3 (1966–7) 24
51 Ibid 37
52 Gascoigne 'The Universities and the Scientific Revolution' 395
53 Ibid 396–7
54 Ibid 398. After the Restoration, the Cambridge medical school continued to act as a catalyst for natural history, where John Ray's influence was felt (400). In botany, at least, by the late seventeenth century, the centre of botanical research resided not in the Royal Society but, for example, in the work of such investigators as Ray, Lister, and Tancred Robinson – some of whom were early members of the Temple Coffee House Botanical Club (formed c 1689), an unofficial offshoot of the Royal Society (401).
55 Sir George Clark *A History of the Royal College of Physicians of London* 2 vols (Oxford 1964–6) 1 (1964) 309–10. A. Rupert Hall 'Medicine and the Royal Society,' in *Medicine in Seventeenth Century England* ed Allen G. Debus (Berkeley and Los Angeles 1974) 424 states that 'the medical men actually formed the strongest group in the Royal Society.' He lists Boyle, Hooke, Locke, Power, and others as society fellows who were not also fellows of the College of Physicians but who may be classified as 'medical.'
56 Ibid 425
57 A.H.T. Robb-Smith 'Cambridge Medicine' in ibid 342–4. Power turned to Harvey's work, and was advised to undertake practical investigations of plants, animals, minerals, and so on; see Webster *Great Instauration* 137.

Power wrote what is considered to be the first work on microscopy published in England, *Experimental Philosophy in Three Books, Microscopical, Mercurial, Magnetical*, published in 1664. In 1646 Browne wrote to Power: 'The knowledge of Plants, Animals, and Minerals, (whence are fetched the Materia Medicamentorum) ... is attainable in gardens, fields, Apothecaries' and Druggists' shops.' He went on: 'See chymical operations in hospitals, private houses ... See what Chymistators do in their officines.' Power's reply included: 'If (after I have finished the Theoreticall part of Physick) you will be pleas'd to induct me into some practicall knowledge, Your commands shall fetch me up any time to Norwich ... ' Two years later Power again wrote to Browne: 'I have ... [walked] in the woods, meadows and Fields instead of Gardens ... I may easyly hereafter bee made a garden Herbalist by any shee-empirick.' The above excerpts are from letters contained in *The Letters of Sir Thomas Browne* ed Geoffrey Keynes (London 1946) 277, 279–80, 282.
58 Robb-Smith 'Cambridge Medicine' 345
59 Ibid 358
60 Rattansi 'Social Interpretation of Science' 24–5. As Rattansi states, for the

Cambridge Platonists 'The goal of knowledge ... was understanding, not power in the vulgar Baconian or Hermetic manner' (25). Bacon, of course, had no use for a Platonism that amalgamated spiritual principles with natural phenomena. Rattansi also gives Robert Boyle credit for introducing a 'cautious' Baconian empiricism into the Continental mechanical philosophy, by accepting 'the Baconian programme of a patient study of "forms" such as fluidity and solidity, heat and cold,' but now being able to 'attribute these much more consistently to the configuration and motion of the small parts of bodies. Bacon had hoped that a knowledge of "forms" would make it possible to "superinduce" them on any piece of matter' (23). According to Kuhn, this was the basis of Boyle's '"novel conception of the chemist as an artificer who fabricates in the microscopic realm as the mechanic does in the macroscopic." The Hermetic conception of a universal science of chemistry was now to be realized in mechanical terms' (ibid).

The natural theology of the Cambridge Platonist Henry More is evidenced in his growing interest in natural history. As Webster states, 'The growing body of evidence accumulated by naturalists provided a substantial basis for the "Universal Principle of Wisdom and Counsel"' (Webster *Great Instauration* 150; Webster's quotation from More is taken from More's 'Antidote against Atheism'.)

61 Hutchison 'Supernaturalism' 297
62 Ibid 298–9. Hutchison noted, however, that Boyle only accepted 'God' as a suitable answer to the question 'what causes this effect' in very special cases, and that generally 'scientific explanations were to avoid invoking God, and here can be found a sense in which Boyle's philosophy marks a continuation of Renaissance naturalism,' in that 'it involved naturalistic *explanations* inside the supernaturalistic *ontology*' (325). In other words, he mixed naturalism with supernaturalism.
63 Ibid 299
64 Ibid 320
65 Hall *Galileo to Newton* 177 states that 'few intelligent men thought that this rather humble level of activity required justification, since they saw the world as there to be described, plants as well as planets.'
66 According to Roy Porter in his review of J.S. Rudwick's *The Meaning of Fossils*, 'The History of Palaeontology' 131, even the speculation of natural historians such as Plot and Lhuyd, say, on the question of whether or not fossils were *sui generis*, could be attributed largely to the fact that there were no easy empirical solutions to this question. Porter states explicitly: 'These fossils contained no residual organic or vegetable matter, were thoroughly mineralized, lay deep in the earth (e.g., in coal shale), and were identical with no living species. There was then no simple scientific solution to the problem; all hinged upon "meanings."'

67 G.A.J. Rogers 'Descartes and the Method of English Science' *Annals of Science* 29 (1972) 244. Rogers, for one, doubts the extent of Descartes's influence on Boyle, even stating (249) that Boyle's methodology was adopted before he was totally aware of Descartes's and Bacon's attitude. If anything, Boyle rejected any a priori method of arriving at hypotheses, preferring empirical evidence to substantiate a hypothesis. Observation and experiment, ie, vital components of the Baconian program, were of considerable value to Boyle. Nevertheless, Boyle placed more confidence in the rational faculty than Bacon; even Bacon stated that his method established 'a true and lawful marriage between the empirical and the rational faculty' (Bacon quoted in Westfall 'Unpublished Boyle Papers Relating to Scientific Method, 1,' 67.) In a more recent study, Spyros Sakellariadis 'Descartes's Use of Empirical Data to Test Hypotheses' *Isis* 73 (1982) 69–76 makes the observation (75) that Descartes's 'extreme interest in observations is consistent with his rejection of the use of specific data, that his rejection of some data can be seen as part of a carefully thought out experimental methodology.' Hence, 'his rejection of those data ought not to be used as evidence that Descartes was a strict a priori deductivist.' What is one to conclude from all this? Perhaps the only conclusion one can reach is that before assessing the impact of Descartes on the English scene, more work has to be done to discover just what his philosophy of science and method were.
68 Hall *Galileo to Newton* 177
69 Evans *Society of Antiquaries* 29. In the next passage Evans says 'There was no break with the school of the past,' citing as one main example Thoroton's *Nottinghamshire*, as carrying on the Elizabethan tradition with little change. This may be true in so far as Thoroton's study happened to be one of the small group of works that carried on in the chorographic tradition, without making the switch to natural history. Unfortunately, Evans fails to notice the main body of regional study that did, in fact, take off in a new direction, thus establishing such a break.
70 Piggott 'Antiquarian Thought' 112–13; see also the same author's *Ruins* 14.
71 See, for example, Gordon K. Chalmers 'Sir Thomas Browne, True Scientist' *Osiris* 2 (1936) 28–79; E.S. Merton *Science and Imagination in Sir Thomas Browne* (New York 1949); and C.A. Patrides ed *Approaches to Sir Thomas Browne* (Columbia and London 1982). Reference material on Browne is contained in Dennis G. Donovan *Sir Thomas Browne and Robert Burton: A Reference Guide* (Boston, c 1981).
72 Evans *Society of Antiquaries* 25–6. Piggott *Ruins* 13–14 notes the even earlier excavation work of John Oglander.
73 Sir Thomas Browne 'Of Artificial Hills, Mounts, or Burrows in Many

Parts of England: what they are, to what end raised, and by what nations' in Sir Thomas Browne *The Works of Sir Thomas Browne* ed Geoffrey Keynes 4 vols 2d ed (London 1964) 3:84–7. Several of the letters between Browne and Dugdale are contained in *Works* 4:299–327, and in *Letters* ed Keynes, which was originally issued in 1931 as volume 6 of Keynes's earlier edition of the *Works*. Sir Thomas acted as a contributor of biographical and other material for John Aubrey. This is evident in one of his letters to Aubrey, dated 24 August 1672: 'There hath been a Roman Castrum by Castor neere Yarmouth, butt plowed up and now nothing or litle discernible thereof, butt I have had many Roman coynes found thereabout,' concluding, 'I hope you proceed in your observations concerning the Druid Stones' (Keynes *Letters* 395–6). Some of the letters illustrate the ties between the chorographer, Dugdale, and the 'scientist' – if that is what Browne may be called. Dugdale, for example, sought information from Browne on where the latter found in Leland 'that expression concerning such buriall of the Saxons, as you mention in your former discourse concerning those raysed heapes of earth' (Keynes *Letters* 338). Browne more than once informed his correspondent of various examples throughout history of the draining of fens (341–4), and also wrote him on a number of topics impinging on chorography, antiquarianism, and natural history, as in the following letter dated October 1660: 'I cannot sufficiently admire the Ingenious industrie of Sr Robert Cotton in preserving so many things of raritie and observation, nor comend yor owne inquiries for the satisfaction of such particulars. The petrified bone you sent mee, wh, with diverse others, was found underground, neare Cunnington, seemes to be the vertebra, spondyle or rackbone of some large fish, and no terrestrious animal, as upon sight conceived, as either of Camel, Rhinoceros, or Elephant, ffor it is not perforated and hollowe, but solid, according to the spine of fishes' (353).

74 This statement requires qualification. The treatise discussed here was apparently written in the late 1650s. It is known that later in his researches Browne went into the field to observe various antiquities, as when he and Robert Plot travelled out to examine the Roman highways (London, BL, Sloane ms 1899 f 3). It is therefore within the realm of possibility that he was also involved in the actual digging for material remains. If that is the case, however, it is just as likely that this was due to the effect of the growing 'scientific' climate of the age.

75 Lester S. King 'The Transformation of Galenism' in Debus *Medicine in Seventeenth Century England* 7. Again, the story is told against the background of the ancients vs moderns issue. A good recent account of Harvey's efforts is contained in Robert G. Frank, Jr *Harvey and the Oxford Physiologists* (Berkeley 1980). Elements of Galenism were retained by some. For example, Robert Sibbald practised a type of Galenic-Hippocratic medicine that followed cer-

tain classical precepts, while at the same time it was also based on modern observation and the study of botany, an early version of what we might call 'holistic' medicine, in which diet, climate, etc were fundamental topics of consideration for the practising medical practitioners.
76 King 'Galenism' 9. As for the Galenists, one has to keep in mind that the key to their whole system of 'form' and 'matter' is derived primarily from Aristotelian philosophy. Also, King, in his discussion, draws a connection between Galenism and the corpuscularians' and Boyle's ideas, the latter two inheriting from the Galenists an 'improved critical attitude' and the 'concept of pattern' (30–1). The Galenists were also criticized by Vesalius for getting away from dissection in anatomy.
77 King 'Galenism' 30
78 Allen G. Debus, 'Paracelsian Medicine: Noah Biggs and the Problem of Medical Reform' in Debus *Medicine in Seventeenth Century England* 36–7
79 Webster *Great Instauration* 120–1
80 Ibid 125–6, 129
81 Ibid 138–41
82 Ray is quoted in ibid 150. At Cambridge, as Webster states (153), the biological subjects predominated, while the same is true of the physical sciences at Oxford.
83 Ibid 249. For Paracelsus, metals and minerals are poison, but with proper usage they are also remedies: 'A hole rotting the skin and eating into the body, what else is it but a mineral? Colcothar – the *caput mortuum* of vitriol – mends the hole. Why? Because Colcothar is the salt that makes the hole' (quoted in Hall *Revolution* 81). As Hall states, for Paracelsus this follows from the analogy between microcosm (internal) and macrocosm (external), wherein 'the pathogenic colcothar or other substance internal to the body must be overcome by introducing the same healing substance from without' (82).
84 Webster *Great Instauration* 255
85 Although the college may have impeded the spread of such knowledge, one may take issue with Webster's contention (ibid 269) that 'the physicians in the seventeenth century played little part in the investigation of the English flora, or in the medical application of these data.' Also, as Hall *Revolution* 234–5 makes clear, in general the college itself 'served in the republican period as a nursery of research in medical science'; see below, in the text.
86 Robert G. Frank, Jr 'The Physician as Virtuoso in Seventeenth-Century England' Shapiro and Frank *English Scientific Virtuosi* 67
87 Ibid 81
88 Ibid 83. Frank observes that the bringing together of medical men and those interested in the physical sciences represents their union in 'the shared per-

ception that a common goal lay before them all' [ie, the promotion of natural philosophy].
89 Walter Charleton, quoted in ibid 90-1
90 Ibid 97
91 Michael Hunter *The Royal Society and Its Fellows* (Preston 1982; reprint ed Chalfont St Giles, Bucks 1985) 24, lists sixteen per cent of the 479 British fellows of the Royal Society elected between 1660 and 1700 as medical practitioners.

As Hall *Revolution* 90 says, Harvey 'was not a critic of traditional medicine,' and for that reason even the hermeticist and occultist London physician Robert Fludd, a critic of Galenic medicine, 'might well have seemed a far more progressive, exciting and active figure than Harvey' to the liberal mind of c 1620. John Aubrey also mentioned that Harvey distrusted chemical cures (ibid 234).
92 Frank, Jr 'Physician as Virtuoso' 102
93 Hall *Revolution* 25
94 Robert Thoroton *The Antiquities of Nottinghamshire* intro by M.W. Barley and K.S.S. Train (East Ardsley, Wakefield 1972) ix. This work, in fact, is a reprint of John Throsby's (1790-6) edition and enlargement of Thoroton's study, and is part of the *Classical County Histories*.
95 Anonymous 'Of the nature of a certaine Stone, found in the Indies, in the Head of a Serpent' *Philosophical Transactions* 1 (1665) 102-3
96 Keith Thomas *Religion and the Decline of Magic* (London 1971) 346. It has been noted, however, that Bacon also believed in certain aspects of astrology. Furthermore, it will also be recalled that in his desire to have any and all phenomena placed under careful examination, he thus called for a history of 'Marvels': 'We have to make a collection or particular natural history of all prodigies and monstrous births of nature; of everything in short that is in nature new, rare, and unusual;' (Bacon *Novum Organum* bk 2 sec 29, quoted in Houghton 'The English Virtuoso' 195. Bacon's warning against including 'fabulous experiments, idle secrets, and frivolous impostures, for pleasure and novelty,' therefore tended sometimes to fall upon deaf ears, even among the fellows of the Royal Society (ibid).
97 For Plot, see F. Sherwood Taylor 'Alchemical Papers of Dr. Robert Plot' *Ambix* 4 (1949) 67-76. It seems that Plot was not the last alchemist, but that that distinction goes to Sir Joseph Banks; see H. Charles Cameron 'The Last of the Alchemists' *Notes and Records of the Royal Society of London* 9 (1951) 109-14.
98 John Ray *The Wisdom of God Manifested in the Works of the Creation* 4th ed (London 1704) 132; on the 'Doctrine of Signatures' see Agnes Arber

Herbals: Their Origin and Evolution. A Chapter in the History of Botany 1470–1670 (Cambridge 1912) chap 8.

99 Robert Sibbald, among others, used the *Philosophical Transactions* as a vehicle for responding to the articles, including Lister's; see Robert Sibbald 'A Letter from Sir Robert Sibbald to Dr. Martin Lister Coll. Med. Lond. and s.r.s. containing an Account of several Shells observed by him in Scotland' *Philosophical Transactions* 19 (1696) 321–5. Also see E.N. Da C. Andrade 'The Birth and Early Days of the *Philosophical Transactions*' *Notes and Records of the Royal Society of London* 20 (1965) 9–27.

100 See Reginald Lennard 'English Agriculture under Charles II: The Evidence of the Royal Society's "Enquiries"' *Economic History Review* 4 (1932) 24.

101 Ibid

102 Ibid 28. Perhaps one reason for the low level of response to the 'Enquiries' lies in their rather complicated nature.

103 Robert Boyle 'General Heads for a Natural History of a Countrey, Great or Small, imparted likewise by Mr. Boyle' *Philosophical Transactions* 1 (1666) 186–9. These were subsequently published in Boyle's *General Heads for the Natural History of a Country ... for the Use of Travellers and Navigators* (London 1692).

104 Ibid 186–7

105 Ibid 188

106 John Hoskins to John Aubrey, nd (1672) Oxford, Bodleian Library, Aubrey ms 12 f 212; quoted in Hunter *John Aubrey* 113.

107 Martin Lister 'An Ingenious proposal for a new sort of Maps of Countrys, together with Tables of Sands and Clays, such chiefly as are found in the North Parts of England' *Philosophical Transactions* 14 (1684) 739–46; parts of Lister's work are quoted in H.C. Darby 'Some Early Ideas on the Agricultural Regions of England' *Agricultural History Review* 2 (1954) 32.

108 Ibid 33; see also Hunter *John Aubrey* 113 and Oxford, Bodleian Library, Aubrey ms 1 f 14.

109 E.G.R. Taylor 'Robert Hooke and the Cartographical Projects of the Late Seventeenth Century' *Geographical Journal* 90 (1937) 532. A copy of the preliminary set of queries survives in Oxford, Bodleian Library, Aubrey ms 4 f 243. Aubrey drafted these in county terms, and several of the questions closely resemble chapter headings in his 'Naturall Historie of Wiltshire.' According to a note in Aubrey's copy of the revised queries (ibid 244), these were discussed at several meetings by Aubrey, Hoskyns, Hooke, and others.

110 See Joseph Ewan and Nesta Ewan *John Banister and His 'Natural History of Virginia, 1678–1692'* (Urbana 1970); Richard Blome *A Description of the Island of Jamaica* (London 1672); and Thomas Glover 'An Account of Virginia,

its Scituation, Temperature, Productions, Inhabitants ... ' *Philosophical Transactions* 11 (1676) 623–36. See Barbara Shapiro 'History and Natural History in Sixteenth- and Seventeenth-Century England: An Essay on the Relationship between Humanism and Science' in Shapiro and Frank *English Scientific Virtuosi* 19.

111 R.W. Frantz *The English Traveller and the Movement of Ideas, 1660–1732* (Lincoln 1934; reprint New York 1968).
112 See Stimson *Scientists and Amateurs* and Martha Ornstein *The Role of Scientific Societies in the Seventeenth Century* 3rd ed (Chicago 1938). Stimson and others have also examined the rapid rise of modern science in mid-seventeenth-century England in relation to Puritanism and Protestantism.
113 Hunter *John Aubrey* 70
114 Childrey *Britannia Baconica* preface
115 Hunter *John Aubrey* 94
116 Joshua Childrey 'An Essay of Dr. John Wallis, exhibiting his Hypothesis about the Flux and Reflux of the Sea' *Philosophical Transactions* 1 (1666) 263–89. The Childrey–Oldenburg correspondence is found in A R. Hall and M.B. Hall trans and eds *The Correspondence of Henry Oldenburg* 13 vols (Madison 1965–86).
117 Hunter *John Aubrey* 94
118 Childrey *Britannia Baconica* 158, 128
119 Ibid 69
120 Ibid preface
121 Ibid, quoted in Hunter *John Aubrey* 94. Hunter perhaps correctly sees this as more of an apology for lapses into theory
122 Childrey *Britannia Baconica* preface, quoted in Hunter *John Aubrey* 94
123 Childrey's account of such 'boisterous exercises' especially reminds one of some of the earlier chorographies.
124 It is difficult to say to what extent Childrey believed these accounts to be true.
125 Joshua Childrey to Henry Oldenburg, 12 July 1669 and 29 March 1670, quoted in Hall *Henry Oldenburg* 6 (1969): 110, 603 respectively
126 See Hunter *John Aubrey* 142; and Joshua Childrey to Henry Oldenburg, 12 July 1669, quoted in Hall *Henry Oldenburg* 6:108.
127 Henry Oldenburg to Joshua Childrey, 24 July 1669, quoted in ibid 6:151, where Oldenburg informs Childrey that 'The encrease of your affection to the Society, and your zeale of promoting their designe ... was represented to them at their last meeting from your late letters, I received of you.' Oldenburg, it may be noted, published in 1671 an English translation of Steno's 1669 treatise on geology, the *Prodromus*.
128 Joshua Childrey to Henry Oldenburg, 14 April 1669, quoted in Hall *Henry Oldenburg* 5 (1968):488–9

129 See Henry Oldenburg to Joshua Childrey, 6 April 1669 and 3 July 1669, quoted in ibid 5:477, and 6:90 respectively; see also Joshua Childrey to Henry Oldenburg, 4 May 1669 and 12 July 1669, quoted in ibid 5:513 and 6:107–8 respectively. Oldenburg took over Hartlib's role as a middleman of scientific interchange; see Marie Boas Hall 'Oldenburg and the Art of Scientific Communication' 277–90.
130 Joshua Childrey to Henry Oldenburg, 12 July 1669, quoted in ibid 6:108 and in Margaret Deacon *Scientists and the Sea, 1650–1900* (London 1971) 72
131 Hall *Henry Oldenburg* 6:108–9. One notebook, the 'Chronologia naturalis,' documented 'the time of all droughts, Comets, Earthquakes, etc.'; the other one, 'Geographia naturalis,' holds observations on the natural rarities of various other countries.
132 See L.R. Cox 'British Palaeontology: A Retrospect and Survey' *Proceedings of the Geologists' Association* 67 (1956) 209–10. Childrey's observations are also discussed in S. Smith 'Seventeenth Century Observations on Rocks [etc.]' *Proceedings of the British Naturalists' Society* 27 (1942) 93–103.
133 Childrey *Britannia Baconica* preface
134 Ibid 137
135 Jones *Ancients and Moderns* 162

CHAPTER NINE

1 John Ray to John Aubrey, 27 October 1691, Oxford, Bodleian Library, Aubrey ms f 13; London, BL, Lansdowne ms 937 f 63
2 Powell *Friends* 106
3 On the question of whether or not Aubrey belongs to the moderns see Hunter *John Aubrey* passim; G.M. Young *Last Essays* (London 1950) 251–2; Lytton Strachey *Portraits in Miniature* (London 1931) 20–5; and Oliver Lawson Dick 'Scholarship and Small Talk' *Listener* 20 November 1947, 904–5. This point is discussed in Jon Bruce Kite 'A Study of the Works and Reputation of John Aubrey, with Emphasis on His *Brief Lives*' (PHD diss, University of California 1962).
4 Oxford, Bodleian Library, Aubrey ms 1 ff 23, 10
5 Hunter *John Aubrey* 119. It may be noted here that because Hunter's study of Aubrey's overall antiquarian and scientific works – though not necessarily his regional work – is quite comprehensive, many of the quotations taken directly from Aubrey's own works are also found, in whole or in part in Hunter's book. In order to eliminate overburdening this study with reference to Hunter, I shall cite the original only, except in cases where references to Hunter would prove

to be especially helpful. This is not to minimize my debt to Hunter.
6 Ibid 106–8. Hunter admits that Aubrey is representative of an age when a 'mixture of mystical science with more modern views on other subjects' was the norm. Hunter writes that this combination of belief and rationalization is not surprising, since any other interpretation of Aubrey's approach would presuppose a split in his intellectual attitudes (198). See Gerald Craven's review of Hunter's book in *Seventeenth Century News* (Winter 1977) 115.
7 Anonymous 'The Oxford Cabinet: Consisting of Engravings from original Pictures in the Ashmolean Museum, and other public and private Collections; with Biographical Anecdotes. By John Aubrey, F.R.S. and other celebrated Writers' *Gentleman's Magazine* 68 (1798) 320. Throughout most of the eighteenth century Aubrey was known mainly for his only published work, the *Miscellanies* (London 1696). His reputation improved as the archaeologists and other antiquaries began to notice his antiquarian efforts.
8 John Aubrey, quoted in John Britton ed '*The Natural History of Wiltshire,' by John Aubrey* (London 1847; reprint Newton Abbot 1969) 11. The introduction to the 1969 reprint edition of *The Natural History*, by K.G. Ponting, has a good general description of Aubrey's life, which I have used. Otherwise, my references to Britton are to the 1847 edition.
9 Ibid 5. Young *Last Essays* 256 states that Aubrey belongs to the moderns because he exhibited 'the disdain of the Augustan for the barbarism of the recent past'; see Kite 'Study' 71.
10 Robert G. Frank, Jr 'John Aubrey, F.R.S., John Lydall, and Science at Commonwealth Oxford' *Notes and Records of the Royal Society of London* 27 (1973) 193–217. Also see Douglas McKie 'The Origins and Foundation of the Royal Society of London' ibid 15 (1960) 1–37 and Ian Anderson 'The Royal Society and Gresham College 1660–1711' ibid 33 (1978) 1–21.
11 Frank, Jr 'Commonwealth Oxford' 201
12 Hunter *John Aubrey* 44. Hartlib recorded meeting Aubrey in these words: 'The 2d of December 52 came to my house of his owne accord the first time Mr. John Aubrey ... He seemed to be a very witty man and a might favourer and promoter of all Ingenious and Verulamian [Baconian] designes ... ' quoted in Frank, Jr 'Commonwealth Oxford' 205.
13 Oxford, Bodleian Library, Aubrey ms 3 f 11. J.E. Jackson *'Wiltshire.' The Topographical Collections of John Aubrey* (London 1862) viii states 'the truth is, that upon the more serious labours of Parochial History – the long investigation of evidences, the thoughtful comparison of them, and the drawing of correct conclusions from them – Aubrey was either unable or unwilling to enter.'
14 Oxford, Bodleian Library, Aubrey ms 1 f 5. Aubrey, it seems, was the first to

vociferously express this desire for what would amount to, in effect, an examination of the entire country through natural histories.

15 Ibid 3 f 10. The manuscript of 'North Division,' called 'Hypomnemata antiquaria A' on the outside of the original parchment, is now Oxford, Bodleian Library, Aubrey ms 3, and was printed by Jackson in his *Topographical Collections*; I have used both the manuscript and Jackson's edition. As it turned out, the others who were to be enlisted in this cooperative effort soon lost interest, so that the project vanished 'in fumo Tabaci' (over their tobacco pipes), leaving Aubrey to go it alone.

16 Ibid 4. Even here, however, Aubrey appears to echo Bacon's statement that antiquities are 'history defaced,' in *De Augmentis Scientiarum*.

17 Jackson *Topographical Collections* 4

18 Ibid 14

19 Ibid 58. Hunter *John Aubrey* 57 states that Aubrey 'advocated a sort of natural theology, considering it "a profound part of Religion to glorifie God in his Workes."' This element also affected the thought and the work of future natural historians.

20 Jackson *Topographical Collections* 411

21 Ibid 241–2

22 Ibid 16

23 This work is contained in Oxford, Bodleian Library, Aubrey ms 4. It was enlarged, corrected, and published by the eighteenth-century antiquary Richard Rawlinson as '*The Natural History and Antiquities of the County of Surrey,*' begun in the year 1673 by John Aubrey 5 vols (London 1718–19). Apparently, Rawlinson did not publish this work as an aid to the better understanding of Aubrey's methodology, but rather as an up-to-date history of Surrey, using Aubrey's collections as an authoritative basis; see Brian Enright 'Richard Rawlinson and the Publication of Aubrey's *Natural History and Antiquities of Surrey*?' *Surrey Archaeological Society Collections* 54 (1955) 126, 128.

24 See Aubrey's preface, found in the first volume of Rawlinson *Surrey*.

25 See Taylor 'Robert Hooke' 529–40; see also Ralph Hyde 'The Ogilby Legacy' *Geographical Magazine* 49 (1976) 115–18.

26 John Aubrey to Anthony à Wood, 11 October 1673, Oxford, Bodleian Library, Wood ms F39 f 231. Ogilby is discussed in J.L. Nevinson's introduction to *The Natural History and Antiquities of the County of Surrey. Volume I, by John Aubrey* (Dorking 1975).

27 Similarly, Plot said of Ogilby that he is a 'cunning Scott, and I must deale warily with him' (Robert Plot to Anthony à Wood. 12 August 1672, Oxford Bodleian Library, Wood ms F39 f 181).

28 Edward Lhuyd to John Aubrey, 12 December 1692, Quoted in Rawlinson *Surrey* 1:4
29 Oxford, Bodleian Library, Aubrey ms 4 f 244. As an example, Query Eight covered 'Springs, Wells, Baths, Cold and Hot Waters, Medicinal, Aluminous, Bituminous, Nitrous, Petrifying, etc'; other queries covered improvements in husbandry, manufacturers, and so on. Most of the natural observation of Surrey is contained in ibid ff 235f, inserted after Aubrey had seen Plot's synopsis of his projected natural history.
30 John Aubrey to Anthony à Wood, 24 October 1674, Oxford, Bodleian Library, Wood ms F39 f 282
31 Oxford, Bodleian Library, Aubrey ms 1 f 6
32 Ibid ff 6–7
33 Ibid f 6
34 Ibid
35 See Hunter *John Aubrey* 138, 192.
36 Ibid 101–2
37 Oxford, Bodleian Library, Aubrey ms 2 f 115
38 Ibid 4 f 7
39 Ibid 1 ff 9–11, 14,19. By 1670 Aubrey could already claim that: 'between S. Wales and the French Sea: I have taken an account of the severall earths, and naturall observables in it, as the nature of the plants in the respective soyles, the nature of the cattle thereon feeding, and the nature of the Indiginae' (John Aubrey to Anthony à Wood, 17 November 1670, Oxford, Bodleian Library, Wood ms F39 f 128).
40 Lister 'An Ingenious proposal' 739–40. In *John Aubrey* 114n1, Hunter notes that there is no reason to assume that Aubrey derived the idea of preparing land-use maps from Lister, even if Britton *Wiltshire* seems to believe Aubrey did just that. It was not until 1815 that the first geological distribution map of England and Wales was made, and, as Young implies, Aubrey's desire for this type of map is evidence of his consciousness of geology's need for a precise codification (Young *Last Essays* 251); see Kite 'Study' 7–8.
41 The quotation is taken from J.E.Jackson 'Memoir of John Aubrey, F.R.S.' *Wiltshire Archaeological and Natural History Magazine* 4 (1858) 94.
42 Oxford, Bodleian Library, Aubrey ms 1 f 9
43 See Britton *Wiltshire* 40–1 and Kite 'Study' 7.
44 On at least one other occasion he verified that an ore deposit that he had found while strolling contained iron; he did this by submitting it to the local forge, where it was examined.

45 Britton *Wiltshire* 9. Cf Oxford, Bodleian Library, Aubrey ms 4 f 511
46 John Harris *Lexicon technicum, or a Universal English Dictionary of Arts and Sciences* 2 vols (London 1704) 'Fossils.' The meaning of the word *fossil* is derived from the Latin *fossiles* ('dug up'), from *fodere* ('to dig').
47 Harris, ibid 'Formed Stones,' applies the term 'formed or figured stones' to rocks that bear a close resemblance to the 'external Figure and Shape of Muscles, Cockles, Periwinkles, and other shells.'
48 See, for example, Britton *Wiltshire* 40, where Aubrey describes some 'iron bulletts, as big as pistoll bulletts' that were often ploughed up out of the earth and that were commented on by Childrey in *Britannia Baconica*.
49 Britton *Wiltshire* 38
50 Ibid 34–5
51 Britton *Wiltshire* 11. Aubrey also noted 'that North Wiltshire is very worme-woodish and more litigious than South Wilts' (ibid 12): see also Kite 'Study' 11–12.
52 Thomas Burnet *Telluris theoria sacra* (London 1681); see also E.G.R. Taylor 'The Origin of Continents and Oceans: A Seventeenth Century Controversy' *Geographical Journal* 116 (1950) 193–8. Many of the theorists based their argument for a flood on the evidence of fossilized marine life. It is interesting that Ray found Aubrey's account of Wiltshire satisfactory except for the latter's 'digression' on geology; Ray thought that it challenged 'the truth of the Letter of the Scripture' (John Ray to John Aubrey, 27 October 1691, Oxford, Bodleian Library, Aubrey ms 1 f 13).
53 Britton *Wiltshire* 47. Into his manuscript Aubrey inserted accounts from the *London Gazette* of three earthquakes that took place in different parts of Italy in 1688 and 1690.
54 Britton *Wiltshire* 47
55 Hunter *John Aubrey* 58–9
56 Britton *Wiltshire* 30. Although this particular idea, according to Aubrey, was first proposed about the year 1626 by Henry Brigges, Savilian Professor of Geometry at Oxford, Brigges was unsuccessful in presenting this 'noble designe' because 'Knowledge of this kind was not at all in fashion' (ibid) until Aubrey's day. Two figures who were especially interested in such schemes were William Petty and Robert Hooke; see also Kite 'Study' 10.
57 Britton *Wiltshire* 71
58 Thomas Fuller *History of the Worthies of England* (London 1662)
59 Oxford, Bodleian Library, Aubrey ms 2 f 90
60 Britton *Wiltshire* 97. H.M. Colvin 'Aubrey's *Chronologia architectonica* in *Concerning Architecture* ed John Summerson (London 1968) 11 states that 'To John Aubrey, therefore, must go the credit for being the first to think historically about medieval English Architecture.'

Notes to pages 182–4 307

61 See Robert W.T. Gunther ed *Further Correspondence of John Ray* (London 1928) 171
62 John Ray to John Aubrey, 21 July 1691, London, BL, Egerton ms 2231 f 11. Cf John Ray to John Aubrey, 22 September 1691, Oxford, Bodleian Library, Aubrey ms 13 f 174.
63 John Britton *Memoir of John Aubrey, F.R.S.* (London 1845) 3–4. Glyn Daniel *A Hundred and Fifty Years of Archaeology* 19 says Aubrey may perhaps 'be called the first English field archaeologist of importance' and, in another place, *The Idea of Prehistory* (Baltimore 1962) 79 he notes that field archaeology 'began with exact observers and painstaking travelers to the past like John Aubrey and [Aubrey's colleague] Edward Lhuyd.' Daniel regards Aubrey as the sort of field archaeologist who excavates sites 'non-scientifically,' but who nevertheless meticulously examines the visible remains of the site. Piggott *Ruins* 16 says that the 'Monumenta' is important because the material in it 'is assembled and set out for its own sake, and not to illustrate a fictitious or quasi-historical narrative,' and adds that Aubrey 'in fact [was] applying to this material the same classificatory method as his colleagues in the Royal Society were using in the natural sciences' (ibid).
64 Walter Charleton *Chorea gigantum* (London 1663) and John Webb *Vindication of Stone-Heng ... Restored* (London 1665); see also R.J.C. Atkinson *Stonehenge* (London 1956) 186ff.
65 See John Aubrey to Anthony à Wood, 8 September 1680, Oxford, Bodleian Library, Wood ms F39 f 348. The chaotic arrangement of the manuscript of the 'Monumenta Britannica,' which now rests at the Bodleian Library (Aubrey mss, TGC24, TGC25), typifies Aubrey's work. Kite 'Study' 74–83 discusses the 'Templa druidum' and other aspects of Aubrey's archaeology.
66 Oxford, Bodleian Library, Aubrey ms, TGC24 f 23. Aubrey returned yearly to Avebury to examine the site.
67 Ibid f 24. For more detail about this see Earl Melton Williams 'John Aubrey's *Templa druidum*: Materials for an Edition' (PH D diss, Florida State University 1978) 4–5.
68 See Jackson *Topographical Collections* 341.
69 Daniel *Prehistory* 79–80. Aubrey attempted to publish his findings thirty years later, at about the same time as his study of the county of Wiltshire.
70 Hunter *John Aubrey* 193, 202
71 Ibid 192
72 Stuart Piggott 'Antiquarian Thought' 108 appears to think otherwise. He says that Aubrey applied to his material 'the same classificatory method and presentation as his colleagues in the Royal Society were using in the natural sciences; the methods of Ray in botany or Lhwyd [Lhuyd] in palaeontology.' This is true only in a general sense.

73 Kite 'Study' 83. Barbara D.Lynch and Thomas F. Lynch 'The Beginnings of a Scientific Approach to Prehistoric Archaeology in 17th and 18th Century Britain' *Southwestern Journal of Anthropology* 24 (1968) passim, cover not only Aubrey's antiquarian-archaeological method and findings, but also those of Dugdale, Plot, Lhuyd, and some of the other regional writers described in the present study. Piggott *Ruins* 9–10 alludes to Aubrey's reaches into comparative ethnography, where the ancient Britons are compared to the newly discovered New World Indians. In this respect Aubrey maintains the same concerns as some of the earlier chorographers, including John Speed (10).

CHAPTER TEN

1 See A.J. Barnouw and B. Landheer eds *The Contribution of Holland to the Sciences* (New York 1943). Ewan *John Banister* 25 has described how England lagged behind in its recognition of the study of botany as a university subject. Also valuable is William T. Stearn 'The Influence of Leyden on Botany in the Seventeenth and Eighteenth Centuries' *British Journal for the History of Science* 1 (1962) 137–58. Apparently, despite their education, the Boates' reputation as medical practitioners was assailed. Clark *History of the Royal College of Physicians of London* 1:262, in discussing the reputation of the Boates, describes their 'contempt for the rules and the College [of Physicians],' difficulties involving charges of malpractice against them, and so on. It did not help matters, apparently, that Gerard 'committed the unique outrage of employing a woman as his apothecary,' and a Dutch one at that!
2 F.V. Emery 'Irish Geography in the Seventeenth Century' *Irish Geography* 3 (1958) 264–5
3 Arnold Boate had a detailed knowledge of Ireland from eight years' service as surgeon-general there. Sir William and Sir Richard Parsons were also major contributors to the book, and these two men probably supplied the author with much of the information that pertained to the rocks and minerals of Ireland.
4 See Arnold Boate's prefatory letter to Samuel Hartlib, in Gerard Boate *Irelands Naturall History* (London 1652).
5 His story in Ireland is documented in Great Britain, Public Record Office *Calendar of State Papers, Domestic, 1649–1650* vol 1 (1649) 66, 588.
6 Boate *Naturall History* prefatory letter
7 Hunter *John Aubrey* 112. Hunter's quotation is taken from the title-page of the *Naturall History*. Both Hartlib's and Boate's concern for the practical application of the new philosophy and its tenets is in part a reflection of Sir Thomas Browne's interest in the practical aspects of the same. For example, 'Browne's interest in plant growth was utilitarian as well as academic,' according to E.S.

Merton 'The Botany of Sir Thomas Browne' *Isis* 47 (1956) 167. Because he had had some experience previously as a fruit grower, Browne saw great potentialities in the art of grafting, which he adumbrated in a long list of experimental combinations (ibid).
8 Arnold Boate to Samuel Hartlib, 16 July 1648; quoted in Emery 'Irish Geography' 265 and in G.H. Turnbull *Hartlib, Dury and Comenius: Gleanings from Hartlib's Papers* (London 1947) 58
9 Hartlib's agricultural ideas are discussed in G.E. Fussell *Old English Farming Books from Fitzherbert to Tull, 1523–1730* (London 1947) chap 4. The effect of these and other ideas of Hartlib on the fellows of the Royal Society is discussed in G.H. Turnbull 'Samuel Hartlib's Influence on the Early History of the Royal Society' *Notes and Records of the Royal Society of London* 10 (1953) 101–30.
10 Samuel Hartlib, quoted in Boate *Naturall History* 'Epistle Dedicatorie'
11 Ibid 9
12 Ibid 10
13 Norden *Cornwall* 3–5
14 Henry Owen ed *'The Description of Penbrokshire' by George Owen of Henllys* Cymmrodorion Record Series no 1, 3 vols (London 1892–1906) 1 (1892) 2
15 Speed *Theatre of the Empire* 117
16 Childrey *Britannia Baconica* 27. For this and the previous three quotations see Davies *Earth in Decay* 55, 57.
17 The quotation is taken from the formulator of the earlier theory, Richard Verstegan (alias Richard Rowlands), in *A Restitution of Decayed Intelligence* (Antwerp 1605) 98. Wallis's stand is found in 'A letter ... Relating to that Isthmus, or Neck of land, which is supposed to have joyned England and France,' *Philosophical Transactions* 22 (1701): 967–79. See Davies *Earth in Decay* chap 2. It is evident that Davies – and in this he is an exception – is not enamoured of the geomorphological content of Boate's book (31).
18 Boate *Naturall History* 49. In a careful analysis of the 'Causes of the loss of such ships as perish Upon this Sea' (49–50), Boate expresses his belief that the major common cause is the long, dark winter nights, when 'some furious storm arising, the ships are dashed upon the rocks.'
19 Ibid 52–3. Boate, however, may have been considerably influenced by Bacon's ideas on the marine environment. Bacon had advocated the collection of information on subjects such as the ebb and flow of the sea, currents, salinity, subterranean physical features, etc. Bacon himself listed several observations on such subjects in *Sylva sylvarum*; see Bacon *Sylva Sylvarum* passim.
20 Boate *Naturall History* 13
21 Ibid 85

22 Other aspects of the bogs merited Boate's attention. At one point (ibid 110) he admires the native Irish for their agility in crossing the deepest bogs from one side to the other over their firm places, 'in which nimble trick, called commonly treading of the bogs, most Irish are very expert, as having been trained up in it from their infancy.' This was, apparently, a lapse from Boate's usual defamatory portrayal of the Irish, a people whom he scurrilously refers to as 'one of the most barbarous Nations of the whole Earth' (124).
23 Ibid 112–13
24 Ibid 71–5
25 Ibid 74
26 Ibid 78–9
27 Ibid 119
28 Taylor *Stuart Geography* 137n
29 Boate *Naturall History* 125
30 He uses the term *mine* as equivalent to *ore*.
31 Charles Webster 'New Light on the Invisible College: The Social Relations of English Science in the Mid-Seventeenth Century' *Transactions of the Royal Historical Society* 5th ser 24 (1974) 34
32 K. Theodore Hoppen 'The Dublin Philosophical Society and the New Learning in Ireland' *Irish Historical Studies* 14 (1964) 100
33 Taylor *Stuart Geography* 132; see also Webster *Great Instauration* 429.
34 Ibid 66
35 T.C. Barnard *Cromwellian Ireland: English Government and Reform in Ireland 1649–1660* (London 1975) 234
36 Hoppen 'New Learning' 101. Molyneux has recently been the focus of J.G. Simm's *William Molyneux of Dublin, 1656–1698* (Dublin 1982). Hoppen 'New Learning' 100 describes *Irelands Naturall History* as 'the only scientific book in the modern manner relating to Ireland, written before the restoration ... Breaking away from the old chorographical method, Boate bases his views on observation and verifiable fact.' See also Hoppen's article 'Some Queries for a Seventeenth Century Natural History of Ireland' *Irish Book* 2 (1963) 60–1.
37 Webster *Great Instauration* 434, 436, 442
38 Barnard *Cromwellian Ireland* 214
39 See Samuel Hartlib *The Advancement of Learning* (London 1653) as quoted in Charles Webster ed *Samuel Hartlib and the 'Advancement of Learning'* (London 1970) 173–4. Boate's interest in economic geography was extended also by natural historians-antiquaries-geographers such as Robert Sibbald in Scotland, who was an avid collector of information on mining, details of mineral springs, meteorological occurrences, etc. Practical information of this type, it was hoped, would lead to a better utilization of natural resources, as recommended by Hartlib.

Notes to pages 193–5 311

CHAPTER ELEVEN

1 R.W.T. Gunther, *Early Science in Oxford* 14 vols (Oxford 1920–45) 12 (1939): preface. Gunther says (333) that it was the publication of Robert Plot's *The Natural History of Oxfordshire* (Oxford 1677) that persuaded Ashmole to donate his collections to Oxford University, and so 'To Plot ... Oxford owes the first public Institution in Britain for the study of Natural History in its widest aspects.' Plot's regional natural histories bore considerable resemblance to Boate's work in Ireland, and to studies done on the Continent. Plot was the first native Briton to fully incorporate Boate's rigidly systematic plan. For more information on the ideas of Boate that influenced Plot, see Webster *Great Instauration* 427–31.
2 Robert Plot 'Plinius Anglicus sive Angliae historia naturalis ac artium' London, Society of Antiquaries, Society of Antiquaries ms 85 f 2
3 Plot *Oxfordshire* 204; Plot *The Natural History of Staffordshire* (Oxford 1686) 329–30
4 Plot 'Plinius' f 2, quoted in Hunter *John Aubrey* 70
5 Wood *Athenae Oxonienses* col 772
6 Francis Bacon 'Parasceve' 2, in Bacon *Works* 4:254
7 See Gunther *Science in Oxford* 12:335–6.
8 The letter to Fell is contained in ibid 12:343–4; Spelman's 'An Interpretation of Villare Anglicum' is also analysed by Aubrey in Oxford, Bodleian Library, Aubrey ms 5.
9 Ibid f 19
10 Gunther *Science in Oxford* 12:344
11 Plot 'Plinius' f 1
12 'Oxford Testimonial to Dr. Plot' 25 July 1674, in Gunther *Science in Oxford* 12:345–6
13 This is especially true of Plot's *Enquiries*, published in 1679.
14 The original manuscript of 'the Naturall History, only,' for 'Surrey,' which Aubrey sent to Plot, survives at Oxford, Bodleian Library, Aubrey ms 4, f 235f, because Plot later returned it to Aubrey at Aubrey's request; see John Aubrey to Anthony à Wood, 3 August 1691, Oxford Bodleian Library, Wood ms F39 f 429 and Hunter *John Aubrey* 72n11.
15 Plot *Oxfordshire* 99; see Hunter *John Aubrey* 73.
16 Aubrey's copy of Plot's *Oxfordshire* is now Oxford, Bodleian Library, Ashmole ms 1722. Aubrey's marginal notes indicate the nature of his assistance to the author. On 336, for example, we learn that Plot derived his information on certain Danish fortifications from a 'note the Dr. had fro J Aubrey.' It appears that Edward Lhuyd came to peruse this same copy; on the title-page is written, in Lhuyd's hand, 'Historiam suam naturalem agri Staffordshire ...'

On Aubrey's assistance to Plot, see also Oxford, Bodleian Library, Rawlinson mss K15281 ff 102–64; K15282 ff 198–294.
17 Robert Plot to John Aubrey, February 1676, London, BL, Egerton ms 2231 ff 100–1; see also Oxford, Bodleian Library, Aubrey ms 13 f 137.
18 Hunter *John Aubrey* 83, 86. Of Plot, Aubrey had this to say: 'I did not think that there had been so much trueth in Mr. R. Sheldons advice to [me] sc: lend not your mss. how ungratefully Dr Plott hath used me!' (Oxford, Bodleian Library, Aubrey ms, TGC25 f 95). This is in contrast to the attitude usually displayed by Aubrey towards Plot. Aubrey wrote to Wood of 'Dr Plott ... whom I shall be able to assist, and he to assist me,' and on yet another occasion he asked Wood: ' ... if you meet with Dr. Plot, pray thank him for his Queres' (John Aubrey to Anthony à Wood, 25 February 1675 and 25 May 1684, Oxford, Bodleian Library, Wood ms F39 ff 292, 372 respectively).
19 An excellent, though brief, article on the scientific nature of the regional work of Plot and that of several other investigators is Frank Emery 'English Regional Studies from Aubrey to Defoe' *Geographical Journal* 124 (1958) 315–25, from which much of the information on Plot's method of dividing the area under study is derived. Additional information on Plot is found in Hunter's *Science and Society* and in Porter's *Making of Geology*; also M.W. Greenslade *The Staffordshire Historians* (Fenton, Stoke-on-Trent 1982), and Michael Pafford's brief 'Robert Plot: A County Historian' *History Today* 20 (1970) 112–17.
20 Plot *Oxfordshire* 'To the Reader'
21 Ibid 69
22 Ibid 'To the Reader'; all the references to the map, which follow immediately, are taken from here.
23 It is amusing that although Plot was generally considered both a man of learning and a man of affairs, some of the Staffordshire gentry to whom he addressed his inquiries used to boast of having 'befooled old Plot.' To such displays, Plot replied in the last sentence of *Staffordshire*: 'I hope all Readers will deale so candidly with me, as only to reprove me calmly, for what is done amiss, which sort of Chastisement I shall cheerfully receive; sincerely promising never to offend in the like manner again ...' Further to this, the publisher of the second edition of *Oxfordshire*, in his remarks to the reader, states that the objections that have been raised against some of Plot's hypotheses have no other foundation than ill nature and censoriousness.
24 Ibid (1667 ed) 1–2
25 He appeared in the *Philosophical Transactions* as the author of commentaries on such topics as the formation of sand and salt from brine, sepulchral lamps, and observations on lead and on electrical bodies; see *Philosophical Transactions*

13 (1683) 96–9; 14 (1684) 806–11; 15 (1685) 930–43; 20 (1698) 384, respectively.
26 When the duke of York visited Oxford with Princess Anne in the spring of 1683, Plot's *Oxfordshire* was presented to him as a gift, together with Wood's history of the university. Plot also had the honour of entertaining the royal party by performing chemical experiments for their satisfaction.
27 Gunther *Science in Oxford* 12:349 contains a list, compiled from Birch's historical notes, of gifts made by Plot in 1683. These included the following:

Jan. 31.	1. Moyra, an earth, wherewith the Turks paint ... walls of their houses.
	2. A depilatory, 2/3 lime, and 1/3 orpiment, made ... in a cataplasm, to take away hair.
Feb. 21.	13. A white earth for polishing silver.
	17. An earth found under Fairy-rings.
June 6.	32. A piece of rock crystal from Madagascar.
June 27.	35. Selenites dodecahedros, mentioned in ... *Natural History of Oxfordshire*.

28 As Gunther *Science in Oxford* 12:353 points out, the fact that the new institution was not as widely known in Oxford as it deserved to be is revealed in the following letter (Edward Lhuyd to John Aubrey, 12 February 1686, London, BL, Egerton ms 2231 f 228): "'Twas well you writ to me of it, for the generality of people at Oxford doe not yet know what ye Musaeum is: for they call ye whole Buylding ye Labradory or Knacecatory and distinguish no farther. That nothing miscarried soe directed to Dr. Plot, was because ye Person was known better than ye place, but things directed to me or Mr. Higgins commonly stayd at the carriers till we fetch'd them.'
   The story of Ashmole, the benefactor, is told by A.L. Humphreys *Elias Ashmole* (Reading 1925). Ashmole's notes on Berkshire were carelessly assembled and printed by E. Churll (see ibid 17–18) under Ashmole's name as *The Antiquities of Berkshire* 3 vols (London 1719). They follow the method of the chorographers, and consequently contain little if any natural history of importance. Also see C.H. Josten *Elias Ashmole (1617–1692)* 5 vols (Oxford 1966) for the Plot-Ashmole connection.
29 See Gunther *Science in Oxford* 12:355–6; for Plot's overall work in chemistry see ibid 1 (1923) 47–50 and 251–61.
30 Taylor 'Alchemical Papers' 69–70. Taylor based his article on the evidence of a volume containing a number of Plot's papers, now London, BL, Sloane ms 3646.
31 Gough *British Topography* 1:xix
32 Martin Lister extended his wishes to Plot for the success of the newly

established Philosophical Society of Oxford, stating that 'Your new Societie will be of great use, it will excite this other here, and emulation is the great promoter of learning'; he went on to observe 'your Methode to be more free and more intent than ours; and I hope you will put us upon new wayes, as well as new matter of Experiments' (Martin Lister to Robert Plot, October 1683, Gunther *Science in Oxford* 12:39).

33 *Dictionary of Natural Biography* credits Chetwynd; but Plot, in *Staffordshire* 61, refers to the 'Right Honourable the Vertuous and most Accomplish't Lady, Jane Lady Gerard Baroness Gerard of Gerards Bromley, *the first actual Encourager of this Designe*' (italics mine). Chetwynd was a distinguished antiquary, elected to the Royal Society in 1678. Among his collections are the papers of William Burton. He was apparently a generous man, for not only did he finance the building of a church at Ingestre (see Plot *Staffordshire* 297–300), but also he acted as a patron to Plot, aiding him financially in the survey of the county. He also assisted Plot by supplying useful answers to Plot's queries; see Gunther *Science in Oxford* 12:215–18.

34 Jacob Wagner 'Historia naturalis Helvetiae curiosa' *Philosophical Transactions* 13 (1683) 268–71. Wagner's study is organized along the same general lines as those of Plot and Aubrey, taking into account the great difference in the topography of the regions under consideration. See also Francis Aston to Robert Plot, 12 July 1683, Gunther *Science in Oxford* 12:36.

35 See S.A.H. Burne 'Early Staffordshire Maps' *Transactions of the North Staffordshire Field Club* 54 (1920) 70. At least one of Plot's correspondents, Charles King, apparently was confident that Plot's second regional natural history 'would be publick' well before the time that it actually was; see Charles King to Robert Plot, 26 March 1684, Gunther *Science in Oxford* 12:216.

36 All of these innovations are discussed in Burne 'Staffordshire Maps' 69.

37 Plot *Staffordshire* 23–4

38 He does the same thing in *Oxfordshire*, where he uses the description of fossils found there as a starting point for a discussion of the origin of fossils in general. He does not, however, in this particular case repeat the origin theory again in *Staffordshire* because of his profound desire to 'avoid all vain repetitions' (2).

39 Ibid 50

40 At one point (ibid 60), Plot alluded to Boate's study of the springs of Ireland in *Irelands Naturall History*, chap 7. Plot, in another work, *De origine fontium, tentamen philosophicum* (Oxford 1685) 7, discussed in greater detail the explanations of the relationship between sea water and spring water as subscribed to by other figures, including Vitruvius, Peter Martyr, Cardano, Molina, Palissy, Gassendi, and Hooke. These usually attributed to Aristotle

the hypothesis of the exchange of water between the oceans and the atmosphere. The theory postulating the subterranean origin of springs they credited to Plato. (These two classical writers made the earliest studies on these subjects that survive complete.)

41 See Emery 'Aubrey to Defoe' 319 for more on Plot's division of the county into three soil types.
42 Plot *Staffordshire* 113; see also his discourse on soil erosion (170).
43 See W.H.B. Court *The Rise of the Midland Industries, 1600–1838* 2d ed (London 1953) 16, 164. Plot *Staffordshire* 110 states that the mountains of the northern part of the county are 'hardly passable, some of them being of so vast a height, that in rainy weather I have frequently seen the tops of them above the Clouds.'
44 This topic comes only after Plot concluded his section on the soil by stating (ibid 125) that: 'reckon'd up by such as have written de Arte combinatoria' there are in total '179001060 different sorts of Earths.'
45 Ibid
46 Plot was the first to use the following geological terms: 'bass,' 'basset,' 'bats,' 'clunch,' 'laming,' and, 'measure.' D.R. Dean in 'The Word "Geology"' *Annals of Science* 36 (1979) 35 states that: 'the earliest direct ancestor of our present word "geology" is the "geologiam" of Richard de Bury, which appears as a deliberate coinage in Chapter II of his *Philobiblon* (written 1344).' The term is here used to denote 'earthly science' in the sense of human laws as contrasted with divine ones. According to Dean (ibid 36), 'the first British work to acknowledge an independently designated science of the earth obviously akin to modern geology' was Daniel Collins's translation of Mickel Pederson Escholt's *Geologia Norvegica* (1657). Plot apparently was aware of Collins's translation, for he cited it (incorrectly) in *Staffordshire* 145. The first book written by an English-speaking person to carry the title 'Geologia' was a cautious criticism of Burnet's *Telluris theoria sacra*, by Erasmus Warren, *Geologia, or a Discourse ... Wherein the Form and Properties Ascribed to It [the earth], in a Book Intituled, 'The Theory of the Earth,' Are Excepted Against* (London 1690). See E.G.R. Taylor 'The English Worldmakers of the Seventeenth Century and Their Influence on the Earth Sciences' *Geographical Review* 38 (1948) 109.
47 Plot *Staffordshire* 131
48 Ibid 170. On the magnetic polarity and the compass, see also Francis Aston to William Musgrave, 24 January 1683, in Gunther *Science in Oxford* 12:50–1.
49 Cox 'British Palaeontology' 210–11
50 H. Hamshaw Thomas 'The Rise of Geology and Its Influence on Contemporary Thought' *Annals of Science* 5 (1947) 328 believes that Plot was 'one of the last champions of the old views in England.'

51 Plot *Oxfordshire* 111; see also another excellent study, Sir Archibald Geikie *The Founders of Geology* 2d ed (New York 1905) 77. The close ties between British naturalists investigating their native land and those studying other lands is revealed, for example, by the fact that – in connection with the examination of fossil remains – Plot credits his 'Ingenious friend,' John Banister, MA, with finding near Oxford an 'Anthropocadites' that he illustrates on plate VIII fig 2 of *Oxfordshire*, commenting that 'I thought its Admittance would be not ungrateful to the reader'; Banister consulted Plot's works for information on various matters; see Ewan *John Banister* 310, 385. Lister, Ray, Lhuyd, and others used Banister's collections to assist their own work (ibid 101, 102).

52 Plot *Oxfordshire* 111–12

53 Ibid 182

54 See ibid 118, 120.

55 Plot's 'petrifying juices' or 'plastic force' was certainly what we now recognize as crystallization, and the workings of the salt principle in the creative plastic virtue were described by Plot in great detail in ibid 121–4. There were certain fossils whose organic nature Plot was prepared to admit, for they possessed not only the outward form of bones but exhibited, though turned to stone, a characteristic bony structure; see W.J. Sollas *The Age of the Earth* (London 1905) 236.

56 Plot *Staffordshire* 392

57 Robert Plot to John Fell, nd (c 1673), Gunther *Science in Oxford* 12:341–2

58 Hunter *John Aubrey* 202

59 Plot *Staffordshire* 396–7. Plot described flints that were 'exactly in the form of a bearded arrow jagg'd at each side, with a larger stemm in the middle.' He concluded therefore that not only are these arrows 'all artificial, whatever is pretended'; but also that they had anciently 'some ways of working of flints by the toole, which may be seen by the marks'; ibid 397. Olao Wormius, in *Museum Wormianum* (Leyden 1655) chap 3:39 had already written that 'some [of these flints] resemble so closely the point of a sword that it is doubtful if they are the work of nature or of art'; see Philip Shorr 'Genesis of Prehistorical Research' *Isis* 23 (1935) 429. According to M.C.W. Hunter 'The Royal Society and the Origins of British Archaeology: I' *Antiquity* 65 (1971) 116 Plot 'insists on the study of archaeological evidence in isolation from the historical sources which were the domain of antiquaries like Camden.' Hunter (ibid II 188–9) seems more sceptical of Aubrey's archaeological interpretations, noting his 'remarkable credulity,' although Hunter also says that on occasion Plot 'was evidently dissatisfied with conclusions based wholly upon things, and therefore linked them to historical sources.'

Notes to pages 204–6   317

60 Plot *Staffordshire* 398
61 R. Pulteney *Historical and Biographical Sketches of the Progress of Botany in England* 2 vols (London 1790) 1:352
62 R. Garner *The Natural History of the County of Stafford* (London 1844) was the first to follow in Plot's footsteps.
63 An account of the antiquities that Plot and Browne examined is contained in a small diary entitled 'Tour in Kent' London, BL, Sloane ms 1899 and also Oxford, Bodleian Library, Rawlinson ms D390 ff 95–6; see Gunther *Science in Oxford* 12:360.
64 Robert Plot to Arthur Charlet, 2 September 1693, ibid 12:396
65 Robert Plot to Arthur Charlet, 1 November 1694, ibid 12:402
66 E.W. Gilbert *Geography as a Humane Study* (Oxford 1955) 4
67 Owen's study is contained in Owen ed '*The Description of Penbrokshire.*'
68 Ibid 1 chap 7. John Leland, in his *Itinerary*, had alluded to geology long before Owen. For example, the first analytical concept recorded in the annals of British geology – that of stratification – came from the researches of Leland: 'The stones ly yn the ground lyke a smothe table: and be bedded one flake under another,' quoted in John Challinor *The History of British Geology: A Bibliographical Study* (Newton Abbot 1971) 59–60.
69 Ibid 60
70 Owen's map of Pembrokeshire was published by William Camden in the 1607 edition of the *Britannia*. According to Karl von Zittel *History of Geology and Palaeontology* trans Maria M. Ogilvie-Gordon (London 1901) 35 it is to Christopher Packe 'that we are indebted for the first geological map of a part of England in his work *A New Philosophical-Chronological Chart of East Kent*, published in 1743.' Also, V.A. Eyles 'Mineralogical Maps as Forerunners of Modern Geological Maps' *Cartographical Journal* 9 (1972) 133 states that 'before 1700 no topographical maps had been published of sufficient accuracy ... to permit the accurate localisation of geological information.'
71 W.H. Fitton 'Notes on the History of English Geology' *Philosophical Magazine* 1 (1832) 443 calls Owen the 'Patriarch of English Geologists.' See also John Challinor 'The Early Progress of Geology' *Annals of Science* 9 (1953) 129.
72 Camden *Britain* 1610 ed 654
73 Frank Emery 'Edward Lhuyd and the 1695 *Britannia*' *Antiquity* 32 (1958) 179–80.
74 Ibid 181
75 Frank Emery 'Edward Lhuyd and Some of his Glamorgan Correspondents: A View of Gower in the 1690s' *Transactions of the Honourable Society of Cymmrodorion* (1965) 68. Davies *Earth in Decay* also refers to Lhuyd's

scientific inquiries, as does B.G. Charles in *George Owen of Henllys: A Welsh Elizabethan* (Aberystwyth 1973). (The title of this last work is indicative of the fact that Owen was not a patriotic Welshman in the narrow sense of the word, but rather, proud that Pembroke men had offered resistance against the Welsh as well as against the English in the past.)

76 Emery *Edward Lhuyd* 39. Lhuyd described palaeontology as 'a new Science in Natural History'; see Emery 'Glamorgan Correspondents' 69.

77 Robert Plot to Edward Lhuyd, 29 January 1691, in Gunther *Science in Oxford* 14 (1945) 133

78 London, BL, Sloane ms 4062 f 262. Lhuyd also referred to Plot as 'a man of as bad morals as ever took a doctors degree. I wish his wife a good bargain of him; and to my self yt I may never meet with ye like again' (Edward Lhuyd to Martin Lister, 17 January 1691, Gunther *Science in Oxford* 14:131.)

79 Oxford, Bodleian Library, Ashmole ms 1722; see Hunter *John Aubrey* 202.

80 Edward Lhuyd to John Lloyd, 16 July 1695, in Gunther *Science in Oxford* 14:279

81 Edward Lhuyd to Richard Richardson, 13 May 1697, ibid 14:335; quoted in Frank Emery '"The Best Naturalist Now in Europe": Edward Lhuyd, F.R.S. (1660–1709)' *Transactions of the Honourable Society of Cymmrodorion* (1969) 57

82 On Lhuyd's remarks on Erdeswicke see *Camden's Britannia* ed and trans Edmund Gibson (London 1695) col 531.

83 Ibid col 638: quoted in Emery 'Best Naturalist' 57

84 For the date of Lhuyd's departure, see Edward Lhuyd to Martin Lister, 4 August 1693, in Gunther *Science in Oxford* 14:197–8; for his return, see Edward Lhuyd to John Lloyd, 10 October 1693, ibid 14:198, where he informs Lloyd that his 'task-masters' 'did not require I should put myselfe to ye trouble and expences of a journey into Wales.'

85 Edward Lhuyd to John Lloyd, 2 May 1695, ibid 14:269–70

86 Nicholas Roberts to Edward Lhuyd, 3 August 1695, Oxford, Bodleian Library, Ashmole ms 1817A f 316

87 Frank Emery 'A Map of Edward Lhuyd's *Parochial Queries in Order to a Geographical Dictionary, etc., of Wales* (1696)' *Transactions of the Honourable Society of Cymmrodorion* (1959) 43–4

88 For Machell, see John Rogan and Eric Birley 'Thomas Machell, The Antiquary' *Transactions of the Cumberland and Westmorland Antiquarian and Archaeological Society* 55 (1956) 132–53. 'Machell's queries were the first to specify the study of geography, history, and antiquities at the parish level,' according to Emery 'A Map' 44.

89 Rupert H. Morris ed *Parochialia* 3 pts (London 1909–11) 1(1909) ix–x

90 Ibid 1: x
91 Ibid 1: xiii
92 They have been published in ibid.
93 For an account of Lhuyd's adventures in Scotland see J. Wreford Watson 'Edward Lhuyd and Scottish Studies' *Scottish Studies* 2 (1958) 117–19, and J.L. Campbell and Derick Thomson *Edward Lhuyd in the Scottish Highlands, 1699-1700* (New York 1963).
94 See Gunther *Science in Oxford* 14: 338–9.
95 Edward Lhuyd to Henry Rowlands, 10 March 1701, ibid 14: 439–40
96 Edward Lhuyd to Richard Rawlinson, 8 June 1701, ibid 14:457. The term 'tories' is used here in its original sense, not its adopted English one.
97 W. Pryce *Archaeologia Cornu-Britannica* (Sherborne 1799) footnote to Lhuyd's first letter
98 See P.W. Carter 'Edward Lhuyd the Scientist' *Transactions of the Honourable Society of Cymmrodorion* (1962) 56. Parts of the *Archaeologia* appeared in the *Philosophical Transactions*, while much of it was published posthumously. *Archaeologia* has been reprinted by the Scolar Press (Menston 1969).
99 Emery 'A Map' 44
100 This work was published in eight volumes as 'a pocket book to be carried into stonepits, etc. such large draughts being folded in, would soon be sullied and torn, abroad in the fields'; Edward Lhuyd to Martin Lister, September 1695, Gunther *Science in Oxford* 14:282. Here again is proclaimed a practical benefit of the new science.
101 Edward Lhuyd to Richard Rawlinson, 17 December 1699, ibid 14:425; see also Piggott 'Antiquarian Thought' 111.
102 For the Aubrey-Lhuyd connection see Glyn Daniel 'Edward Lhuyd: Antiquary and Archaeologist' *Welsh History Review* 3 (1967) 345–59. Richard Ellis 'Some Incidents in the Life of Edward Lhuyd' *Transactions of the Honourable Society of Cymmrodorion* (1907) 1–51 contains useful biographical data (reprinted in Gunther *Science in Oxford* 14). Daniel 'Antiquary' 352 notes how Lhuyd's thought reflects the transition from 'mediaeval antiquarianism' to 'modern archaeology,' and that he 'recognized the myth and legend element in the writers that had gone before.'

CHAPTER TWELVE

1 See [Sir Robert Sibbald] *The Autobiography of Sir Robert Sibbald* (Edinburgh 1833) 37–8.
2 Frank Emery 'The Geography of Robert Gordon, 1580–1661, and Sir Robert Sibbald, 1641–1722' *Scottish Geographical Magazine* 74 (1958) 3–12. Emery is an authority on the Scottish regional natural historians, as on the Welsh.

3 Robert Sibbald *An Account of the Writers Ancient and Modern ... of North-Britain called Scotland ...* (Edinburgh 1710).
4 Ibid 3. Here Sibbald is referring to the works of Ptolemy, Tacitus, and others.
5 Ibid 17–24. For Pont, see C.G. Cash 'The First Topographical Survey of Scotland' *Scottish Geographical Magazine* 17 (1901) 399–414; Ian A.G. Kinniburgh 'A Note on Timothy Pont's Survey of Scotland' *Scottish Studies* 12 (1968) 187–9; and D.G. Moir and R.A. Skelton 'New Light on the First Atlas of Scotland' *Scottish Geographical Magazine* 84 (1968) 149–59.
6 Robert Gordon to Sir John Scotstarvet, 24 January 1648, ibid 149
7 Emery 'Geography' 5
8 Robert Gordon, quoted in ibid 4–5
9 See ibid 6.
10 Whether in their original state or transcribed, much of this and other such material collected by Sibbald eventually found its way into the Advocate's Library; see Mitchell and Clark eds 'What Sibbald received from James Gordon' *Geographical Collections* 2 (1908) xix–xx.
11 The quotation regarding the number of Scotsmen at Leyden is taken from L.W. Sharp *Early Letters of Robert Wodrow, 1698–1709* (Edinburgh 1937) xli–xliii. E.W. Gudger, in 'George Marcgrave, The First Student of American Natural History' *Popular Science Monthly* 81 (1912) 255 states that Marcgraf 'would certainly have raised himself to the rank of the first natural historian of his time, and possibly that of greatest since Aristotle,' had he lived long enough to properly organize his Brazilian collections. Marcgraf's *Historia Naturalis Brasiliae* comprised 303 folio pages and was illustrated by 429 figures. It described such items as plants and animals, the diseases of the country, the uses of herbs as remedies, as well as the aboriginal inhabitants and their character and customs (only one section is devoted to the native population); see Gudger 260–1. On Leyden's influence also see Rev Dr Stokes 'Dudley Loftus: A Dublin Antiquary of the Seventeenth Century' *Proceedings and Papers of the Royal Society of the Antiquaries of Ireland* 21 (1890–1) 19 (which deals with Leyden as a centre of oriental study), and Stearn 'The Influence of Leyden on Botany,' which discusses the reasons for Leyden's international predominance (142–3) as well as Linnaeus's ties with Leyden.
12 Robert Sibbald, quoted in F. Paget Hett ed *The Memoirs of Sir Robert Sibbald, 1641–1722* (London 1932) 64
13 [Sibbald] *Autobiography* 28
14 The second volume, the 'Descriptions of Ancient and Modern Scotland, with a Scottish Atlas,' was to have contained the material from Gordon, and from his own correspondents.

15 Robert Sibbald *Scotia illustrata, sive Prodromus historiae naturalis* ... (Edinburgh 1684) pt 2 bk 4, 48–9, 55
16 Robert Sibbald *The History, Ancient and Modern, of the Sheriffdoms of Fife and Kinross* ... (Edinburgh 1710) 25; see Piggott 'Antiquarian Thought' 112.
17 Ibid. Sibbald did not have much to say on the origin of these; see Stuart Piggott 'The Ancestors of Jonathon Oldbuck' *Antiquity* 29 (1955) 151.
18 Robert Sibbald *Historical Enquiries, Concerning the Roman Monuments and Antiquities in the North-Part of Britain Called Scotland* ... (Edinburgh 1707) preface
19 This quotation is taken from Sibbald's *History and Description of Stirling-Shire, Ancient and Modern* (Stirling 1892) 16–17, a reprint of part of the original work, which is dated there as 1707.
20 *Camden's Britannia*, col 883. Sibbald's contribution to this work included a 'Disclosure Concerning the Thule of the Ancients.'
21 Symson revised this work in 1692; it was published under the title *A Large 'Description of Galloway' by Andrew Symson* (Edinburgh 1823) and was edited by Thomas Maitland, according to Mitchell and Clark eds *Geographical Collections* 2:xxviii; it is also contained in ibid 2:51–99. For the life of Symson see Maitland v–xvi.
22 See Emery 'Geography' 8–9.
23 Andrew Symson to Edward Lhuyd, May 1708, Oxford, Bodleian Library, Ashmole ms 1817A, f 499. This work represents one of the earliest parish-by-parish surveys of any county, although it appears that Symson never had it published.
24 Martin Martin, quoted in Frantz *The English Traveller* 21
25 Martin Martin, quoted in ibid 29
26 Philip Falle *An Account of the Island of Jersey* (London 1694) quoted in Emery 'Aubrey to Defoe' 323; Martin Martin *A Late Voyage to St Kilda, The Remotest of all Hebrides, or Western Isles of Scotland* (London 1698) preface
27 Ibid 51, 271. The results of Martin's travels had previously appeared in 'Several Observations in the North Islands of Scotland. Communicated to the Royal Society by Mr. Martin Martin' *Philosophical Transactions* 19 (1697) 727–9.
28 Donald J. Macleod ed *'A Description of the Western Islands of Scotland' by Martin Martin* (Stirling 1934) 64–5
29 Ibid 63. Many other seventeenth-century natural historians and scientists, such as William Molyneux, also held the notion of an affinity between the study of natural history and religion; see K. Theodore Hoppen 'The Royal Society and Ireland: William Molyneux, F.R.S. (1656–1698)' *Notes and Records of the Royal Society of London* 18 (1963) 129. In many respects this outlook can be

322 Notes to pages 220-2

traced to Bacon himself, who, according to Moody E. Prior 'Bacon's Man of Science' *Journal of the History of Ideas* 15 (1954) 362 'represents his scientist as a religious man.' Although to some extent, perhaps, Bacon considered it necessary to 'defend the new learning against the charge that it leads to atheism,' conceding that 'a little natural philosophy inclineth the mind to atheism,' a further consideration 'bringeth the mind back to religion' (ibid 362–3). For Bacon, then, the glory of God is inspired by the (scientific) investigation of His works, and therefore he claimed that the new philosophy does a service to religion. Nevertheless, according to Prior (ibid 363), the true basis of religion for Bacon was 'the knowledge of God's will and law, matters which lay beyond man and hence were knowable only through divine revelation,' so that in the long run Bacon did not presume to be able to attain to the mysteries of God solely through the contemplation of nature. Such matters were apparently knowable only through divine revelation.

30 See Mitchell and Clark eds *Geographical Collections* 2:219–23.
31 Ibid 2:221
32 Emery 'Geography' 11
33 Edition of 1716, published in London (BL copy C45 C1)
34 Ibid, opposite the title-page. Toland, perhaps best known for his inquiries into comparative religion, was also supposed to serve as Gibson's consultant on Ireland when the latter was putting together the revised *Britannia*. However, a quarrel between the two men soon put an end to this short-lived association.
35 *Camden's Britannia* col 1073
36 This quotation is taken from ibid col 1075, which I have used instead of the original *A Description of the Isles of Orkney* ... (Edinburgh 1693).
37 James Wallace *An Account of the Islands of Orkney* ... (London 1700); see also 'An Abstract of a Book, viz. *An Account of the Islands of Orkney*. By James Wallace, M.D. and Fellow of the Royal Society. To which is added, an *Essay concerning the Thule of the Ancients*' *Philosophical Transactions* 22 (1700) 543–6.
38 See Emery 'Geography' 11. The extent of the interaction among the natural historians is further exemplified by Nicolson's correspondence with Lhuyd, which ranged over subjects such as the comparison of Cumbria and North Wales, described here as 'much of a piece' in their natural history (Oxford, Bodleian Library, Ashmole ms 1816 f 454f).
39 See the comments on Leigh in the letter of Richard Rawlinson to Edward Lhuyd, 14 September 1701, Oxford, Bodleian Library, Ashmole ms 1817A f 278; quoted in Emery 'Aubrey to Defoe' 321.
40 For example, see Charles Leigh to William Musgrave, 5 January 1685, Gunther *Science in Oxford* 12:250–1, which contains references by Leigh to Plot's lectures and experiments.

Notes to pages 222–8 323

41 Charles Leigh *The Natural History of Lancashire, Cheshire and the Peak in Derbyshire* (Oxford 1700) 65, 75, 80
42 Ibid 62, 100–1. Leigh also describes the head of a 'Stag of Canada' found eight yards within Marle in Lancashire, stating that 'these Creatures being Foreign to this Island, I think, sufficiently demonstrate the Universality of the Deluge' (184).
43 Ibid 99, 119–20
44 Ibid preface
45 Ibid 64–5
46 Lhuyd possessed, if not the actual book, a copy of Leigh's proposals for his *Lancashire*, on which he jotted a drawing of a marine plant, with some notes. Leigh's biographer in the *Dictionary of National Biography*, C.W.S., claims that all of Leigh's writings are of little value.
47 The correspondence between the two men is contained in Oxford, Bodleian Library, Ashmole ms 1816 ff 397–443, among other places; see also Emery 'Aubrey to Defoe' 321.
48 Edward Lhuyd to John Morton, 15 May 1700, London, BL, Sloane ms 4063 f 25
49 Edward Lhuyd to John Morton, 2 April 1704, ibid f 243. For the Lhuyd–Morton tie see Emery 'Aubrey to Defoe' 322.
50 John Morton 'Proposals for Subscription to the Natural History of Northamptonshire' Oxford, Bodleian Library, Ashmole ms 1820 f 79
51 Ibid
52 John Morton to Hans Sloane, 22 June 1706, London, BL, Sloane ms 4040 ff 183–4
53 John Morton *The Natural History of Northamptonshire* (London 1712) 14–17
54 Ibid 19
55 Ibid 14–15
56 Ibid 26
57 Ibid 97
58 Ibid 345
59 Ibid 538
60 See Oxford, Bodleian Library, Ashmole ms 1816 f 442; also quoted in Emery, 'Aubrey to Defoe' 322.
61 Morton *Northamptonshire* 97; quoted in Porter *Making of Geology* 39
62 Ibid 39–40
63 Ibid 38. Here Porter is referring specifically to developments in the study of the mineral kingdom, stating that 'Without the ... [work] of Plot, Lister, Ray, Webster, Lhwyd and Woodward, future generations would have built on sand' (note the names of the regional writers in this select group).
64 See Gunther *Science in Oxford* 14:270.

65 Sibbald's first proposals focused on Italian and French academies as models, and included literary, religious, and historical (civil) interests, which were dropped in favour of the narrower scientific, medical, and practical concerns of the type upheld by the Royal Society of London. Sibbald's plan was probably drafted in its revised form in about 1701 or 1702; unfortunately, he did not live to see his plan fulfilled. In the 1680s Sibbald also attempted to establish medical training at the university.

CHAPTER THIRTEEN

1 Joseph Glanvill *Plus ultra: Or, the Progress and Advancement of Knowledge Since the Days of Aristotle. In an Account of some of the most Remarkable Late Improvements of Practical, Useful Learning* (London 1668) 75
2 G.V. Scammell 'The New Worlds and Europe in the Sixteenth Century' *Historical Journal* 12 (1969) 389–412 argues that in general the discoveries – by Europeans – were slow to generate interest in the geography of the New World. But at least 'as the discoveries unfolded it was increasingly clear how much there was of which the [ancient] Greeks and Romans were either ignorant or misinformed,' and the discoveries also helped to stimulate 'rational speculation or impartial, factual description' (ibid 393, 397). On this subject see also Lucien Febvre and Henri-Jean Martin *The Coming of the Book* trans David Gerard (new ed London 1976) 272–82 and P. Marshall and Glyndwyr Williams *Great Map of Mankind* (London 1982).
3 Nathaniel Salmon *Antiquities of Surrey* (London 1736) preface
4 Fussner's work, however, still holds considerable influence.
5 The possible exception to this statement might be the political and legal bodies of writing, areas in which much research has been conducted and which, of course, relate to the topic at hand.
6 Lily B. Campbell *Tudor Conceptions of History and Tragedy in 'A Mirror for Magistrates'* (Berkeley 1936) 2
7 F. Smith Fussner *Tudor History and the Historians* (New York 1970) 277
8 Douglas *English Scholars* 1951 ed 166
9 In Philipot's *Kent Surveyed*, the author purported to rely on 'common Records ... [on material] shut up in the private Muniments, Escripts, and Registers of particular Families ... ' ('To the Reader').
10 John Earle *Microcosmographie* ed A.S. West (London 1897) 57–8. Much of the discussion here of the merits of the chorographers is based on similar remarks made by Strauss about the German topographer-historians.
11 On the relationship – or in some cases on the lack of relationship – between

Notes to pages 232–6 325

science and the humanistic disciplines see Douglas Bush *Science and English Poetry: A Historical Sketch, 1590–1950* (New York 1950); Herschel Baker *The Wars of Truth: Studies in the Decay of Christian Humanism in the Earlier Seventeenth Century* (Cambridge 1952); C.P. Snow *The Two Cultures and the Scientific Revolution* (Cambridge 1959); and Barbara Shapiro *Probability and Certainty,* not only the most recent work on the subject but also probably the best.

12 Taylor *Stuart Geography* 85–6. Taylor says that the more popular earlier standpoint 'was that Nature's mysteries were not intended to be unravelled, a knowledge of them being a prerogative of the Deity' (86).
13 Herbert Butterfield *Man on His Past* (Cambridge 1955) 32
14 See Bacon *Works* 4:114.
15 These developments and currents are succinctly described in the works of Hunter, Emery, and Piggott, and admirably summarized in Lynch 'Prehistoric Archaeology' passim.
16 See Sir Francis Bacon *The Advancement of Learning* (London 1605; reprint ed London 1915) 95
17 Morton 'Proposals' f 79
18 Ibid
19 Richard Westfall *Science and Religion in Seventeenth Century England* (Cambridge 1958; reprint ed Ann Arbor 1973) 31. Sir Thomas Browne was one of the first to consider such intellectual endeavours as a form of repaying God for having endowed man with reason.
20 Morton *Northamptonshire* ii
21 Ibid; the note is located immediately before the title-page.
22 Ray *Wisdom* 3rd ed (London 1701) 199–206; see also Gordon L. Davies 'The Concept of Denudation in Seventeenth-Century England' *Journal of the History of Ideas* 27 (1966) 281. In the preface of the fourth edition of Ray's *Wisdom* Ray had written: 'Note that by the Works of the Creation in the Title, I mean the Works created by God at first, and by him conserved to this Day in the same State and Condition in which they were at first made; for Conservation, according to the Judgment both of Philosophers and Divines, is a continued creation.'
23 See Lankester *The Correspondence of John Ray* 242. It is interesting to note that Strabo's work had its teleological aspects – with obvious differences. For him, 'regions are laid out, not in a fortuitous way, but as though in accordance with some calculated plan'; quoted in Paassen *Classical Tradition* 30.
24 Britton *Wiltshire* 53
25 Thomas *Decline of Magic* passim

26 Thomas Hobbes *Opera* 2 (1839) 127, quoted in Lynn Thorndike *A History of Magic and Experimental Science* 8 vols (New York 1923–58) 7 (1958) 74–5; Kite 'Study' 18
27 Stuart Piggott *British Prehistory* (London 1949; reprint ed London 1955) 10
28 William Stukeley *Palaeographia Britannica: or Discourses on Antiquities in Britain* (London 1743) 15. Stukeley's interest in some of these sites was probably aroused by his reading, in 1719, a transcript of Aubrey's 'Monumenta Britannica': see Stuart Piggott *William Stukeley* (Oxford 1950) 45.
29 Piggott 'Archaeological Draughtsmanship' 171
30 See John Tabor 'An accurate Account of a tessellated Pavement, Bath, and other Roman Antiquities, lately discover'd near East Bourne in Sussex. Being part of a Letter of January 26, 1717, from the learned Dr John Tabor of Lewis, to Dr John Thorpe, R.S.S. and by him communicated to the Royal Society' *Philosophical Transactions* 30 (1717) 563; referred to in Hunter 'Origins of British Archaeology: 1' 116.
31 Plot *Staffordshire* 397; see Lynch 'Prehistoric Archaeology' 46.
32 Douglas *English Scholars* passim
33 Stuart Piggott *The Druids* (London 1968) 143–5
34 Edward Lhuyd to William Nicolson, 20 April 1698, quoted in Gunther *Science in Oxford* 14:362
35 Margaret C. Jacob *The Newtonians and the English Revolution, 1689–1720* (Ithaca 1976) has more to say on this and related subjects; also refer to Roy Porter's, Michael Hunter's and Charles Webster's works.
36 Fussner *Tudor History* 245 was among the first to speculate that there might be a 'connection between the methods of history and of science,' noting that 'this is still a dark plain, on which anyone may stumble.' For more on this subject, of course, see Shapiro *Probability and Certainty* passim.
37 See Simon Schaffer 'Natural Philosophy and Public Spectacle in the Eighteenth Century' *History of Science* 21 (1983) 2.
38 R.S. Porter 'Science, Provincial Culture and Public Opinion in Enlightenment England' *British Journal for Eighteenth-Century Studies* 3 (1980) 20
39 Ibid 32
40 Ibid 21. At least, this was the case by the Enlightenment.
41 Sergio Moravia 'The Enlightenment and the Sciences of Man' *History of Science* 18 (1980) 248. Moravia states that the 'mainstream' of seventeenth-century thought makes this idenification, one that spilled over into the next century.
42 Schaffer 'Public Spectacle' 2. Schaffer states (ibid) that the love of the marvellous was claimed as a major strength of natural philosophy, but that it was also a threat: 'if in the earlier part of the century, wonder could be profitably exploited by successful entrepreneurs, by the later eighteenth century it could also be

displayed as a threat to social order,' and 'it was too easy to use the audience-relation as a resource against natural philosophers.'
43 Margaret Espinasse 'Decline and Fall of Restoration Science' 72
44 Quoted in ibid
45 Quoted in ibid 76. As Epinasse states (77), acquaintance with both types of science was considered desirable as long as 'there was no danger of its being taken seriously or being in any serious degree useful.'
46 Ibid 75. Espinasse points out that 'even the prestige of Newton could not obviate the contempt which had come to be the stock attitude of many influential writers by the early eighteenth century.'
47 Quoted in Houghton 'English Virtuosi' 196. Casaubon, critic of the Royal Society, also criticized, among other things, the criterion of utility adopted by society members. Not everyone, of course, thought in exactly these terms. Burnet, Whiston, Woodward, and other 'world makers,' for example, kept the scriptural-natural historical analogy intact. But since their cosmogonies centred more on theory than on collection and observation, it is uncertain whether Casaubon would have placed them in the category of 'sick brains,' although many others would have.
48 Unfortunately, many modern historians of science take on this line of thinking and extend it backwards in time, as being representative of the second half of the seventeenth century.
49 Robert A. Aubin 'Grottoes, Geology, and the Gothic Revival' *Studies in Philology* 31 (1934) 408–16
50 Lhuyd was one of the first to foresee the impending decline in natural history. On 15 October 1695, he complained to Lister that none of the divines and few masters of colleges at Oxford were 'sensible of the value' of natural history (Oxford, Bodleian Library, Lister ms 36 f 133).
51 W.P. Jones 'The Vogue of Natural History in England, 1750–1770' *Annals of Science* 2 (1937) 345
52 Ibid. Hankins *Science and the Enlightenment* passim also presents a strong argument for a thriving natural history among scientific circles from the 1670s on.
53 Jones 'Vogue' 346–7
54 Allen *Naturalist in Britain* 18. It is not surprising that Allen believes (17) that perhaps 'the very brilliance of the Royal Society under Newton helped to ensure an unusual violence in the inevitable reaction,' ie, there had occurred a swing away from empirical thought altogether after Newton's death in 1727. As far as British natural history as a whole is concerned, Allen identifies the period 1725–60 as 'largely a blank,' except for 'signs of energy in entomology' (15–16). (As noted, Jones sees the take-off point for natural history as c 1750.)

55 William Wotton *Reflections upon Ancient and Modern Learning* 2d ed (London 1697)
56 See Stephen Briggs 'Thomas Hearne, Richard Rawlinson, and the Osmondthick Hoard' *Antiquaries Journal* 58 (1978) 256. Briggs credits the later, more bizarre antiquarian activities of William Stukeley as being partly responsible for ending the rationalistic, pioneering spirit displayed by Plot or Lhuyd, so that there was a 'relapse into vagery of speculation' (255).
57 It would not be inappropriate to apply Allen's statement (*Naturalist in Britain* 17–18) that 'scientific genius tends to display itself in sudden, brief bursts of magnificent intensity, followed by long periods of comparative darkness,' to the particular context of the history of British regional study.
58 Ibid 20–1. In his conclusion to *Science and Society*, Hunter comments on the continuity between the seventeenth and eighteenth centuries as symbolized by the Royal Society, noting, however, that 'the Society's renown still co-existed with tension and decay at its London home' (189). This sort of situation did little to encourage fellows to undertake serious natural historical study; the English government's ineffectual role in the promotion of science did not help matters (190).
59 These were the comments of Thomas Tanner (Oxford, Bodleian Library, Ashmole ms 1817B f 9), and Tancred Robinson (ibid ms 1817A f 344), respectively; quoted in Emery 'Aubrey to Defoe' 322–3. Piggott *Ruins* 21 says that as far as historical and antiquarian scholarship is concerned, there was a general lapse of standards after 1730 or so, when 'the necessity of an empirical approach was forgotten.' Levine *Humanism and History* chap 4 notes the Augustan contribution to the transformation of antiquarianism into archaeology.
60 Martin *Islands* preface; also quoted in Emery 'Aubrey to Defoe' 323
61 Allen *Naturalist in Britain* 54
62 Davies *Earth in Decay* 96
63 Ibid. Davies (31) refers to the death of Ray in 1705 as 'a convenient terminus to that first phase in the history of British geomorphology.'
64 Stuart Piggott 'Prehistory and the Romantic Movement' *Antiquity* 11 (1937) 35; see also Piggott's 'The Origins of the English County Archaeological Societies' *Transactions of the Birmingham and Warwickshire Archaeological Society* 86 (1974) 1–15. Of added interest, Hunter *Science and Society* 191 notes that the contributions made by scientists to the field of technology were also disappointing.
65 See G.S. Rousseau's review of Joseph M. Levine's *Dr. Woodward's Shield: History, Science, and Satire in Augustan England* (Berkeley and Los Angeles 1977), in *History of Science* 17 (1979) 143. A scientific dimension, as Levine states, and as this study has shown, to historical and antiquarian study had been devel-

oped by Woodward's day (c 1700). But it was neither firm enough, nor were there enough practitioners of it, to enable Woodward's famous bronze shield to be conclusively disclosed as a fake, and not the rare antiquity of great historical value that Woodward claimed it was. The classifiers and collectors, however, were eventually to accrue enough information, especially for purposes of comparison, to enable such evaluations to be accurately made. This was to take time, so that the process was not firmly entrenched until well into the Enlightenment. Piggott *Ruins* 20 notes that as far as antiquities are concerned, Aubrey, Lhuyd, and others appreciated the need for recording and collating as preliminaries to interpretation.

66 Levine 'Natural History' 61
67 Ibid 61–3. Levine (62) states: 'Unfortunately, nothing that Ray did and only a little of what the members of the Royal Society did collectively, really measures up [as far as modern historians of science are concerned]. What they did, therefore (so the argument would seem to go), was dubious as science, which it merely anticipates.' This has meant that 'the revolution that brings about modernity must be deferred in those places to a later date, if at all' (ibid).
68 Ibid 67–8
69 William Wotton, quoted in ibid 68
70 Ibid 69: see also 72 n37.
71 This assumes that paradigms cover more than just scientific thought.
72 Levine *Humanism and History* 176

Select Bibliography

This bibliography focuses primarily, although not exclusively, on titles not listed in the text or in the Notes.

BIBLIOGRAPHICAL REFERENCE

Anderson, John P. *The Book of British Topography* (London 1881)
Bandinel, Bulkeley *A Catalogue of the Books, relating to British Topography, and Saxon and Northern Literature, bequeathed to the Bodleian Library, in the Year MDCCXCIX, by Richard Gough* (Oxford 1914)
Batten, John *Historical and Topographical Collections relating to the Early History of Parts of South Somerset* (Yeovil 1894)
British Museum *General Catalogue of Printed Books* 261 vols (London 1931)
Challinor, John *The History of British Geology: A Bibliographical Study* (Newton Abbot 1971)
Chubb, Thomas *The Printed Maps in the Atlases of Great Britain and Ireland. A Bibliography 1579–1870* (London 1927)
Davies, Godfrey *Bibliography of British History: Stuart Period, 1603–1714* (Oxford 1928)
Dictionary of National Biography ed Leslie Stephen and Sidney Lee 22 vols (London 1937–8)
Donovan, Dennis G. *Sir Thomas Browne and Robert Burton: A Reference Guide* (Boston c 1981)
Fordham, Sir George Herbert *The Road-Books and Itineraries of Great Britain 1570–1850* (Cambridge 1924)
Freeman, R.B. *British Natural History Books, 1495–1900; A Handlist* (Folkestone 1980)

332 Select Bibliography

Fussell, George E. and V.G.B. Atwater 'Travel and Topography in Seventeenth-Century England: A Bibliography of Sources for Social and Economic History' *Library* 4th ser 13 (1933) 292–311

Gomme, G.L. *Index of Archaeological Papers, 1665–1890* (London 1907)

Gough, Richard *British Topography* 2 vols (London 1780)

Henrey, Blanche *British Botanical and Horticultural Literature before 1800* (Oxford 1975)

Hoare, Sir Richard Colt *A Catalogue of Books relating to the History and Topography of England, Wales, Scotland, Ireland* (London 1815)

Humphreys, A. *A Hand-book to County Bibliographies* (London 1917)

Jayawardene, S.A. 'Western Scientific Manuscripts before 1600: a Checklist of Published Catalogues' *Annals of Science* 35 (1978) 143–72

Keynes, Geoffrey *A Bibliography of Sir William Petty F.R.S.* (Oxford 1971)

Knight, David *Sources for the History of Science 1660–1914* (Ithaca 1975)

Levien, Edward 'On Unpublished Devonshire Manuscripts in the British Museum' *Journal of the British Archaeological Association* 18 (1862) 134–45

Lewis, S. *A Topographical Dictionary of England* 5 vols (London 1831)

London, William *Catalogue: The Most Venible Books in England, Orderly and Alphabetically Digested* (London 1657)

Lowndes, William Thomas *The Bibliographer's Manual of English Literature* 4 vols new ed (London and New York 1858)

Maddison, Francis, Dorothy Styles, and Anthony Wood *Sir William Dugdale, 1605–1686: A List of His Printed Works* (Warwick 1953)

Malcolm, James Pellar *Lives of Topographers and Antiquaries who have written concerning the Antiquities of England* (London 1815)

Mitchell, Sir Arthur and James Tosach Clark eds *Geographical Collections Relating to Scotland, made by Walter Macfarlane* Scottish History Society Publications 3 vols (Edinburgh 1907–9)

Nichols, John *Bibliotheca topographica Britannica* 8 vols (London 1780–90)

Nickson, M.A.E. *The British Library: Guide to the Catalogues and Indexes of the Department of Manuscripts* (London 1978)

Nicolson, William *The English, Scotch, and Irish Historical Libraries* 3rd ed rev (London 1736)

Paget, H.F. *The Early Maps of Scotland* 3rd ed rev (Edinburgh 1973)

Pollard, A.W. and G.R. Redgrave *A Short-Title Catalogue of Books Printed in England, Scotland and Ireland, And of English Books Printed Abroad: 1475–1640* (London 1926, reprint ed London 1964)

Prince, John *Danmonii orientales illustres; or, The Worthies of Devon* (Exeter 1701)

Richardson, John *The Local Historian's Encyclopedia* (New Barnet, Herts 1974)

Rawlinson, Richard *The English Topographer* (London 1720)

Sarjeant, William A.S. *Geologists and the History of Geology. An International Bibliography from the Origins to 1978* 5 vols (London 1980)
Skelton, R.A. *County Atlases of the British Isles, 1579–1850; a Bibliography* 5 pts (London 1964–70)
Taylor, E.G.R. *Late Tudor and Early Stuart Geography 1583–1650* (London 1934, reprint ed New York 1968)
– *Tudor Geography, 1485–1583* (London 1930)
Thompson, J.W. *A History of Historical Writing* 2 vols (New York 1942)
Upcott, William *A Bibliographical Account of the Principal Works relating to English Topography* 3 vols (London 1818)
Ward, A.W. and A.R. Waller eds *The Cambridge History of English Literature* 15 vols (Cambridge 1907–27)
Wells, J.E. *Manual of Writings in Middle English* (London 1916)
Wing, D.G. *A Short-Title Catalogue of Books Printed in England, Scotland, Ireland, Wales, and British America, and of English Books Printed in Other Countries: 1641–1700* 3 vols (New York 1945–51, 2d ed rev (New York 1972)
Worrall, John *Bibliotheca topographica Anglicana* (London 1736)

PRIMARY AND SECONDARY

Primary

Aubrey, John *'Monumenta Britannica,' parts I–III, by John Aubrey* ed John Fowles, ann Rodney Legg (Sherborne 1980–2)
Beaumont, John *An Historical, Physiological, and Theological Treatise of Spirits, Apparitions, Witchcrafts, and Other Magical Practices* (London 1705)
Bede, The Venerable *Historical Works* ed J.A. Giles 2 vols (London 1843–5)
Bellenden, John *The Works of John Bellenden* ed Thomas Maitland 2 vols (Edinburgh 1822)
Biondo, Flavio *Italia Illustrata* (Basel, 1531)
Birch, Thomas *The History of the Royal Society of London* 4 vols (London 1756–7)
Blith, Walter *The English Improver, Or a New Survey of Husbandry* (London 1649)
Bodin, Jean *Method for the Easy Comprehension of History* trans Beatrice Reynolds (New York 1969)
Boethius, Hector *Scotorum historiae* (Paris 1574)
Borlase, William *The Natural History of Cornwall* (Oxford 1758)
Boyle, Robert *General Heads for the Natural History of a Countrey ... for the Use of Travellers and Navigators* (London 1692)
Brand, John *A Brief Description of Orkney, Zetland, Pightland-Firth and Caithness* (Edinburgh 1701)

Browne, Sir Thomas *Pseudodoxia Epidemica: or Enquiries into Very many commonly received Tenents, And commonly presumed Truths* (London 1646)
– *Religio Medici* (London 1642)
Buchanan, George *Rerum Scoticarum historia* ... (Edinburgh 1583)
Burnet, Thomas *Telluris theoria sacra* (London 1681)
Burton, William *A Commentary on Antoninus his Itinerary* (London 1658)
Butcher, Richard *The Survey and Antiquities of the Towne of Stamford in the County of Lincolne* (London 1646)
Chapman, Henry *Thermae Redivivae: The City of Bath Described* (London 1673)
Charleton, Walter *Chorea Gigantum* (London 1663)
Cunningham, William *The Cosmographical Glasse* (London 1559)
Dale, Samuel *Pharmacologia* (London 1693)
– *The History and Antiquities of Harwich and Dovercourt, Topographical, Dynastical and Political. First Collected by Silas Taylor, alias Domville, Gent* (London 1730)
Dallaway, James *Antiquities of Bristow* (Bristol 1834)
Dallington, R. *The View of Fraunce* (London 1604)
Dick, Hugh G. 'Thomas Blundeville's "The True order and Methode of wryting and reading Hystories (1574)"' *Huntington Library Quarterly* 3 (1939–40) 149–70
Digges, Leonard *A Boke named Tectonicon* (London 1562)
– *A geometrical practise, named Pantometria* (London 1571)
Dodridge, Sir John *The Lawyers Light* (London 1629)
Dunstar, Samuel *Anglia rediviva* (London 1699)
Eliot, John *The Survey, or Topographical description of France* (London 1592)
Falle, Philip *An Account of the Island of Jersey* (London 1694)
Folkingham, William *Feudigraphia* (London 1610)
Fuller, Thomas *History of the Worthies of England* (London 1662)
Gerard, John *The Herball, or General Historie of Plantes* (London 1597)
Gildas *The Epistle of Gildas* (London 1638)
Godwyn, T. *Romanae historiae anthologia. An English Exposition of Romane Antiquities* (London 1614)
Guidott, Thomas *A Discourse of Bathe and the Hot Waters there* (London 1676)
Habington, Thomas *A Survey of Worcestershire by Thomas Habington* ed John Amphlett. Worcestershire Historical Society Publications, 2 vols (Oxford 1895–9)
Hakluyt, Richard *The Principall Navigations, Voiages, Traffiques and Discoueries of the English Nation* (London 1589)
Harris, John *Lexicon technicum, or a Universal English Dictionary of Arts and Sciences* 2 vols (London 1704)
Henry of Huntingdon 'Historiarum' *Rerum Anglicarum scriptores post Bedam praecipui* ed Sir Henry Savile (London 1596)

Select Bibliography 335

Holland, Philemon *The Historie of the World: commonly called 'The Natural Historie of C. Plinius Secundus'* (London 1601)
John of Fordun *Chronica gentis Scotorum* ed William F. Skene. Historians of Scotland, vol 1 (Edinburgh 1871)
Johnston, Nathaniel *Enquiries for Information towards the Illustrating and Compleating the Antiquities and Natural History of Yorkshire* (London 1683?)
– *The Excellency of Monarchical Government* (London 1686)
Jones, Inigo *The Most Notable Antiquity of Great Britain Vulgarly Called Stonehenge ... Restored* (London 1655)
Jones, John *Bathes of The Bathes ayde* 4 pts (London 1572–4)
Kennett, White *Life of Mr. Somner* (Oxford 1693)
King, Daniel *The Vale-Royale of England, or County Palatine of Chester* (London 1656)
Leycester, Sir Peter *Historical Antiquities in Two Books* (London 1673)
Llwyd, Humphrey *The Breviary of Britayne* trans Thomas Twyne (London 1573)
Merrick, Rice *A Booke of Glamorganshire Antiquities* ed James Andrew Corbett (London 1857)
Nashe, Thomas *Nashe's Lenten Stuffe* (London 1599)
Norden, John *John Norden's Survey of Barley, Hertfordshire 1593–1603* ed. J.C. Wilkerson (Cambridge 1974)
Ogilby, John *Britannia* (London 1675)
Ormerod, George *The History of the County Palatine and City of Chester* 3 vols (London 1819)
Ortelius, Abraham *Theatrum orbis terrarum* (London 1606)
Pemble, William *A Briefe Introduction to Geography* (Oxford 1630)
Petty, Sir William *Petty-Southwell Correspondence* ed H. Lansdowne 2 vols (London 1927)
Pilkington, Gilbert *The Turnament of Tottenham* (London 1631)
Pinkerton, John, ed *A General Collection of the Best and Most Intersting Voyages and Travels in All Parts of the World* 17 vols (London 1808–14)
Raleigh, Sir Walter *The Historie of the World* (London 1614)
Rawlinson, Richard *The Life of Mr. Anthony à Wood* (London 1711)
Recorde, Robert *The Castle of Knowledge* (London 1556)
Roe, William J. *Ancient Tottenham* (Tottenham 1950)
Rudder, Samuel *A New History of Gloucestershire* (Cirencester 1779)
Sacheverell, William *An Account of the Isle of Man* (London 1702)
– *An Account of the Isle of Man* ed Rev. J.G. Cumming (Douglas 1859)
Salmon, Nathaniel *Antiquities of Surrey* (London 1736)
Sammes, Aylett *Britannia antiqua illustrata: or, The Antiquities of Ancient Britain, Derived from the Phoenicians* (London 1676)
Spelman, Sir Henry *Archaeologus in modum Glossarii* (London 1626)

- 'Icenia: sive Norfolciae descriptio topographica.' In *The English Works of Sir Henry Spelman ... together with his Posthumous Works* ed 'H.S.' 2d ed (London 1727)
Stillingfleet, Edward *Origines Britannicae* (London 1685)
Stukeley, William *Palaeographia Britannica: or Discourses on Antiquities in Britain* (London 1743)
Tanner, Thomas *Bibliotheca Britannico-Hibernica* (London 1748)
Todd, Hugh *Account of the City and Diocese of Carlisle* ed Richard Saul Ferguson. Cumberland and Westmorland Antiquarian and Archaeological Society, Tract Series 5 (London 1890)
Vergil, Polydore *Anglica historia* (London 1534)
Warner, Erasmus *Geologia, or a Discourse* (London 1690)
Webb, John *Vindication of Stone-Heng Restored* (London 1665)
Worsop, Edward *A discouerie of Sundrie Errours* (London 1582)
Wood, Anthony à *'Survey of the Antiquities of the City of Oxford'* composed in 1661–6 ed Andrew Clark 3 vols (Oxford 1889–99)
Woodward, John *Brief Instructions for Making Observations in all Parts of the World* (London 1696)

Secondary

Adams, Eleanor *Old English Scholarship in England from 1566 to 1800* Yale Studies in English no 55 (New Haven 1917)
Alexander, Peter *Ideas, Qualities and Corpuscles: Locke and Boyle on the External World* (Cambridge 1985)
Andrews, John *Ireland in Maps* (Dublin 1961)
Arber, Agnes 'A Seventeenth-Century Naturalist: John Ray' *Isis* 34 (1943) 319–24)
- *Herbals: Their Origin and Evolution. A Chapter in the History of Botany 1470–1670* (Cambridge 1912)
Atkinson, R.J.C. *Stonehenge* (London 1956)
Aubin, Robert A. 'Grottoes, Geology and the Gothic Revival' *Studies in Philology* 31 (1934) 408–16
Austin, W.H. *The Relevance of Natural Science to Theology* (London 1975)
Bernard, T.C. *Cromwellian Ireland* (London 1975)
Beazley, Charles R. *The Dawn of Modern Geography* (London 1905)
Birken, William Joseph 'The Royal College of Physicians of London and Its Support of the Parliamentary Cause in the English Civil War' *Journal of British Studies* 23 (1983) 47–62
Bottigheimer, Karl S. *English Money and Irish Land* (Oxford 1971)
Boud, R.C. 'The Early Development of British Geological Maps' *Imago Mundi* 2d ser 1 (1975) 73–95

Bowle, John *John Evelyn and His World: A Biography* (London 1981)
Boyes, J. 'Sir Robert Sibbald: A Neglected Scholar' *The Early Years of the Edinburgh Medical School* ed R.G.W. Anderson and A.D.C. Simson (Edinburgh 1976)
Brett-James, N.G. *The Growth of Stuart London* (London 1935)
Bromehead, C.E.N. 'Geology in Embryo up to 1600 A.D.' *Proceedings of the Geologists' Association* 56 (1945) 89–134
Burke, P. 'A Survey of the Popularity of Ancient Historians 1450–1700' *History and Theory* 5 (1966) 135–52
Burl, Aubrey 'Geoffrey of Monmouth and the Stonehenge Bluestones' *Wiltshire Archaeological and Natural History Magazine* 79 (1985) 178–83
– 'John Aubrey's *Monumenta Britannica*' *Wiltshire Archaeological and Natural History Magazine* 77 (1983) 163–6
Burne, S.A.H. 'Early Staffordshire Maps' *Transactions of the North Staffordshire Field Club* 54 (1920) 54–87
Burton, Edward *The Life of John Leland* (London 1896)
Bury, J.B. *The Idea of Progress* (New York 1955)
Butt, J. 'The Facilities for Antiquarian Studies in the 17th Century' *Essays and Studies* 24 (1938) 64–80
Cameron, H. Charles 'The Last of the Alchemists' *Notes and Records of the Royal Society of London* 9 (1951) 109–14
Campbell, E.M.J. 'The Beginnings of the Characteristic Sheet to English Maps' *Geographical Journal* 128 (1969) 411–15
Carozzi, Albert V. 'Robert Hooke, Rudolph Erich Raspe, and the Concept of "Earthquakes"' *Isis* 61 (1970) 85–91
Chamberlin, Russell *The Idea of England* (London 1986)
Clark, Peter and Paul Slack eds *Crisis and Order in English Towns 1500–1700* (Toronto 1972)
Clarke, Archibald L. 'John Leland and King Henry VII' *Library* 3rd ser 2 (1911) 132–49
Close, Charles 'The Old English Mile' *Geographical Journal* 76 (1930) 338–42
Cochrane, Eric 'Science and Humanism in the Italian Renaissance' *American Historical Association Review* 81 (1976) 1039–57
Cohen, I. Bernard 'The Eighteenth Century Origins of the Concept of Scientific Revolution' *Journal of the History of Ideas* 37 (1976) 257–88
Corbett, W.J. 'Elizabethan Village Surveys' *Transactions of the Royal Historical Society* 2d ser 11 (1897) 67–87
Cragg, G.R. *From Puritanism to the Age of Reason* (Cambridge 1950)
Craig, W.S. *History of the Royal College of Physicians of Edinburgh* (Edinburgh 1976)

Cranston, Maurice 'John Locke and John Aubrey' *Notes and Queries* 195 (1950) 552–4; 197 (1952) 383–4
Crawford, O.G.S. 'Primitive English Land-Marks and Maps' *Empire Survey* 1 (1931) 3–12
Crone, G.R. 'Early Mapping of the British Isles' *Scottish Geographical Magazine* 78 (1962) 73–80
Cross, W. Redmond 'Dutch Cartographers of the Seventeenth Century' *Geographical Review* 6 (1918) 66–70
Daly, James *Sir Robert Filmer and English Political Thought* (Toronto 1979)
Darby, H.C. 'The Agrarian Contribution to Surveying in England' *Geographical Journal* 82 (1933) 529–35
Deacon, Margaret *Scientists and the Sea, 1650–1900* (London 1971)
Dean, D.R. 'The Age of the Earth Controversy: Beginnings to Hutton' *Annals of Science* 38 (1981) 435–56
Dean, Leonard F. 'Bodin's *Methodus* in England before 1625' *Studies in Philology* 39 (1942) 160–6
Denholm-Young, N. and H.H.E. Craster 'Roger Dodsworth and his Circle' *Yorkshire Archaeological Journal* 32 (1934) 5–32
Dijksterhuis, E.J. *The Mechanisation of the World Picture* (Oxford 1961)
Dodd, A.H. 'The Early Days of Edward Lhuyd' *National Library of Wales Journal* 6 (1950) 301–6
Dunlop, R. 'Sixteenth-Century Maps of Ireland' *English Historical Review* 20 (1905) 309–39
Earwaker, J.P. *East Cheshire: Past and Present* (London 1877)
Edelen, Georges 'William Harrison (1535–1593)' *Studies in the Renaissance* 9 (1962) 256–72
Edwards, W.N. *The Early History of Palaeontology* (London 1967)
Eyles, V.A. 'Mineralogical Maps as Forerunners of Modern Geological Maps' *Cartographical Journal* 9 (1972) 133–5
– 'The History of Geology: Suggestions for Further Research' *History of Science* 5 (1966) 77–86
Ferguson, Wallace 'Humanist Views of the Renaissance' *American Historical Review* 45 (1939) 1–28
Finlayson, Michael G. *Historians, Puritanism, and the English Revolution* (Toronto 1983)
Fisch, Harold 'The Scientist as Priest: A Note on Robert Boyle's Natural Theology' *Isis* 44 (1953) 252–65
Fletcher, Harold and William H. Brown *The Royal Botanic Garden in Edinburgh 1670–1970* (Edinburgh 1970)
Flint, Valerie I.J. 'The *Historia regnum Britanniae* of Geoffrey of Monmouth: Parody and Its Purpose. A Suggestion' *Speculum* 54 (1979) 447–68

French, Peter *John Dee: The World of an Elizabethan Magus* (London 1972)
Fryde, E.B. *Humanism and Renaissance Historiography* (London 1983)
Fuller, Jean Overton *Francis Bacon: A Biography* (London 1981)
Fussell, G.E. 'Crop Nutrition in the Late Stuart Age (1660–1714)' *Annals of Science* 14 (1958) 173–84
Galbraith, Vivian H. 'An Autograph MS of Ranulph Higden's *Polychronicon*' *Huntington Library Quarterly* 23 (1959) 1–18
Gardner, R.C.B. 'Sir Robert Cotton, A Great Collector' *Chambers Journal* 7th ser 14 (1924) 782–4
Geikie, Sir Archibald *The Founders of Geology* 2d ed (London 1905)
Gerish, William *Sir Henry Chauncey, Kt.: A Biography* (London 1907)
Glacken, Clarence J. *Traces on the Rhodian Shore* (Berkeley 1967)
Gottfried, Rudolph 'Antiquarians at Work' *Renaissance News* 11 (1958) 114–20
Gregory, J.C. 'Chemistry and Alchemy in the Natural Philosophy of Sir Francis Bacon' *Ambix* 2 (1938) 93–111
Hall, D.H. 'History of the Earth Sciences' *History of Science* 14 (1976) 149–95
Hall, Marie Boas 'Oldenburg, the *Philosophical Transactions*, and Technology' *The Uses of Science in the Age of Newton* ed John G. Burke (Berkeley and Los Angeles 1983)
Hare, J.E. 'Aristotle and the Definition of Natural Things' *Phronesis* 24 (1979) 168–79
Harley, J.B. and William Ravenhill 'The Gift of a Saxton Atlas to the University of Exeter' *Devon and Cornwall Notes and Queries* 34 (1980) 194–201
Harlow, C.G. 'Robert Ryece of Preston 1555–1638' *Proceedings of the Suffolk Institute of Archaeology* 32 (1973) 43–70
Hatfield, Gary C. 'Force (God) in Descartes' Physics' *Studies in History and Philosophy of Science* 10 (1979) 113–40
Hay, Denys 'The Historiographers Royal in England and Scotland' *Scottish Historical Review* 29 (1950) 15–29
Heawood, Edward *A History of Geographical Discovery in the Seventeenth and Eighteenth Centuries* (New York 1965)
Heninger, S.K. Jr *The Cosmographical Glass: Renaissance Diagrams of the Universe* (San Marino 1977)
Henry, John 'Occult Qualities and the Experimental Philosophy: Active Principles in Pre-Newtonian Matter Theory' *History of Science* 24 (1986) 335–81
Hill, Christopher *Intellectual Origins of the English Revolution* (Oxford 1965)
Hinch, J. de W. *Notes on Boate's 'Naturall History of Ireland' 1652* (Wexford 1928)
Hollaender, A.E.J. and William Kellaway eds *Studies in London History* (London 1969)

Hoppen, K. Theodore 'Some Queries for a Seventeenth Century Natural History of Ireland' *Irish Book* 2 (1963) 60–1

Horton, Mary 'In Defense of Francis Bacon: A Criticism of the Critics of the Inductive Method' *Studies in History and Philosophy of Science* 4 (1973) 241–78

Hoskins, W.G. ed *Mirror of Britain* (London 1957)

Hoskins, W.G. and H.P.R. Finberg *Devonshire Studies* (London 1952)

Housman, John E. 'Higden, Trevisa, Caxton, and the Beginnings of Arthurian Criticism' *Review of English Studies* 23 (1947) 209–17

Howell, A.C. 'Sir Thomas Browne and 17th-Century Scientific Thought' *Studies in Philosophy* 22 (1925) 61–80

Huddesford, William *The Lives of those Eminent Antiquaries John Leland, Thomas Hearne and Anthony Wood* 2 vols (Oxford, 1772)

Hutchieson, Alexander R. 'Timothy Pont, Scotland's Pioneer Cartographer' *The Book of the Society of the Friends of Dunblane Cathedral* 6 (1951) 47–9

Hyde, Ralph 'The Ogilby Legacy' *Geographical Magazine* 49 (1976) 115–18

Isard, Walter 'The Scope and Nature of Regional Science' *Proceedings of the Regional Science Association* 2 (1956) 13–26

Jackson, J.E. *The History of the Parish of Grittleton, in the County of Wiltshire* (London 1843)

Jacob, J.R. 'Restoration, Reformation and the Origins of the Royal Society' *History of Science* 13 (1975) 155–76

Jahn, Melvin E. 'The Old Ashmolean Museum and the Lhwyd Collections' *Journal of the Society for the Bibliography for Natural History* 4 (1966) 244–8

James, Francis Godwin *North Country Bishop: A Biography of William Nicolson* (New Haven 1956)

Jones, Charles 'Bede as Early Medieval Historian' *Medievalia et Humanistica* 4 (1946) 26–36

Jones, W.P. 'The Vogue of Natural History in England, 1750–1770' *Annals of Science* 2 (1937) 345–57

Jordan, W.K. *Philanthropy in England, 1480–1660* (London 1959)

Keller, Abraham C. 'Ancients and Moderns in the Early 17th Century' *Modern Language Quarterly* 11 (1950) 79–82

Kerridge, Eric *The Agricultural Revolution* (London 1967)

Kiely, E.R. *Surveying Instruments* (New York 1947)

Laudan, Laurens 'The Clock Metaphor and Probabilism: The Impact of Descartes on English Methodological Thought, 1650–65' *Annals of Science* 22 (1966) 73–104

Levine, Joseph M. 'From Caxton to Camden: The Quest for Historical Truth in Sixteenth Century England' (PH D diss Columbia University 1965)

– *Dr. Woodward's Shield: History, Science, and Satire in Augustan England* (Berkeley and Los Angeles 1977)

Select Bibliography 341

Levine, Philippa *The Amateur and the Professional: Antiquarians, Historians and Archaeologists in Victorian England, 1838–1886* (Cambridge 1986)
Liddell, J.R. 'Leland's Lists of Manuscripts in Lincolnshire Monasteries' *English Historical Review* 54 (1939) 88–95
Lloyd, John E. 'Geoffrey of Monmouth' *English Historical Review* 62 (1942) 460–8
Lukermann, F. 'The Concept of Location in Classical Geography' *Annals, Association of American Geographers* 51 (1961) 194–210
MacGillivray, Royce 'Local History as a Form of Popular Culture in Ontario' *New York History* 65 (1984) 367–76
Manuel, F.E. *The Religion of Isaac Newton* (Oxford 1974)
Marckwardt, A.H. 'Nowell's *Vocabularium Saxonicum* and Somner's *Dictionarium*' *Philological Quarterly* 26 (1947) 345–51
Mattingly, Garrett *The Armada* (Boston 1959)
Mendyk, Stan E. 'Robert Plot: Britain's "Genial Father of County Natural Histories"' *Notes and Records of the Royal Society of London* 39 (1985) 159–77
– 'Scottish Regional Natural Historians and the *Britannia* Project' *Scottish Geographical Magazine* 101 (1986) 165–73
Merton, R.K. 'Science, Technology and Science in Seventeenth Century England' *Osiris* 4 (1938) 360–632
Minshull, Roger *Regional Geography: Theory and Practice* (Chicago 1967)
Moir, D.G. and R.A. Skelton 'New Light on the First Atlas of Scotland' *Scottish Geographical Magazine* 84 (1968) 149–59
Moore, John M. 'Manuscript Charts by John Adair: A Further Discovery' *Scottish Geographical Magazine* 101 (1986) 105–10
Moravia, Sergio 'The Enlightenment and the Sciences of Man' *History of Science* 18 (1980) 247–68
Morgan, John 'Puritanism and Science: A Reinterpretation' *Historical Journal* 22 (1979) 535–60
Morrison, James C. 'Philosophy and History in Bacon' *Journal of the History of Ideas* 38 (1977) 585–606
Mulligan, Lotte and Glenn Mulligan 'Reconstructing Restoration Science: Styles of Leadership and Social Composition of the Early Royal Society' *Social Studies of Science* 11 (1981) 327–64
Murphy, Michael 'Methods in the Study of Old English in the Sixteenth and Seventeenth Centuries' *Medieval Studies* 30 (1968) 344–50
Nauert, Charles G. 'Humanists, Scientists, and Pliny: Changing Approaches to a Classical Author' *American Historical Review* 84 (1979) 72–85
North, F.J. *Humphrey Lhuyd's Maps of England and Wales* (Cardiff 1937)
Norton, William *Historical Analysis in Geography* (London and New York 1984)

O'Domhnaill, Sean 'The Maps of the Down Survey' *Irish Historical Studies* 3 (1943) 381-92
Ogilvie, M.H. *Historical Introduction to Legal Studies* (Toronto 1982)
Pace, K. Claire '"Strong Contraries ... Happy Discord": Some Eighteenth-Century Discussions about Landscape' *Journal of the History of Ideas* 40 (1979) 141-55
Parsons, A.E. 'The Trojan Legend in England' *Modern Language Review* 24 (1929) 253-64, 494-508
Patrides, C.A. *Approaches to Sir Thomas Browne* (Columbia and London 1982)
Patterson, Louise D. 'Recorde's Cosmography, 1556' *Isis* 42 (1951) 208-18
Pearce, E.C. 'Matthew Parker' *Library* 6 (1925) 209-28
Peshall, Sir John ed *The Ancient and Present State of the City of Oxford* (London 1773)
Phythian-Adams, Charles *Re-thinking English Local History* (Leicester 1987)
Piggott, Stuart 'The Sources of Geoffrey of Monmouth II: The Stonehenge Story' *Antiquity* 15 (1941) 305-19
Porter, Roy 'Creation and Credence: The Career of Theories of the Earth in Britain, 1660-1820' *Natural Order: Historical Studies of Scientific Culture* ed Barry Barnes and Steven Shapin (Beverly Hills and London 1979)
Powell, Ken and Chris Cook *English Historical Facts 1485-1603* (London 1977)
Preston, Joseph H. 'Was There an Historical Revolution?' *Journal of the History of Ideas* 38 (1977) 353-64
Pryce, W. *Archaeologia Cornu-Britannica* (Sherborne 1790)
Pulteney, R. *Historical and Biographical Sketches of the Progress of Botany in England* 2 vols (London 1790)
Puschmann, T. *A History of Medical Education* (London 1891)
Putnam, B.H. 'The Earliest Form of Lambarde's *Eirenarcha* and a Kent Wage Assessment of 1563' *English Historical Review* 41 (1926) 260-73
Quinn, Anthony *Francis Bacon* (Oxford 1980)
Quinn, David *The Elizabethans and the Irish* (Ithaca 1966)
Richardson, John *The Local Historian's Encyclopedia* (New Barnet, Herts 1974)
Ritchie, James 'Natural History and the Emergence of Geology in the Scottish Universities' *Transactions of the Edinburgh Geological Society* 15 (1952) 297-316
Roncaglia, Alessandro *Petty: The Origins of Political Economy* trans Isabella Cherubini (New York 1985)
Roots, Ivan *The Great Rebellion 1642-1660* (London 1966)
Rowe, John Howard 'The Renaissance Foundations of Anthropology' *American Anthropologist* 67 (1965) 1-20
Rowse, A.L. *Reflections on the Puritan Revolution* (London 1986)
Ruestow, Edward G. *Physics at Seventeenth and Eighteenth Century Leiden: Philosophy and the New Sciences in the University* (The Hague 1973)

Rye, W.E. *England as Seen by Foreigners* (London 1865)
Schnapper, Antoine 'The King of France as Collector in the Seventeenth Century' *Journal of Interdisciplinary History* 17 (1986) 185–202
Scroggs, E.S. 'Sir William Dugdale' *Journal of the British Archaeological Association* 2d ser 2 (1937) 1–6
Seaton, Ethel *Literary Relations of England and Scandinavia in the 17th Century* (Oxford 1935)
Semple, G. *The Geography of the Mediterranean Region: Its Relation to Ancient History* (New York 1931, reprint ed New York 1971)
Shorr, Philip 'Genesis of Prehistorical Research' *Isis* 23 (1935) 425–41
Shumaker, Wayne *The Occult Sciences in the Renaissance: A Study in Intellectual Patterns* (Berkeley 1972)
Simcock, A.V. *The Ashmolean Museum and Oxford Science 1683–1983* (Oxford 1983)
Simmons, Jack ed *English County Historians* (East Yardsley, Wakefield 1978)
– *Parish and Empire* (London 1952)
Skeat, T.C. 'Two "Lost" Works by John Leland' *English Historical Review* 65 (1950) 505–8
Skinner, Quentin 'History and Ideology in the English Revolution' *Historical Journal* 8 (1965) 151–78
Sollas, W.J. *The Age of the Earth and other Geological Studies* (London 1905)
Spiller, Michael R.G. *'Concerning Natural Experimental Philosophie': Meric Casaubon and the Royal Society* (The Hague 1980)
Stephens, James *Francis Bacon and the Style of Science* (Chicago 1975)
Stokes, E. 'The Six Days and the Deluge: Some Ideas on Earth History in the Royal Society of London, 1660–1775' *Earth Science Journal* 3 (1969) 13–39
Thomas, Keith *Man and the Natural World* (London 1983)
Tillyard, E.M.W. *The Elizabethan World Picture* (New York 1943)
Toulmin, Stephen and June Goodfield *The Discovery of Time* (London 1965)
Trevor-Roper, Hugh *The Romantic Movement and the Study of History* (London 1969)
Turnbull, G.H. 'Samuel Hartlib's Influence on the Early History of the Royal Society' *Notes and Records of the Royal Society of London* 10 (1953) 101–30
Turner, A.J. '"A World of Wonders in One Closet Shut"' *History of Science* 24 (1986) 209–15
– 'Mathematical Instruments and the Education of Gentlemen' *Annals of Science* 30 (1973) 51–88
Tyacke, Sarah ed *English Map-Making 1500–1650* (London 1983)
Underdown, David *Pride's Purge* (Oxford 1971)
Vasoli, Cesare 'The Contribution of Humanism to the Birth of Modern Science' *Renaissance et Reforme* new ser 15 (1979) 1–15

Vickers, Brian, ed *Occult and Scientific Mentalities in the Renaissance* (Cambridge 1984)
Wallace, Karl R. *Francis Bacon on the Nature of Man* (Urbana 1967)
Walters, Gwyn and Frank Emery 'Edward Lhuyd, Edmund Gibson, and the Printing of Camden's *Britannia*, 1695' *Library* 5th ser 32 (1977) 109–37
Waters, D.W. 'Science and the Techniques of Navigation in the Renaissance' *Art, Science, and History in the Renaissance* ed Charles S. Singleton (Baltimore 1967)
Webster, Charles 'Henry Power's Experimental Philosophy' *Ambix* 15 (1967) 150–78
Wilson, C.H. *England's Apprenticeship, 1603–1713* (London 1965)
Woolf, D.R. 'Erudition and the Idea of History in Renaissance England' *Renaissance Quarterly* 40 (1987) 11–48
Wright, John Kirkland *The Geographical Lore of the Time of the Crusades* (New York 1925)
Wright, Peter 'Astrology and Science in Seventeenth-Century England' *Social Studies of Science* 5 (1975) 399–422
Wright, Thomas 'On Antiquarian Excavations and Researches in the Middle Ages' *Archaeologia* 30 (1844) 438–57
Yates, Frances A. 'Essay Reviews: Science in Its Context' *History of Science* 11 (1973) 286–91
Yeo, Richard 'An Idol of the Market-Place: Baconism in Nineteenth Century Britain' *History of Science* 23 (1985) 251–92
Zagorin, Perez *A History of Political Thought in the English Revolution* (London 1954)
Zittel, Karl von *History of Geology and Palaeontology* trans Maria Ogilvie-Gordon (London 1901)

Index

Agricola, Georg 26, 28, 33, 75; *De natura fossilium* 255n.149
Agriculture, Board of 241, 243
alchemy and alchemists 131–2, 163, 198, 284nn.69, 80, 299n.97
Aldrovandi, Ulisse 28, 33
Allen, David Elliston 36, 327n.54, 328n.57; *Naturalist in Britain* 36
ancients and moderns 4, 114, 119, 147, 159, 171, 184, 246, 290n.6, 302n.3, 303n.9
Anderson, F.H. 117, 123
Anglo-Saxons and Anglo-Saxon period 10, 20, 50, 52, 55, 81, 108, 241; antiquary and 9; heptarchy 41, 48; institutions 126; language 48; laws 47; place-names 48; pots 159
Annales school x
Anne I, queen 256n.157
Antoninus 39; *Itinerary* 39, 46
Archer, Simon 104–6, 110; *see also* Dugdale, William
Aristotelianism 27–8, 30, 128, 130–1, 152, 155, 156, 157, 298n.76; neo-Aristotelianism 147; pseudo-Aristotelianism 30
Aristotle 28, 119, 314n.40

Aston, Margaret 8, 277n.35
astrology 131, 163, 171, 299n.96
Aubrey, John xiii, 8, 24, 43, 126–7, 168, 170–84, 205, 230, 290n.6, 299n.91; appreciation of 171, 182, 184, 245; 'Aubrey Holes' 183; and Avebury 171, 183–4, 236; on Bacon 128, 180, 283n.56; as Baconian 171, 173; biographical details of 171–2; on Burnet 180; reliance on Camden 174, 178; and Charles II 171; respect for chorographers 178; on court of James I 175; draining schemes of 142; and Dugdale 173, 183; environmental determinism of 171; and field-work 46, 182, 184, 236; on fossils 165, 177, 179; fascination with freaks 181; use of Fuller 182; on Garsdon 175; and geological survey 165, 178–80, 201, 237; and Hartlib 136, 148, 172–3, 303n.12; on Harvey 172; and husbandry 164 (Awbrey), 172; Kennett on 170; on learning 175; and Lhuyd 176, 182, 207; and medical folklore 161; and Ogilby 176; and Petty 144, 172; and Plot 25, 176–7, 193, 195, 312n.18; polymathical interests 163; as popu-

larizer of science 150, 212; pride in own work 195; Ray on 170, 182; reflections on the past 5, 104, 172–4; and the Royal Society 166, 170, 172, 177, 182–3; scientific antiquarianism of 171, 179, 182–4, 236; 'secret call' 177; and Stonehenge 183; supernaturalism of 131–2, 171, 181; works: *Brief Lives* 171; 'Chorographia super-et subterranea naturalis' 178–9; 'Essay towards the Description of the North Division of Wiltshire' 173–5; 'Hypothesis of the Terraqueous Globe. A Digression' 180; *Miscellanies* 171, 181; 'Monumenta Britannica' 171, 182–3, 326n.28; 'Naturall Historie of Wiltshire' 171, 177–82, 300n.109; *Perambulation of Surrey* 175–7; 'Queries In Order to the Description of Britannia' 176; 'Some Excerpta, out of John Norden's Dialogues' 182; 'Templa Druidum' 183

Austen, Ralph 140; *Treatise of Fruit-Trees* 140

Avebury 171, 183, 236; *see also* Aubrey, John

Bacon, Francis xi, xiii, 35, 114, 130–1, 140–1, 145–7, 157, 159, 197, 229, 232–3, 238, 246, 255n.141, 255–6n.154, 280nn.8, 11, 14; 281n.31, 282nn.37, 49, 283n.55, 284nn.65, 73, 78, 288–9n.41, 289n.45 passim; on ancients 34, 119, 121, 132; and antiquarianism 126; and civil history 115–18, 127–8; and Copernicanism 70, 134, 266n.57; death of xii; division between science and religion 149–50; influence of 11, 118, 128, 134, 137, 140–1, 143, 145–8, 150–5, 157–8, 161, 167–9, 191; and intellectual climate 6, 34; and knowledge 27, 123, 134; and magic, supernatural 34, 131–3, 135, 151; and natural history 25–6, 115–25, 127–8, 144; and the Royal Society 25, 134, 137, 146–8, 151–2, 154–5, 235; 'Solomon's House' 144, 161; works: 'Catalogue of Particular Histories by Titles' 123–4; *De Augmentis Scientiarum* 117, 161, 266n.57, 281n.21, 304n.16; *History of Henry VII* 117, 125; *New Atlantis* 132, 137; *Novum Organum* (*Organon*) 114, 117, 136–7, 167, 169; 'Parasceve' 122; *Sylva Sylvarum* 137

Baconians 115–16, 135, 144, 148, 150, 153, 155, 157, 166; Aubrey as 171, 173; influence of 137, 146, 150–5, 159, 171, 235; Molyneux as 192; as natural historians 18, 108, 149

Baconism xii–xiii, 31–2, 34, 108, 114–15, 118, 121, 127, 134, 137, 146–7, 151–6, 158, 245, 279n.2, 283n.64, 286nn.15,20, 288n.41, 303n.12; in natural history 141, 145; and science xii, 131, 136, 139, 146, 245, 292n.42; 'vulgar' 132, 152–3, 155

Bainbrigg, Reginald 54

Bale, John 45–6; *Laboryouse Journey and Serche of Johan Leylande, for Englandes Antiquitiees* 45, 260n.32; *Scriptorum illustrium majoris Britanniae ... catalogus* 48; *see also* Leland, John

Balfour, Andrew 215

Banister, John 165, 316n.51; 'Natural History of Virginia' 165

Bateman, Christopher 73

Beale, John 132, 138–40; *Herefordshire Orchards* 138–41, 286n.17

Bede, the Venerable 10, 40, 48, 79, 258n.9; *Historia abbatum* 40; *Historia ecclesiastica gentis anglorum* 40; *De natura rerum liber* 258n.9; *De temporum ratione* 258n.9; *Works* 258n.9
Bedwell, William 275n.74
biblical (Noachian) flood 29–30, 75, 77, 180, 202, 222, 239, 306n.52
Biondo, Flavio 8, 18, 50–1, 248n.6; *Italia Illustrata* 50–1; *Roma Instaurata* 8; *Roma Triumphans* 50
Blith, Walter 142, 288n.30
Blome, Richard 165; *Description of the Island of Jamaica* 165
Blundeville, Thomas 23; *His Exercises* 23
Boate (Boet), Arnold 140–1, 186, 190, 192; 'Interrogatorie' ('Interrogatory') 140, 308nn.1, 3; *see also* Boate, Gerard; Invisible College
Boate (Boet), Gerard 139, 141, 143, 148, 158, 162, 185–92, 221, 289n.45, 308nn.1, 7, 309nn.18–19, 22, 310nn.22, 36, 39, 311n.1; *Irelands Naturall History* 139, 145, 158, 166, 185–92; *see also* Boate, Arnold; Invisible College
Bodley, Thomas, library of xiv, 9, 13, 37, 88, 104, 196
Boodt, Boethius de 33; *Gemmarum et lapidum historia* 33
'Book of Nature' 11
Borlase, William 242; *Natural History of Cornwall* 242
Bourne, William 26, 75; *Booke called the Treasure for traveilers* 26, 252n.90
Bowen, Margarita 35–6; *Empiricism and Geographical Thought* 36, 267n.78
Boyle, Robert 147–9, 151, 162,

284n.71, 288nn.39, 41, 290nn.18, 19, 293n.43, 294n.55, 295nn.60, 62, 296n.67, 298n.76, 300n.103; as Baconian 143–4; and Hartlib 140, 143–4, 192; on the Oxford Experimental School 156; and Petty 143; and pseudo-science 131; and the Royal Society 152; and religion 291nn.18–19; and supernaturalism 149–50, 157, 163; and theorizing 153–4; *Sceptical Chymist* 131
Brazil 185; *see also* Marcgraf, George
Bridges, John 242; *History and Antiquities of the County of Dorset* 242
Bridgman, Orlando 20, 252n.79; and strict settlement 20
Brigges, Henry 306n.56
Brooke, Ralph 262n.65; *Certaine Errours* 262n.65
Brooke, Robert 19
Browne, Thomas 32, 149, 156, 159–60, 172, 185, 271n.2, 294n.57, 296–7nn.73, 74, 308–9n.7, 325n.19; works: *Hydriotaphia* 159; *Religio Medici* 156, 159, 172; *see also* Power, Henry
Burckhardt, Jacob 31
Burghley, Lord 60–1, 63, 65–7, 264n.14; *see also* Norden, John
Burnet, Thomas 180, 269n.90, 327n.47; *Telluris theoria Sacra* 180, 315n.46
Burton, Robert 88; *Anatomy of Melancholy* 88
Burton, Thomas 12–13
Burton, William 82, 84, 88–90, 93, 99, 104–7, 250n.45, 276n.17, 278n.48, 314n.33; *Particular Description of Leicester Shire* 82, 104
Butterfield, Herbert xii, 232

Cambrensis, Giraldus *see* Gerald of Wales

Cambridge Platonists 133, 156–7, 160, 294–5n.60; *see also* neo-Platonism and neo-Platonists

Camden, William 6–7, 35, 44, 46, 49–55, 57, 73, 86, 110, 166, 169–70, 177, 213, 225, 230–1, 256n.157, 260n.26, 262n.65, 274n.67, 316n.59 passim; used by Aubrey 174, 178; used by Boate 188; 'British Strabo' 49; and British tradition 51; on Brut 50; influence on Burton 88; and Carew 54, 77, 270nn.93–4; influence on Childrey 167; and Cotton collection 262n.75; and European correspondents 50; influence on Gerard 99; and Hakluyt 51; influence of 61, 74, 77, 79, 81–2, 87–8, 91, 93, 99, 111, 167, 174, 194, 223, 225; influenced by Lambarde 49, 52, 261n.57; used by Leigh 223; and Norden 62, 67, 72; influence on Plot 194; purpose of research 38; and Society of Antiquaries 13; and Speed 78–9, 81; Stow on 263n.83; works: *Annales ... regnante Elizabetha* 51; *Britain (Britannia* 1610) 38, 51; *Britannia* 6, 16, 23–4, 49–55, 57–8, 62, 72, 77, 82, 87, 96, 98, 111, 169, 206, 264n.6, 270nn.93–4, 317n.70; *Camden's Britannia* 206, 208, 213, 217, 221–2, 246, 318n.82, 321n.20, 322n.34; *Remaines of a greater worke Concerning Britaine* vi

Carew, Richard 13, 25, 54, 68, 73, 77–8, 82, 100–1, 107, 270nn.93–4, 278n.48, 281n.16; *Survey of Cornwall* 68, 77–9, 82, 85, 87, 90–5, 97, 118, 178

Carpenter, Nathanael 74, 93, 252n.92, 267n.78, 268n.79, 269n.90; *Geographie [Geography] Delineated Forth in Two Books* 74, 252n.92, 269n.90

Carroll, Lewis 3
Cartesianism 135, 147, 152, 155–6, 293
Casaubon, Meric xi, 50, 240, 327n.47
Celtis, Konrad 50, 249n.18; *Germania Illustrata* 50
Cesalpino, Andrea 26, 28
Charles I, king 104, 276n.13
Charles II, king 171, 183, 215
Chauncey, Henry 66, 242; *Historical Antiquities of Hertfordshire* 242, 278n.54
Chetwynd, Walter 193, 202, 272n.20, 314n.33
Childrey, Joshua 166–70, 185, 188, 227, 290n.6, 292n.34, 301nn.116, 123, 124, 127, 302nn.129–32; *Britannia Baconica* 166–9, 185, 227, 269n.88, 306n.48
civil war (1642–6, 1648) 24, 103–4, 138, 142, 149, 160, 232
Clark, Stuart 117
classicism 241
Coke, Edward 83
College of Antiquaries 13, 24; *see also* Society of Antiquaries
College of Physicians of Edinburgh 215
College of Physicians of London 156, 161, 294n.55, 308n.1
Collins, Daniel 315n.46
Colonna, Fabio 30; *De glossopetris dissertatio* 30
Comenius, John Amos (Komenský, Jan Amos) 130, 283n.64
Coote, Charles H. 58
Copernicanism 69, 134, 244
Copernicus, Nicolaus 69
Cotton, Robert 13, 77–9, 84, 88, 243, 270n.105, 297n.73; library of 13, 100, 109, 250n.36, 262n.75; politicization of 18

Court, Inns of *see* Inns of Court
Crane, Robert 84–5; *see also* Reyce, Robert
Cudworth, Ralph 157

Dale, Samuel 279n.54; *Harwich and Dovercourt* 279n.54
Daniel, Glyn 37, 170, 184; *Hundred and Fifty Years of Archaeology* 170, 307n.63; *Idea of Prehistory* 307n.63
Dee, John 21, 23, 54, 130, 252n.90, 284nn.69, 73
Descartes, René xii, 114, 155–8, 292n.40, 296n.67; *Discourse on Method* 115, 135; *see also* Cartesianism
Dioscorides 32, 255n.141; *Materia medica* 255n.141
Dodridge (Doddridge), John 100–1; *History of ... Chester* 100
Dodsworth, Roger 105; *see also* Dugdale, William
Douglas, David C. x, 6, 102, 237; *English Scholars, 1660–1730* 35
Down Survey 144, 192, 289nn.44, 46; *see also* Petty, William
Drayton, Michael 88, 225, 273n.36, 278n.48; *Poly-olbion* 273n.36
Druids 184, 212, 237, 297n.160
Dublin Philosophical Society 145, 192
Dugdale, William 8, 24, 102–13, 117–18, 126, 136, 166, 169, 183, 297n.73; and Archer 105–6; and Aubrey 173, 183; and Burton 105–6; biographical details of 103; as chorographer 106–7; and Dodsworth 105, 109; draining schemes of 142; influence of 111–13, 162, 173–4, 178, 278nn.48, 49, 54; politicization of 18; research of 102–3; anticipated by Rous 13; as royalist 104; and scientific antiquarianism 110, 159, 277n.40; and Spelman 105; works of: *Antiquities of Warwickshire* 14, 24, 102–3, 105–12, 158, 169, 173, 276n.27; *Baronage of England* 110, 'Directions' 277n.38; *History of Imbanking* 110; *History of St Paul's Cathedral* 110; *Monasticon Anglicanum* 14, 105, 109, 172, 277n.37; *Origines Juridiciales* 110; *View of the Late Troubles in England* 104
Dury, John 141, 283n.64
Dymock, Cressy 138

Earle, John 231; *Microcosmographie* 231
Edward I, king 20, 101
Edward II, king 111
Elizabeth I, queen 6, 49, 54, 60–1, 63–5, 69, 85, 91, 231, 264n.14, 265n.40
Emery, Frank V. ix, 36, 206, 209, 214; 'Aubrey to Defoe' ix, 312n.19, 315n.41, 319–20n.2
Enlightenment 56, 263n.85, 282n.31
Erasmus, Desiderius 20, 175
Erdeswicke, Sampson 54, 90, 107, 203, 208, 269–70n.91, 318n.82
Escholt, Mickel Pederson 315n.46
Evans, Joan 158–9, 296n.69
Evelyn, John 140, 143, 151; *Sculptura* 143–4

Falle, Philip 219
Fell, John 194, 204
Ferguson, Arthur 42
Ferrers, Henry 90
feudum 18
Ficino, Marsilio 129
Fitzherbert, Anthony 19; *Boke of Husbandrie* 26

Fitzherbert, John 268n.80
Fitzstephen, William 41
flood, Noachian *see* biblical flood
Fludd, Robert 132, 299n.91
Fordun, John of 259n.21
fossils ('figured stones') 25, 295n.66, 306nn.46–7; and Aubrey 165, 177, 179; cataloguing of 237; and Childrey 168; and earth chronology 30–1; and Colonna 30; and Harris 306nn.46, 47; and Hooke 29, 202–3, 206, 254n.124, 268n.85; interpretation of 25, 27–30, 76, 119, 179, 201–3, 206, 223, 226, 238, 295n.66, 306n.46; and Leigh 222–3; and Lhuyd 20, 206, 211, 238, 295n.66; and Lister 202, 206, 224, 238; and Morton 223–6; and Plot 29, 197, 201–3, 205–6, 295n.66, 316nn.51, 55; and Ray 30; and Sibbald 216; and Xenophanes 119
Fothergill, John 37
Foucault, Michel 32, 34; *Order of Things* 255n.145
Frank, Robert G., Jr 161
Frantz, R.W. 165
Fuller, Thomas 182; *Worthies of England* 182
Fussner, Frank Smith x, 15, 35, 230–1; *Historical Revolution* 230, 250n.36, 256n.156, 280n.8; *Tudor History* 326n.36

Galen 32, 128, 155, 160
Galenism 129, 131, 160, 297–8nn.75, 76
Galilei, Galileo 155, 292n.40, 293n.47
Gassendi, Pierre 151, 155, 314n.40
gavelkind 47
Genesis, book of 29
Geoffrey of Monmouth 8, 41, 53, 63, 248n.6, 258n.20

George III, king 243
Georgical Committee 163, 165, 195
Gerald of Wales 41–2, 47, 100, 167, 188, 190, 200; *Descriptio Cambriae* 41; *Itinerarium Cambriae* 41; *Topographica Hibernica* 41
Gerard, John 26, 280n.15; *Herball* 26
Gerard, Thomas 98–100; *Somerset* 98–9
Gervase of Canterbury 10
Gervase of Tilbury 5–6; *Otia Imperialia* 5–6
Gesner, Conrad 26, 28, 33
Gibbon, Edward 18
Gibson, Edmund 51, 127, 206, 213–14, 217, 322n.34; *see also* Camden, William
Gilbert, William 155, 178, 267n.78, 284n.73
Gildas, Saint 10, 39, 61; *De Excidio et conquestu Britanniae* 39
Glanvill, Joseph 131, 229; *Plus ultra* 229
Glover, Thomas 165
Godfrey, John T. 112; *see also* Thoroton, Robert
Gordon, Charles 214–15
Gordon, James 214–15
Gordon, Robert 214
Goscelin 40
Gothic revival 241
Gough, Richard 5, 9, 37, 51, 73, 198, 278n.49
Gransden, Antonia 10, 12, 14, 40, 42
Greenslade, M.W. 36
Grey, William 275n.74
Guillim, John 84; *Display of Heraldrie* 84, 271n.2
Gunther, R.W.T. 193

Haak, Theodore 141
Habington, Thomas 105–6, 200
Hakluyt, Richard 51, 86, 267n.73; *Voyages* 86
Hall, A. Rupert xi–xii, 162, 279n.2, 293n.42, 294n.55, 298nn.83, 85
Hall, Marie Boas 152
Halley, Edmund 240
Halliday, F.E. 77; see also Carew, Richard
Harris, John 242; *History of Kent* 242
Harrison, William 46
Hartlib, Samuel 121, 132, 136–44, 148, 156, 162, 164, 170, 172, 186, 192, 232, 283n.64, 286n.11, 287nn.22, 24, 288n.41, 289n.42, 308n.7, 309n.9 passim; circle of 136, 138, 140, 143, 146, 286nn.11, 15, 289n.45; *Essay for the Advancement of Husbandry-Learning* 186; *Legacy [Legacie] of Husbandry* 139, 143, 164
Harvey, William 155, 160–3, 172, 246, 294n.57, 296–7n.73, 297n.74, 299n.91
Hasted, Edward 242; *Survey of the County of Kent* 242
Hatton, Christopher 104, 106
Hay, Denys 51, 54; on Camden 262n.62
'Heads of Enquiries' 163–4; see also Georgical Committee; Royal Society
Hearne, Thomas 45, 260n.32
Helmont, Joan Baptista van 131–2, 160, 284n.71
Henry II, king 89
Henry VII, king 8
Henry VIII, king 45, 175
Henry of Huntingdon 41; *Historia Anglorum* 41
Hermes Trismegistus 130
Herodotus 38

Heylyn, Peter 22, 274n.74; *Microcosmus* 252n.92; *Two Journeys* 274n.74
Hickes, George 246
Higden, Ranulph 41–2, 259nn.20–1; *Polychronicon* 42, 259n.18; see also Trevisa, John
Hill, Thomas 26, 75; *Contemplation of Mysteries* 26
Hippocrates 160
historicism 17
Hoare, Colt 170
Hobbes, Thomas 143, 172, 236
Hoeniger, F.D. and J.F.M. 36
Holland, Philemon 51
Hooke, Robert 27, 29–31, 119, 131, 150–3, 197–8, 202–3, 206, 238, 254n.124, 255n.135, 268n.85, 269n.90, 278n.2, 293n.43, 294n.55, 300n.109, 314n.40
Hooker, Richard 93, 95
Hoppen, K. Theodore ix, 137, 146–7, 284n.80, 286n.15
Hopton, Arthur 22
Hoskins, W.G. 36
Hoskyns, John 165, 300n.109
Houghton, Walter E., Jr xi, 150–1
Hunter, Michael ix, 37, 242, 291n.13, 293n.47, 299n.91, 305n.40, 316n.59, 328nn.58, 64 passim; on ancients and moderns xi; on Aubrey 132, 171, 181, 184, 195, 302nn.5, 6, 305n.40; on Childrey 166; on Hartlib 186; on the Royal Society 147–8; on science 148, 150, 184; *John Aubrey and the Realm of Learning* 36; *Science and Society* 290nn.5, 10, 312n.19

Inns of Court 19, 60, 87
Interregnum 134, 136–7, 139–40, 142, 148, 285n.9

Invisible College 141, 143–4, 192, 287n.24; *see also* Boyle, Robert
Ives, J. 277n.38; *Select Papers* 277n.38

James I, king 54, 67–8, 71, 78, 94, 100, 190, 231, 274n.69
Johnson, Thomas 280n.15
Jones, Inigo 183; *Stone-Heng ... Restored* 183
Jones, R.F. 137, 169, 281n.19; *see also* ancients and moderns
Jonson, Ben 77; *Execration* 77
Jonston, John (Jonstonus, Joannes) 34; *Historia naturalis de quadripedibus* 34

Kendrick, T.D. 9–10, 14, 46, 90, 103, 166, 262n.73, 270n.106; *British Antiquity* 35
Kennett, White 170, 226, 278n.42; *Life of Mr Somner* 278n.42; *Parochial Antiquities* 253n.111
Kilbourne, Richard 111; *Survey of the County of Kent* 111, 277n.42, 278n.43
King, Daniel 169, 278n.48; *Vale-Royal* 169
Kircher, Athanasius 131, 284n.68
Kite, Jon Bruce 184
Knyvet, Lady Anne 58, 262n.3
Kuhn, Thomas xii–xiii, 27, 115, 239, 244–5, 254n.116

Lambarde, William 6, 20–1, 44, 46–9, 51–2, 54–5, 58, 62, 67, 73, 77, 85, 166, 169 (Lambert), 261nn.56–7, 264n.7, 278n.48; and Anglo-Saxon studies 20, 47–8; used by Camden 49, 52; used by Carew 77–8; use of documents 48; interests of 47; politicization of 18; purpose of research 23; religious bias of 48–9, 81; used by Speed 79; works: *Archaionomia* 20, 261n.43; *Archeion* 21; *Eirenarcha* 252n.86; *Perambulation of Kent* 6, 14, 21, 24, 47–9, 52–3, 61, 79, 169, 263n.83; *Topographical Dictionary* 62, 261nn.44–5
Latimer, Robert 172
latitudinarianism 34, 148–9, 286–7n.20
Leibniz, Gottfried Wilhelm 154
Leigh, Charles 222–3, 322nn.39–40, 323nn.42, 46; *Natural History of Lancashire, Cheshire, and the Peak in Derbyshire* 222–3
Leigh, Edward 111; *England Described* 111, 278n.48
Leland, John 6, 9, 13, 35, 39, 44–7, 51, 53, 78–80, 170, 173, 176, 229, 246, 248n.3, 260nn.26, 33, 262n.65, 278n.48 passim; work edited by Bale 45; influence of 46, 183, 194, 225, 229; as 'King's Antiquary' 44; manuscripts of 88, 99; 'New Year's Gift' (*Itinerary*) 45–6, 225, 317n.68; research plan of 45; used by Speed 79–80; *see also* Bale, John
Lennard, Reginald 164
Levine, Joseph M. ix, 245–6, 328nn.59, 65, 329n.67; 'Natural History' 257n.171, 260n.33, 284n.78
Levy, Fritz J. x, 35, 45, 73
Leycester, Peter 278n.54; *Historical Antiquities* 278n.54
Leyden, university of 185, 215, 218–19, 320n.11
Lhuyd (Lhwyd), Edward xiii, 9, 145, 147, 152, 153, 184, 246 (Lhwyd), 295n.66, 327n.50; and Aubrey 176, 182, 207; and *Britannia* project 206, 208–14 passim; 'Camden Tour' 208–10; and earth chronology 76, 157; on denudation 235; and field study 170 (Lhwyd) 207–8, 227–8; and fossils 20,

206, 211, 238, 295n.66; hardships of travel 210; and Leigh 223; and Lloyd 208; and Martin 219; and Morton 224, 226; and Plot 207, 211–12; queries 209, 218–19; research on 36; entry into the Royal Society 211; scientific antiquarianism of 211–212, 237–8, 241, 307nn.63,72; and Sibbald 213–14, 221; speculations of 157; works: 'Archaeologia Britannica' 208–10; 'Glossography' 210–11; *Parochial Queries* 209–10
Lhwyd *see* Lhuyd; Llwyd
Linnaeus, Thomas 33, 241; his system 33, 243
Linnaean Society 242
Lister, Martin 153, 163, 165, 178–9, 184, 202, 205–6, 216, 224, 226, 238, 294n.54, 300nn.99, 107, 305n.40, 313–14n.32, 323n.63, 327n.50
Lithgow, William 8, 275n.74
Little Ice Age 210
Littleton, Thomas 19, 103; *Tenures in Englysshe* 19, 103
Livy 107, 251n.61
Lloyd, John 208, 318n.84
Llwyd (Lhwyd) Humphrey 81, 85, 100; *Angliae Regni Florentissimi Nova Descriptio* 85
local history, defined 247n.1, 248n.17
Locke, John 240, 294n.55
Lyly, William 260n.33
Lynam, Edward 60, 263–4n.6, 264n.13

MacGillivray, Royce 104
Machell, Thomas 209; 'Queries' 209, 318n.88
McKisack, May 9, 100
Malmesbury, William of 41; *Gesta pontificum anglorum* 41

Malpighi, Marcello 245
Marcgraf (Marcgrave), Christian 215
Marcgraf (Marcgrave), George 215; *Historia Naturalis Brasiliae* 215, 320n.11
Martin, Martin 213, 219–20, 243; *Description of Skye* 220; *Late Voyage to St Kilda* 213, 218–20; *Western Islands* 220
Medici family 262n.75
Mela, Pomponius 38
Mercator, Gerardus 50
Merrett, Christopher 161
microcosm and macrocosm 129, 298n.83
Mirandola, Pico della 129
Molyneux, William 192, 209, 263n.85, 321n.29
Momigliano, Arnaldo 9–11, 14, 16, 25, 248n.30
Montague, Charles 219
Moray, Robert 131
More, Henry 156, 295n.60
More, Thomas 136–7, 285n.3
Morello, Nicoletta 29
Morton, John 223–7, 234; *Natural History of Northamptonshire* 223–6, 234, 263n.85; 'Proposals for Subscriptions' 234
myths: Arthurian 8, 15, 41–2; Trojan (Brut) 8, 15, 50, 53, 67, 78, 81, 91, 126, 238; Phoenician 223

Nennius 10, 39, 79, 81; *Historia Brittonum* 39
neo-Platonism and neo-Platonists 28, 129–30, 134, 156–7; *see also* Cambridge Platonists; Plato; Platonism
Newton, Isaac xii, 116, 132–3, 147, 150–2, 154–5, 157, 238–9, 244, 327nn.46, 54

Niccoli, Niccolò 262n.75
Nichols, John 49, 272n.20
Nicolson, William 206, 221, 224, 322n.38
Norden, John 3, 35, 57–74, 81, 90, 138, 187; accuracy of work 73; biographical details of 58; and *Britannia* 51, 72; on Brut 62, 67; and Burghley 60, 63; on Camden 264n.6, 278n.48; and Cecil 67, 71; and Copernicanism 69–70; and James I 67; on London 62–3, 65; emulated by Plot 194; research plan of 58, 60, 86; used by Speed 80; as surveyor 13, 70–1, 74; on Verulamium 66; and Webb 86; works: 'Chorographical discription of the Severall Shires' 64; *Christian Familiar Comfort* 64, 69; *Cornwall* 67, 71, 73; *England, An Intended Guyde* 72; *Essex* 64; *Eye to Heaven in Earth* 59, 70; *Good Companion for a Christian* 59; *Hartfordshire* 65–6; *Middlesex* 3, 57, 61–4, 66; *Mirror of Honor* 58, 64, 265n.32; *Pensive Soules Delight* 69; *Prayer for the Prosperous Proceedings of the Earle of Essex* 66; *Preparatiue to His Speculum Britanniae* 58, 65; *Progresse of Pietie* 64; *Sinfull Mans Solace* 59; 'Speculum Britanniae' 3, 58, 61, 65; *Store-House of Varieties* 69; *Surveyors Dialogue* 71–2; *Vicissitudo Rerum* 69–70
Nowell, Laurence 47, 264n.7; *Dictionarium Angliae topographicum et historicum (Topographical Dictionary)* 47, 62; *see also* Lambarde, William

Office of Address 140–1, 287n.22, 288n.41; *see also* Hartlib, Samuel

Ogilby, John 165, 176; 'Description of the Whole World' 176, 195, 304nn.26–7
Oldenburg, Henry 131, 136, 140, 147–9, 151, 163, 166–8, 232, 301n.127, 302n.129
Ortelius, Abraham 50
Oruch, Jack 90, 253n.92
Owen, George 54, 78, 178–9, 187, 191, 201, 205–6, 208, 268n.78, 269n.90, 317nn.68, 70, 318n.75; *Penbrokshire* 205

Packe, Christopher 317n.70
Paracelsianism 129, 131–2, 160
Paracelsus, P.A. 34, 128–31, 134, 149, 283nn.61, 63, 298n.83
Paris, Matthew 10, 100
Parker, Matthew 13, 20
Pasquier, Estienne 17–18, 20, 251n.61; *Recherches* 17
Pausanias 38
Pemble, William 22; *Briefe Introduction to Geography* 252n.92
Petty, William 140–1, 143–5, 156, 160, 162, 164, 172, 192, 289nn.45–6, 306n.56; *Advice of W.P.* 144; *see also* Down Survey; Dublin Philosophical Society
Phaer, Thomas 19; *Presidentes* 19
Philipot, John 111, 166, 169; *Villare Cantianum* 111, 169
Philosophical Society of Oxford 198, 211, 214, 313–14n.32
Philosophical Transactions 143, 163–4, 166, 197–9, 232, 237, 300n.99, 312–13n.25
Piggott, Stuart ix, 54, 158, 236–7, 249n.18, 277n.40, 296n.72, 307nn.63, 72, 308n.73, 328n.59

Pitt, Moses 192; *English Atlas* 192
Plato 11, 119, 130
Platonism 28, 130, 133, 160, 314–15n.40; *see also* Cambridge Platonists; neo-Platonism
Pliny 23, 32, 38, 42, 119, 255n.141; *Historia naturalis* (*Natural History*) 32, 258n.9
Plot, Robert xiii, 25, 113, 145, 148, 153, 158, 184, 193–206, 216, 227, 230, 234, 235, 237, 245, 284n.80, 295n.66, 297n.74, 299n.97; and alchemy 132, 163, 198; and Aubrey 25, 176–7, 193, 195; biographical details of 193; and Dugdale 277n.38; field-work of 195–6; and fossils 29, 197, 201–3, 205–6, 295n.66, 314n.38; as geologist 200–1; and the Gordons 214; influence of 165, 223–7, 242; and Lhuyd 207, 211–12; research method of 196–7; and Morton 224–7, 234; on Ogilby 304n.27; and the Philosophical Society of Oxford 198; research on 36; and the Royal Society 197–8; scientific antiquarianism of 157, 202–3; interest in supernatural 132; works: 'Natural History of England' 176; *Natural History of Oxfordshire* 25, 29, 113, 158, 165–6, 177, 193, 195–7, 199, 202, 204, 207; *Natural History of Staffordshire* 197, 199–200, 202–4
Pocock, John G.A. xi, 16–17
Pole, William 90–1, 93–5
Pollard, A.W. 59, 264nn.7, 13, 265n.18
Polybius 5, 107
Pont, Timothy 214
Porter, Roy ix, 227, 239, 280n.15, 295n.66, 312n.19, 323n.63
post-Restoration period 144
Powel, David 100

Power, Henry 32, 156, 294nn.55, 57; *see also* Browne, Thomas
printing, impact of 15
Prior, Moody 322n.29; 'Bacon's Man of Science' 322n.29
Ptolemaic system 23, 50, 70, 141
Ptolemy 21–2, 38–9, 50, 61, 70, 80, 91, 229, 320n.4; *Geographia* (*Geography*) 21, 39, 53, 257–8n.5; 'Tables' 50
puritanism 37, 139, 148, 285n.7, 287n.20
Puritans 75, 84, 90–1, 137, 138, 140, 141, 173, 221, 279n.2, 285n.7, 286nn.15, 20, 287nn.20, 21, 22 passim
Purver, Margery 125, 282n.33, 290n.2; *Royal Society* 282n.33
Pythagoras 34, 119

Raleigh, Walter 183
Rastell, William 19; *Entrees* 19
Rattansi, P.M. 130, 132, 294–5n.60
Raven, Charles 281n.16; *English Naturalists* 281n.16
Rawlinson, Richard 222, 304nn.23–4, 328n.56
Ray, John 30–1, 76, 119, 151, 153, 156, 160, 163, 170–1, 182, 184, 206, 235, 245, 294n.54, 298n.82, 306n.52, 307n.72, 323n.63, 325n.22, 328n.63, 329n.67
real-property law 19, 87
Reeds, Karen Meier 32
Restoration 24, 27, 110, 112, 139–42, 149–50, 160, 166, 181, 294n.54, 310n.36
Reyce (Ryece), Robert 84–6, 93; *Breviary of Suffolk* 84–6
Richard III, king 101
Risdon, Tristram 90–1; *Devon* 91, 93
Roberts, Nicholas 206, 209

Rogers, G.A.J. 157
Rollright stones 204
romanticism 241
Rossi, Paolo 27; *Dark Abyss of Time* 27, 125, 133, 254n.116, 281–2n.31
Rous, John 13, 90; collection of 109; *see also* Dugdale, William
Rowse, A.L. 35; *England of Elizabeth* 35
Royal Society xi, 35–7, 75, 110, 127–8, 130, 136, 145, 146, 160–1, 167–8, 177–8, 181, 192, 220, 232, 282n.31, 286n.15, 287nn.21, 27 passim; and agriculture 143, 164, 168; aims of 25, 137, 144, 146, 158; Aubrey and 166, 170, 172, 177, 182–3; Baconism of 25, 34, 115, 134, 137, 140, 146–8, 150–5, 158–9, 166, 235, 279n.2, 283n.64, 288n.41; decline of 240–1; empiricism of 151–2, 154; and field studies 26, 158; as gentleman's club 27, 147; Georgical Committee of 163, 165, 195; and Hartlib circle 136–7, 140–1, 146–8; 'Heads of Enquiries' 163–4; latitudinarianism of 148–9; Leigh and 222; Lhuyd and 211; and medicine 160, 162; diverse membership of 139, 147–8, 150; political associations of membership 148; models of 147; Montague and 219; Morton and 224; and natural history 34, 146–7, 163–6, 241, 245 passim; as organizer of science 155; philosophical characterization of 149, 152, 155; *Philosophical Transactions* of 143, 163–4, 166, 197–9, 232, 237, 300n.99; and physicians 152, 162; Plot and 197–8; and pseudo-science 131, 150–1, 163; Puritan associations of 136–7, 139–40, 148; and religion 136–40 passim, 148–50, 286n.20; Sprat on 146; its involvement in theorizing 152–5; utilitarianism of 114, 148, 154, 279n.2; Wotton on 245
Royal Society (Academy) for Scotland 228
Rudder, Samuel 242; *New History of Gloucestershire* 242
Rudwick, Martin J.S. 27–8; *Meaning of Fossils* 27; *see also* Porter, Roy

Sacheverell, William 242; *Isle of Man* 242
Salmon, Nathaniel 229, 242; *Antiquities of Surrey* 324n.3
Sammes, Aylett 223
Saxton, Christopher 51, 60, 65, 78–9, 196; *Atlas* 58, 60
Schiffman, Zachary Sayre 17
scientific antiquarianism 25, 182, 212, 217, 238 passim
Selden, John 77, 83–4
Shadwell, Thomas 125, 151; *Virtuoso* 151
Shapiro, Barbara ix, 31, 37, 126–7, 153, 155, 232, 281n.15; *English Scientific Virtuosi* 37; *Probability and Certainty* 37, 279n.8, 280n.16, 283n.52, 326n.36
Sharpe, Kevin 35; *Sir Robert Cotton 1586–1631* 250n.36, 252n.88
Sibbald, Robert 9, 152, 162, 185, 213–19, 221–2, 245, 263n.85, 297–8n.75, 300n.9, 310n.39; *Fife and Kinross* 216–18; *Linlithgow and Stirling* 217; *Nuncius Scoto-Britannus* 215–16; *Scotia Illustrata* 216
'signatures' 129; doctrine of 163, 299n.98
Smyth, John 274–5n.74; *Lives of the Berkeleys* 275n.74

Society of Antiquaries: Elizabethan 9, 13, 24, 51, 73–4, 77, 84, 158, 242, 250n.36, 262n.67, 271n.1; Georgian 156, 242
'Solomon's House' 144, 161
Somner, William 88, 109–11, 275n.74, 278nn.42, 48; *Antiquities of Canterbury* 275n.74
Southern, Richard 10, 41, 47, 55
'Speculum Britanniae' 3, 58, 61, 65; see also Norden, John
Speed, John 78–82, 86, 88, 96, 183, 187–8, 194, 196, 225, 270nn.98, 106; *Genealogies recorded in the Sacred Scriptures* 81; *History of Great Britaine* 79, 81, 270n.101; *Theatre of the Empire of Great Britaine* 78–80, 82, 88, 178, 270nn.98, 101
Spelman, Henry 18, 77–8, 84, 109, 194, 250n.36, 273–4n.60, 276–7n.34; *Archaeologus in modum Glossarii* 274n.60, 277n.34; *Epistle on Tithes* 77, 105; 'Icenia' 273–4n.60; 'Interpretation of Villare Anglicum' 194, 311n.8
Spinoza, Benedictus de 154
Sprat, Thomas 131, 146, 148–9, 163, 290n.2; *History of the Royal Society* 146
Stafforde, Robert 21; *Geographicall and Anthologicall description of all the Empires and Kingdomes* 21
Steno 269n.89, 301n.127; *Prodromus* 269n.89, 301n.127
Stimson, Dorothy 27, 301n.112
Stonehenge 183, 204, 236; see also Aubrey, John
Stow, John 24, 46, 54, 79, 87, 178, 183, 203, 260n.26, 262n.65, 263n.83, 278n.48; *Survay of London* 24, 86–7, 263n.83

Strabo 21–2, 38–9, 50, 52–3, 119, 229, 257–8n.5, 325n.23; 'British Strabo' (Camden) 49
Strachey, Lytton 25
Strangman, James 84, 271n.1
Stukeley, William 156, 170, 236, 326n.28, 328n.56
Suetonius 107
surveying 70–1, 74, 80, 182, 268n.83, 289n.45
Symond, Père Richard 180
Symson, Andrew 218, 321n.23; 'Description of Galloway' 218; 'Villare Scoticum' 218

Tacitus 23, 80, 107, 320n.4; *Germania* 50, 52–3
Talbot, Robert 46
Taylor, E.G.R. 35, 325n.12
Theophrastus 255n.141
Thirsk, Joan 138, 141
Thomas, Keith 236
Thompson, J.W. 35; *History of Historical Writing* 35
Thoroton, Robert 112–13, 162; *Antiquities of Nottinghamshire* 112, 278n.49, 269n.69, 299n.94; see also Godfrey, John T.
Throsby, John 278n.53, 299n.94
title search 19, 87
Toland, John 220–1, 322n.34
Trevisa, John 42, 259n.20; see also Higden, Ranulph
Turner, William 26, 131, 280n.15
Twysden, Roger 277n.37

Vadian (Vadianus), Joachim 39
Vaughan, Thomas 132
Verenius, Bernhard 269n.89
Vergil, Polydore 61, 262n.73
Verulamium 66

Vesalius, Andreas 149, 160, 298n.76
Vico, Giambattista 27, 31
Victoria County Series 49
virtuosi x, 27, 125, 127, 134, 140, 144, 149, 150–2, 155, 158, 206, 224, 232, 234, 236, 238, 240, 292nn.27, 28; *see also* Shadwell, Thomas

Wagner, Jacob (Wagnero, Jacobo) 199, 314n.34
Wallace, James (the elder) 221; *Description of the Isles of Orkney* 221; *see also* Wallace, James (the younger)
Wallace, James (the younger) 221; *see also* Wallace, James (the elder)
Wallis, John 156, 161, 188
Walters, H.B. 44, 260n.26
Warren, Erasmus 315n.46
Webb, William 86–7; *Description of the City and County Palatine of Chester* 86–7
Webster, Charles ix, 37, 134, 138, 140, 145, 160, 285nn.7–8, 86; *Great Instauration* 37, 137, 254n.113, 298nn.82–5; *Paracelsus to Newton* 285n.86
Weever, John 84, 100, 178, 194; *Ancient Funerall Monuments* 84, 99, 100, 178, 194
Weinberger, Jerry 114

Westcote, Thomas 83, 90–1, 93–9; *View of Devonshire in 1630* 95–6, 98
Westfall, Richard xi
Weston, Richard 138, 286n.11
Whitehead, A.N. 6
Wilkins, John 151
William I, king 20
Williams, William 210
Willis, Thomas 153, 160
Wodrow, Robert 320n.11
Wood, Anthony à 13, 45, 58, 91, 110, 177, 182, 203, 222, 275n.6, 313n.26; *Athenae Oxonienses* 36
Woodward, John 165, 226, 327n.47, 328–9n.65; *Brief Instructions* 165
Worcester, William of 43, 46; *Itinerary (Itineraria)* 43, 46, 259–60n.23
Wotton, William 14, 240–1, 245
Wren, Christopher 172, 176
Wright, James 278n.54; *Rutlandshire* 278n.54

Xanthius 119
Xenophanes 119

Yates, Frances A. 133, 284n.73, 285n.84
Yorkshire Philosophical Society 239

Zouche, Richard vi; *Dove* vi

www.ingramcontent.com/pod-product-compliance
Lightning Source LLC
Chambersburg PA
CBHW020350080526
44584CB00014B/962